PROBLEMS AND SOLUTIONS IN REAL ANALYSIS

Second Edition

Series on Number Theory and Its Applications   Vol. 14

PROBLEMS AND SOLUTIONS IN REAL ANALYSIS

Second Edition

Masayoshi Hata
Kyoto University, Japan

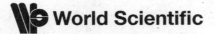
World Scientific

EW JERSEY · LONDON · SINGAPORE · BEIJING · SHANGHAI · HONG KONG · TAIPEI · CHENNAI · TOKYO

Published by

World Scientific Publishing Co. Pte. Ltd.

5 Toh Tuck Link, Singapore 596224

USA office: 27 Warren Street, Suite 401-402, Hackensack, NJ 07601

UK office: 57 Shelton Street, Covent Garden, London WC2H 9HE

Library of Congress Cataloging-in-Publication Data

Names: Hata, Masayoshi.

Title: Problems and solutions in real analysis / by Masayoshi Hata (Kyoto University, Japan).

Description: Second edition. | New Jersey : World Scientific, 2016. | Series:
    Series on number theory and its applications ; volume 14 | Includes
    bibliographical references and index.

Identifiers: LCCN 2016040761| ISBN 9789813142817 (hardcover : alk. paper) |
    ISBN 9789813142824 (pbk. : alk. paper)

Subjects: LCSH: Mathematical analysis--Problems, exercises, etc. | Number
    theory--Problems, exercises, etc.

Classification: LCC QA301 .H37 2016 | DDC 515.076--dc23

LC record available at https://lccn.loc.gov/2016040761

British Library Cataloguing-in-Publication Data

A catalogue record for this book is available from the British Library.

Printed in Singapore

Preface to the First Edition

Rome was not built in a day...

There is no shortcut to good scholarship. To learn mathematics you are to solve many 'good' problems without haste. Mathematics is not only for persons of talent. Tackling difficult problems is like challenging yourself. Even if you do not make a success in solving a problem, you may set some new knowledge or technique you lacked.

This book contains more than one hundred and fifty mathematical problems and their detailed solutions related mainly to Real Analysis. Many problems are selected carefully both for students who are presently learning or those who have just finished their courses in Calculus and Linear Algebra, or for any person who wants to review and improve his or her skill in Real Analysis and, moreover, to make a step forward, for example, to Complex Analysis, Fourier Analysis, or Lebesgue Integration, etc. The solutions to all problems are supplied in detail, which should compete well with the famous books written by Pólya and Szegö more than thirty-five years ago.

Some problems are taken from Analytic Number Theory; for example, the uniform distribution (Chapter 12) and the prime number theorem (Chapter 17). The latter is treated in a slightly different way. They may be useful for an introduction to Analytic Number Theory.

Nevertheless the reader should notice that all solutions are not short and elegant. It may always be possible for the reader to find better and more elementary solutions. The problems are merely numbered for convenience' sake and so the reader should grapple with them using any tools, which makes a difference from the usual exercises in Calculus. One may use integration for problems on series, for example. The author must confess that there are some problems expressed in an elementary way, whose simple and elementary proof could not be found by

v

the author. The reason why he dared to include such problems and the solutions beyond the limits of Calculus is to urge the reader to find better ones.

The author wishes to take this opportunity to thank Professor S. Kanemitsu for invaluable help in the preparation of the manuscript.

Enjoy mathematics with a pen!

Kyoto, JAPAN *M. Hata*

Preface to the Second Edition

This second edition gave me the opportunity of correcting a number of misprints, improving some problems and solutions to clarify essential meaning and including new materials by appending three chapters: Bernoulli Numbers, Metric Spaces and Differential Equations. Note that the numbering and the signs of the Bernoulli numbers are defined differently from the first edition. Several figures are also included to help the reader's understanding, which are drawn by using 'Grapher 2.5' bundled with Mac OSX.

The author expresses his deep gratitude to Dr. Vladimir Lucic, Prof. Elton P. Hsu and Prof. Hisashi Okamoto for their beneficial comments. Also thanks are due to the staff of World Scientific Publishing Co. for the excellent help and cooperation.

Kyoto, JAPAN   *M. Hata*

Notations

1. The largest integer not exceeding a real number x is called the integer part of x and denoted by $[x]$.

2. Let $f(x)$ be a real-valued function defined on an open interval (a, b) and let c be any point in $[a, b)$. The right-sided limit of $f(x)$ as x approaches to c satisfying $x > c$ is denoted by $f(c+)$ or

$$\lim_{x \to c+} f(x)$$

if it exists. Note that in some other books the notation '$x \to c + 0$' instead of '$x \to c+$' are used. Similarly the left-sided limit $f(c-)$ is defined for $c \in (1, b]$ if it exists.

3. The right derivative of $f(x)$ at c is denoted by $f'_+(x)$ if it exists. We similarly define the left derivative $f'_-(x)$ if it exists.

4. Given two sequences $\{a_n\}$ and $\{b_n\}$ with $b_n > 0$ for all $n \geq 1$, we write

$$a_n = O(b_n)$$

if $|a_n| \leq C b_n$ holds for all $n \geq 1$ with some positive constant C. In particular, $a_n = O(1)$ means merely that $\{a_n\}$ is a bounded sequence. This is a convenient way of expressing inequalities, known as Landau notation. Note that the symbol $O(b_n)$ does not mean a specific sequence. For example, one can write $O(1) + O(1) = O(1)$. Usually we use the big O notation to describe the asymptotic behavior of $\{a_n\}$ and the majorant sequence $\{b_n\}$ is chosen among standard positive sequences such as n^α, $(\log n)^\beta$, $e^{\delta n}$, etc., where

$$e = \lim_{n \to \infty} \left(1 + \frac{1}{n}\right)^n = \sum_{n=0}^{\infty} \frac{1}{n!}$$

is the base to the natural logarithm.

5. If the ratio a_n/b_n converges to 0 as $n \to \infty$, then we write

$$a_n = o(b_n).$$

In particular, $a_n = o(1)$ means simply that a_n converges to 0 as $n \to \infty$.

6. If the ratio a_n/b_n converges to 1 as $n \to \infty$, then we write

$$a_n \sim b_n$$

and we say that $\{a_n\}$ and $\{b_n\}$ are asymptotically equivalent or that $\{a_n\}$ is asymptotic to $\{b_n\}$. This gives clearly an equivalence relation. The sequence $\{b_n\}$ is also said to be the principal part of $\{a_n\}$. Note that $a_n \sim b_n$ if and only if

$$a_n = b_n + o(b_n).$$

7. Landau notations can also be used to describe the asymptotic behavior of a function $f(x)$ as $x \to c$. If $|f(x)| \le Cg(x)$ holds on a sufficiently small neighborhood of the point c with some positive function $g(x)$ and positive constant C, then we write

$$f(x) = O(g(x))$$

as $x \to c$. We can also define $f(x) = o(g(x))$ and $f(x) \sim g(x)$ as $x \to c$ in the same manner as sequences, even when $x \to \infty$ or $x \to -\infty$.

8. The n times derivative of a function $f(x)$ defined on an interval I is denoted by $f^{(n)}(x)$ if it exists. The set of all functions on I possessing the continuous n times derivative is denoted by $\mathscr{C}^n(I)$. If the interval I contains an end point, the derivative at this point should be regarded as the one-sided derivative. In particular, the set of all continuous functions defined on I is denoted by $\mathscr{C}(I)$.

9. The *sign* function or signum function $\operatorname{sgn}(x)$ is defined by

$$\operatorname{sgn}(x) = \begin{cases} 1 & (x > 0) \\ 0 & (x = 0). \\ -1 & (x < 0) \end{cases}$$

10. We will reasonably suppress or abbreviate parentheses used in some cases. For example, we write $\sin n\theta$ and $\sin^2 x$ instead of $\sin(n\theta)$ and $(\sin x)^2$ respectively.

11. The real part and imaginary part of a complex number z are denoted by $\operatorname{Re} z$ and $\operatorname{Im} z$ respectively.

12. For any subset E in a linear space we denote by $E - E$ the set of all points $x \in E$ which can be written as $x = u - v$ for some $u, v \in E$.

Contents

Chapter 1

Sequences and Limits

[Summary of Basic Points]

1. Let $\{a_n\}$ be a sequence of real or complex numbers. A necessary and sufficient condition for the sequence to converge is that for any $\epsilon > 0$ there exists an integer $N > 0$ such that

$$|a_p - a_q| < \epsilon$$

holds for all integers p and q greater than N. This is called the Cauchy criterion.

2. Any monotone bounded sequence is convergent.

3. For any sequence $\{a_n\}$ the *inferior limit* and the *superior limit* are defined by the limits of monotone sequences

$$\liminf_{n \to \infty} a_n = \lim_{n \to \infty} \inf\{a_n, a_{n+1}, \dots\}$$

and

$$\limsup_{n \to \infty} a_n = \lim_{n \to \infty} \sup\{a_n, a_{n+1}, \dots\}$$

respectively. Note that the inferior and superior limits always exist if we adopt $\pm\infty$ as limits.

4. A bounded sequence $\{a_n\}$ converges if and only if the inferior limit coincides with the superior limit.

1

PROBLEM 1.1

Prove that $n \sin(2n! e\pi)$ converges to 2π as $n \to \infty$.

PROBLEM 1.2

Prove that

$$\left(\frac{1}{n}\right)^n + \left(\frac{2}{n}\right)^n + \cdots + \left(\frac{n}{n}\right)^n$$

converges to $e/(e-1)$ as $n \to \infty$.

PROBLEM 1.3

Prove that

$$e^{n/4} n^{-(n+1)/2} \left(1^1 2^2 \cdots n^n\right)^{1/n}$$

converges to 1 as $n \to \infty$.

This was proposed by Cesàro (1888) and solved by Pólya (1911).

PROBLEM 1.4

Suppose that a_n and b_n converge to α and β as $n \to \infty$ respectively. Show that

$$\frac{a_0 b_n + a_1 b_{n-1} + \cdots + a_n b_0}{n}$$

converges to $\alpha\beta$ as $n \to \infty$.

PROBLEM 1.5

Suppose that a sequence $\{a_n\}_{n \geq 1}$ satisfies

$$a_{m+n} \leq a_m + a_n$$

for all integers $m, n \geq 1$. Show that a_n/n either converges or diverges to $-\infty$ as $n \to \infty$.

This is essentially due to Fekete (1923). In various places we may benefit from this useful lemma in deducing the existence of limit.

PROBLEM 1.6

Suppose that a sequence $\{a_n\}_{n \geq 0}$ satisfies

$$\frac{a_{m+n} + a_{|m-n|}}{2} \leq a_m + a_n$$

for all non-negative integers m, n. Show that a_n/n^2 either converges or diverges to $-\infty$ as $n \to \infty$.

This is a quadratic version of the previous problem. The form of the inequality originates from the formula: $(x + y)^2 + (x - y)^2 = 2(x^2 + y^2)$.

PROBLEM 1.7

For any $0 < \theta < \pi$ and integer $n \geq 1$ show that

$$\sin \theta + \frac{\sin 2\theta}{2} + \cdots + \frac{\sin n\theta}{n} > 0.$$

This inequality was conjectured by Fejér and proved by Jackson(1911) and by Gronwall(1912) independently. Landau(1934) gave a shorter (maybe the shortest) elegant proof. See also **PROBLEM 5.8**. Note that

$$\sum_{n=1}^{\infty} \frac{\sin n\theta}{n} = \frac{\pi - \theta}{2}$$

for $0 < \theta < 2\pi$, which will be shown in **SOLUTION 7.12**.

PROBLEM 1.8

For any $\theta \in \mathbb{R}$ and any integer $n \geq 1$ show that

$$\frac{\cos \theta}{2} + \frac{\cos 2\theta}{3} + \cdots + \frac{\cos n\theta}{n + 1} \geq -\frac{1}{2}.$$

This was shown by Rogosinski and Szegö(1928). Verblunsky(1945) gave another proof. Koumandos(2001) obtained the lower bound $-41/96$ for $n \geq 2$. Note that

$$\sum_{n=0}^{\infty} \frac{\cos n\theta}{n + 1} = \frac{\pi - \theta}{2} \sin \theta - \cos \theta \log \left(2 \sin \frac{\theta}{2} \right)$$

for $0 < \theta < 2\pi$.

For a simpler cosine sum Young(1912) showed that

$$\sum_{n=1}^{m} \frac{\cos n\theta}{n} > -1$$

for any $\theta \in \mathbb{R}$ and integer $m \geq 2$. Brown and Koumandos(1997) improved the right-hand side by replacing -1 by $-5/6$.

PROBLEM 1.9

Given a positive sequence $\{a_n\}_{n \geq 0}$ satisfying $\sqrt{a_1} \geq \sqrt{a_0} + 1$ and

$$\left| a_{n+1} - \frac{a_n^2}{a_{n-1}} \right| \leq 1$$

for all integer $n \geq 1$, show that

$$\frac{a_{n+1}}{a_n}$$

converges as $n \to \infty$. Show moreover that $a_n \theta^{-n}$ converges as $n \to \infty$, where θ is the limit of a_{n+1}/a_n.

This is due to Boyd (1969).

PROBLEM 1.10

Let E be a bounded closed set in the complex plane containing an infinite number of points, and let M_n be the maximum of $|V(x_1, \ldots, x_n)|$ as n points x_1, $\ldots$, x_n run through the set E, where

$$V(x_1, \ldots, x_n) = \prod_{1 \leq i < j \leq n} (x_i - x_j)$$

is the Vandermonde determinant. Show that $M_n^{2/(n(n-1))}$ converges as $n \to \infty$.

This is due to Fekete (1923) and the limit
$$\tau(E) = \lim_{n \to \infty} M_n^{2/(n(n-1))}$$
is called the transfinite diameter of the set E. See **PROBLEM 15.10**.

PROBLEM 1.11

Define a sequence $\{a_n\}$ inductively by

$$a_n + \frac{a_{n-1}}{2} + \frac{a_{n-2}}{3} + \cdots + \frac{a_1}{n} = \frac{1}{n+1}$$

for $n \geq 1$. Show that $\{a_n\}$ is monotonously decreasing and converges to 0 as $n \to \infty$.

PROBLEM 1.12

Put $c_1 = 1/2$ and define a sequence $\{c_n\}$ inductively by

$$(n+2)c_{n+1} = (n-1)c_n + \sum_{k=1}^{n} c_k c_{n+1-k}$$

for $n \geq 1$. Show that $\{c_n\}$ coincides with the sequence $\{a_n\}$ defined in the previous problem.

PROBLEM 1.13

For any positive sequence $\{a_n\}_{n \geq 1}$ show that

$$\left(\frac{a_1 + a_{n+1}}{a_n}\right)^n > e$$

holds for infinitely many n's, where e is base of the natural logarithm. Prove also that the constant e cannot be replaced by any larger number in general.

PROBLEM 1.14

Let $f(x)$ be a function defined on $I = (0, a]$ satisfying $0 < f(x) < x$ and

$$f(x) = x - \alpha x^{\beta+1} + o(x^{\beta+1})$$

as $x \to 0+$ for some constants $\alpha, \beta > 0$. For a given $x_0 \in I$ define a sequence $\{x_n\}$ by $x_{n+1} = f(x_n)$ for $n \geq 0$. Show then that

$$\lim_{n \to \infty} n^{1/\beta} x_n = \frac{1}{(\alpha\beta)^{1/\beta}}.$$

Note that the limit is independent of the choice of x_0. For example, for the sequence defined by

$$x_n = \underbrace{\sin \circ \sin \circ \cdots \circ \sin}_{n}(x_0)$$

with $0 < x_0 < \pi$, we have $x_n \sim \sqrt{3/n}$ as $n \to \infty$, because $\alpha = 1/6$ and $\beta = 2$.

Solutions for Chapter 1

SOLUTION 1.1

Let r_n and ϵ_n be the integral and fractional parts of $n!e$ respectively. Using the expansion

$$e = 1 + \frac{1}{1!} + \frac{1}{2!} + \cdots + \frac{1}{n!} + \cdots,$$

we have

$$r_n = n!\left(1 + \frac{1}{1!} + \frac{1}{2!} + \cdots + \frac{1}{n!}\right) \quad \text{and} \quad \epsilon_n = \frac{1}{n+1} + \frac{1}{(n+1)(n+2)} + \cdots,$$

because

$$\frac{1}{n+1} < \epsilon_n < \frac{1}{n+1} + \frac{1}{(n+1)^2} + \cdots = \frac{1}{n}.$$

Thus we get $\sin(2n!e\pi) = \sin(2\pi\epsilon_n)$. Note that $\epsilon_n \to 0$ implies the irrationality of e. Since $n\epsilon_n$ converges to 1 as $n \to \infty$, we have

$$\lim_{n\to\infty} n\sin(2\pi\epsilon_n) = \lim_{n\to\infty} \frac{\sin(2\pi\epsilon_n)}{\epsilon_n} = 2\pi;$$

hence $n\sin(2n!e\pi)$ converges to 2π as $n \to \infty$. $\qquad\qquad\square$

REMARK. More precisely one gets

$$\epsilon_n = \frac{1}{n} - \frac{1}{n^3} + O\left(\frac{1}{n^4}\right);$$

so we have

$$n\sin(2n!e\pi) = 2\pi n\epsilon_n + \frac{4\pi^3}{3}n\epsilon_n^3 + O(n\epsilon_n^5)$$

$$= 2\pi + \frac{2\pi(2\pi^2 - 3)}{3n^2} + O\left(\frac{1}{n^3}\right)$$

as $n \to \infty$.

Let $\{d_n\}$ be any monotone increasing sequence of positive integers diverging to ∞ and satisfying $d_n < n$ for $n > 1$. We divide the sum into two parts as follows.

$$a_n = \left(\frac{1}{n}\right)^n + \left(\frac{2}{n}\right)^n + \cdots + \left(\frac{n-1-d_n}{n}\right)^n,$$

$$b_n = \left(\frac{n-d_n}{n}\right)^n + \cdots + \left(\frac{n}{n}\right)^n.$$

First the sum a_n is roughly estimated above by

$$\frac{1}{n^n} \int_0^{n-d_n} x^n\, dx = \frac{(n-d_n)^{n+1}}{(n+1)n^n} < \left(1 - \frac{d_n}{n}\right)^n.$$

Now using an inequality $\log(1-x) + x < 0$ valid for $0 < x < 1$ we obtain

$$0 < a_n < e^{n\log(1-d_n/n)} < e^{-d_n},$$

which converges to 0 as $n \to \infty$.

Next by Taylor's formula for $\log(1-x)$ we can take a positive constant c_1 such that

$$\left|\log(1-x) + x\right| \le c_1 x^2$$

holds for $|x| \le 1/2$. Thus for any integer n satisfying $d_n/n \le 1/2$ we get

$$\left| n\log\left(1 - \frac{k}{n}\right) + k \right| \le \frac{c_1 k^2}{n} \le \frac{c_1 d_n^2}{n}$$

for $0 \le k \le d_n$. Suppose further that d_n^2/n converges to 0 as $n \to \infty$. For example $d_n = [n^{1/3}]$ satisfies all the conditions imposed above. Next take a positive constant c_2 satisfying

$$|e^x - 1| \le c_2|x|$$

for any $|x| \le 1$. Since $c_1 d_n^2/n \le 1$ for all sufficiently large n, we have

$$\left| e^k\left(1 - \frac{k}{n}\right)^n - 1 \right| = \left| e^{n\log(1-k/n)+k} - 1 \right|$$

$$\le \frac{c_1 c_2 d_n^2}{n}.$$

Dividing both sides by e^k and summing from $k = 0$ to d_n, we get

$$\sum_{k=0}^{d_n} \left| \left(1 - \frac{k}{n}\right)^n - e^{-k} \right| \le \frac{c_1 c_2 d_n^2}{n} \sum_{k=0}^{d_n} e^{-k}.$$

Hence

$$\left| b_n - \sum_{k=0}^{d_n} e^{-k} \right| < \frac{e c_1 c_2 d_n^2}{(e-1)n},$$

which implies that

$$\left| b_n - \frac{e}{e-1} \right| \le \frac{e}{e-1} \left(\frac{c_1 c_2 d_n^2}{n} + e^{-d_n} \right).$$

Therefore $a_n + b_n$ converges to $e/(e-1)$ as $n \to \infty$. □

SOLUTION 1.3

The logarithm of the given sequence can be written as

$$\sum_{k=1}^{n} f\left(\frac{k}{n}\right) - n \int_0^1 f(x)\, dx,$$

where $f(x) = x \log x$. Since f' is integrable over $(0, 1)$, it follows from the remark after SOLUTION 5.7 that the above sum converges to

$$\left(1 - \frac{1}{2} \right)(f(1) - f(0+)) = 0;$$

hence the limit is 1. □

SOLUTION 1.4

Let M be an upper bound of the two convergent sequences $|a_n|$ and $|b_n|$. For any $\epsilon > 0$ we can take a positive integer N satisfying $|a_n - \alpha| < \epsilon$ and $|b_n - \beta| < \epsilon$ for all integers $n > N$. If $n > N^2$, then

$$|a_k b_{n-k} - \alpha\beta| \le |(a_k - \alpha)b_{n-k} + \alpha(b_{n-k} - \beta)|$$
$$\le (M + |\alpha|)\epsilon$$

for any integer $k \in [\sqrt{n}, n - \sqrt{n}]$. Therefore

$$\left| \frac{1}{n} \sum_{k=0}^{n} a_k b_{n-k} - \alpha\beta \right| \le \frac{1}{n} \sum_{\sqrt{n} \le k \le n - \sqrt{n}} |a_k b_{n-k} - \alpha\beta|$$
$$+ 2\left(|\alpha\beta| + M^2\right) \frac{[\sqrt{n}] + 1}{n}$$
$$\le (M + |\alpha|)\epsilon + 2\left(|\alpha\beta| + M^2\right) \frac{\sqrt{n} + 1}{n}.$$

We can take n so large that the last expression is less than $(M + |\alpha| + 1)\epsilon$. □

ANOTHER SOLUTION. Put $\tilde{a}_n = a_n - \alpha$. Then we have

$$\frac{a_0 b_n + a_1 b_{n-1} + \cdots + a_n b_0}{n} = \frac{\tilde{a}_0 b_n + \tilde{a}_1 b_{n-1} + \cdots + \tilde{a}_n b_0}{n}$$
$$+ \alpha \frac{b_0 + b_1 + \cdots + b_n}{n}.$$

The first term on the right-hand side converges to 0 as $n \to \infty$, because

$$\frac{|\tilde{a}_0 b_n + \tilde{a}_1 b_{n-1} + \cdots + \tilde{a}_n b_0|}{n} \leq M \frac{|\tilde{a}_0| + |\tilde{a}_1| + \cdots + |\tilde{a}_n|}{n}$$

and the arithmetic mean of a convergent sequence converges to the same limit. By the same reason the second term converges to $\alpha\beta$ as $n \to \infty$.

SOLUTION 1.5

For an arbitrarily fixed positive integer k we put $n = qk + r$ with $0 \leq r < k$. Applying the given inequality for q times we get $a_n = a_{qk+r} \leq qa_k + a_r$; so,

$$\frac{a_n}{n} \leq \frac{a_k}{k} + \frac{a_r}{n}.$$

Taking the limit as $n \to \infty$, we get

$$\limsup_{n\to\infty} \frac{a_n}{n} \leq \frac{a_k}{k}.$$

The sequence a_n/n is therefore bounded above. Since k is arbitrary, we conclude that

$$\limsup_{n\to\infty} \frac{a_n}{n} \leq \inf_{k\geq 1} \frac{a_k}{k} \leq \liminf_{k\to\infty} \frac{a_k}{k},$$

which completes the proof. $\qquad\qquad\qquad\qquad\qquad\qquad\qquad\square$

SOLUTION 1.6

For an arbitrarily fixed positive integer k we put $n = qk + r$ with $0 \leq r < k$. Put $c_\ell = a_{k\ell+r}$ for brevity. Adding $m - 1$ inequalities

$$c_{\ell+1} + c_{\ell-1} \leq 2(c_\ell + a_k)$$

for $\ell = 1, \ldots, m - 1$ one after another, we get

$$c_0 + c_m \leq c_1 + c_{m-1} + 2(m-1)a_k.$$

Adding again the above inequalities for $m = 2, \ldots, M$, we thus obtain

$$c_M \leq Mc_1 - (M-1)c_0 + M(M-1)a_k.$$

Therefore

$$\limsup_{n\to\infty} \frac{a_n}{n^2} \le \frac{a_k}{k^2}$$

and the sequence a_n/n^2 is bounded above. Since k is arbitrary, we conclude that

$$\limsup_{n\to\infty} \frac{a_n}{n^2} \le \inf_{k\ge 1} \frac{a_k}{k^2} \le \liminf_{k\to\infty} \frac{a_k}{k^2},$$

which completes the proof. □

SOLUTION 1.7

Denote by $s_n(\theta)$ the left-hand side of the inequality to be shown. Put $\vartheta = \theta/2$ for brevity. Since

$$s_n'(\theta) = \operatorname{Re}\left(e^{i\theta} + e^{2i\theta} + \cdots + e^{ni\theta}\right) = \frac{\cos(n+1)\vartheta \sin n\vartheta}{\sin \vartheta},$$

we obtain the candidates of extreme points of $s_n(\theta)$ over the interval $(0, \pi]$ by solving the equations $\cos(n+1)\vartheta = 0$ and $\sin n\vartheta = 0$; that is,

$$\frac{\pi}{n+1}, \ \frac{2\pi}{n}, \ \frac{3\pi}{n+1}, \ \frac{4\pi}{n}, \ \ldots.$$

Note that the last two candidates are $(n-1)\pi/(n+1)$ and π if n is even, and $(n-1)\pi/n$ and $n\pi/(n+1)$ if n is odd. In any case $s_n'(\theta)$ vanishes at least at n points in $(0, \pi)$.

Since $s_n'(\theta)$ can be expressed as a polynomial in $\cos\theta$ of degree n and $\cos\theta$ maps the interval $[0, \pi]$ onto $[-1, 1]$ homeomorphically, this polynomial possesses at most n real roots in $[-1, 1]$. Hence all these roots must be simple and give the actual extreme points of $s_n(\theta)$ except for $\theta = \pi$. Clearly $s_n(\theta)$ is positive in a right neighborhood of $\theta = 0$, and the maximal and minimal points stand in line alternately from left to right. Thus $s_n(\theta)$ attains its minimal values at the points $2\ell\pi/n$ in $(0, \pi)$ when $n \ge 3$. In the cases $n = 1$ and $n = 2$, however, $s_n(\theta)$ has no minimal points in $(0, \pi)$.

Now we will show that $s_n(\theta)$ is positive on $(0, \pi)$ by induction on n. It is clear when $n = 1$ and $n = 2$, because $s_1(\theta) = \sin\theta$ and $s_2(\theta) = (1+\cos\theta)\sin\theta$. Suppose that $s_{n-1}(\theta) > 0$ for $n \ge 3$. Then the minimal values of $s_n(\theta)$ are certainly attained at $2\ell\pi/n$ in $(0, \pi)$, whose values are

$$s_n\left(\frac{2\ell\pi}{n}\right) = s_{n-1}\left(\frac{2\ell\pi}{n}\right) + \frac{\sin 2\ell\pi}{n} = s_{n-1}\left(\frac{2\ell\pi}{n}\right) > 0.$$

Therefore $s_n(\theta) > 0$ on $(0, \pi)$ and this completes the proof. □

REMARK. Landau(1934) gave the following elegant shorter proof using mathematical induction on n. Suppose that $s_{n-1}(\theta) > 0$ on $(0, \pi)$. If $s_n(\theta)$ attains the non-positive minimum at some point, say θ^*, then $s_n'(\theta^*) = 0$ implies

$$\sin\left(n + \frac{1}{2}\right)\theta^* = \sin\frac{\theta^*}{2},$$

and hence

$$\cos\left(n + \frac{1}{2}\right)\theta^* = \pm\cos\frac{\theta^*}{2}.$$

We thus have

$$\sin n\theta^* = \sin\left(n + \frac{1}{2}\right)\theta^* \cos\frac{\theta^*}{2} - \cos\left(n + \frac{1}{2}\right)\theta^* \sin\frac{\theta^*}{2}$$
$$= \sin\frac{\theta^*}{2}\cos\frac{\theta^*}{2} \pm \cos\frac{\theta^*}{2}\sin\frac{\theta^*}{2},$$

being equal either to 0 or $\sin\theta^* \geq 0$ according to the sign. We are led to a contradiction, because

$$s_n(\theta^*) = s_{n-1}(\theta^*) + \frac{\sin n\theta^*}{n} \geq s_{n-1}(\theta^*) > 0.$$

SOLUTION 1.8

The proof is substantially based on Verblunsky(1945). Put $\vartheta = \theta/2$ for brevity. Let $c_n(\vartheta)$ be the left-hand side of the inequality to be shown. It suffices to show this on $[0, \pi/2]$. Since $c_1(\vartheta) = \cos\vartheta/2 \geq -1/2$ and

$$c_2(\vartheta) = \frac{2}{3}\cos^2\vartheta + \frac{1}{2}\cos\vartheta - \frac{1}{3} \geq -\frac{41}{96},$$

we can assume that $n \geq 3$. Note that

$$\cos n\theta = \frac{\sin(2n+1)\vartheta - \sin(2n-1)\vartheta}{2\sin\vartheta}$$
$$= \frac{\sin^2(n+1)\vartheta - 2\sin^2 n\vartheta + \sin^2(n-1)\vartheta}{2\sin^2\vartheta},$$

whose numerator is the second difference of the non-negative sequence $\{\sin^2 n\vartheta\}$. Using this formula we get

$$c_n(\vartheta) = \frac{1}{2\sin^2\vartheta}\sum_{k=1}^{n}\frac{\sin^2(k+1)\vartheta - 2\sin^2 k\vartheta + \sin^2(k-1)\vartheta}{k+1},$$

which can be written as

$$\frac{1}{2\sin^2\vartheta}\left(-\frac{2\sin^2\vartheta}{3}+\frac{\sin^2 2\vartheta}{12}+\cdots+\frac{2\sin^2(n-1)\vartheta}{n(n^2-1)}\right.$$
$$\left.-\frac{(n-1)\sin^2 n\vartheta}{n(n+1)}+\frac{\sin^2(n+1)\vartheta}{n+1}\right).$$

Hence we obtain

$$c_n(\vartheta) \geq -\frac{1}{3}+\frac{\cos^2\vartheta}{6}+\frac{\sin^2(n+1)\vartheta-\sin^2 n\vartheta}{2(n+1)\sin^2\vartheta}$$
$$=-\frac{1}{6}-\frac{\sin^2\vartheta}{6}+\frac{\sin(2n+1)\vartheta}{2(n+1)\sin\vartheta}.$$

For any ϑ satisfying $\sin(2n+1)\vartheta \geq 0$ we obviously have $c_n(\vartheta) \geq -1/3$. Moreover if ϑ belongs to the interval $(3\pi/(2n+1), \pi/2)$, then using Jordan's inequality $\sin\vartheta \geq 2\vartheta/\pi$,

$$c_n(\vartheta) \geq -\frac{1}{3}-\frac{1}{2(n+1)\sin(3\pi/(2n+1))}$$
$$\geq -\frac{1}{3}-\frac{2n+1}{12(n+1)} > -\frac{1}{2}.$$

Thus it suffices to consider on the interval $[\pi/(2n+1), 2\pi/(2n+1)]$.

In general, we consider an interval in the form

$$\left[\frac{\alpha\pi}{2n+1}, \frac{\beta\pi}{2n+1}\right].$$

For any ϑ satisfying $\sin(2n+1)\vartheta \leq c$ on this interval it follows that

$$c_n(\vartheta) \geq -\frac{1}{6}-\frac{\sin^2\vartheta}{6}-\frac{c}{2(n+1)\sin\vartheta}.$$

Now the right-hand side can be written as $-1/6-\varphi(\sin\vartheta)$, where $\varphi(x)$ is a concave function; hence, the maximum of φ is attained at an end point of the interval. By using the inequality

$$\alpha\pi\sin\vartheta \geq 7\vartheta\sin\frac{\alpha\pi}{7},$$

we get

$$\varphi\left(\sin\frac{\alpha\pi}{2n+1}\right) = \frac{1}{6}\sin^2\frac{\alpha\pi}{2n+1}+\frac{c}{2(n+1)\sin(\alpha\pi/(2n+1))}$$
$$\leq \frac{(\alpha\pi)^2}{6(2n+1)^2}+\frac{c}{2(n+1)}\cdot\frac{2n+1}{7\sin(\alpha\pi/7)}.$$

Since $n \geq 3$, the last expression is less than

$$\frac{(\alpha\pi)^2}{294} + \frac{c}{7\sin(\alpha\pi/7)}.$$

Similarly we get an estimate for another end point.

For $\alpha = 1$ and $\beta = 4/3$ we can take $c = \sqrt{3}/2$ so that the values of φ at the corresponding end points are less than 0.319 and 0.28 respectively. Similarly for $\alpha = 4/3$ and $\beta = 2$ we can take $c = 1$ so that the values of φ at the end points are less than 0.314 and 0.318 respectively. Therefore the maximum of φ on the interval $[\pi/(2n+1), 2\pi/(2n+1)]$ is less than $1/3$, which implies that $c_n(\vartheta) > -1/2$. $\qquad\qquad\square$

| **SOLUTION 1.9** |

We first show that

$$\frac{a_{n+1}}{a_n} > 1 + \frac{1}{\sqrt{a_0}} \qquad\qquad (1.1)$$

by induction on n. When $n = 0$ this holds by the assumption. Put $\alpha = 1 + 1/\sqrt{a_0}$. Suppose that (1.1) holds for $n \leq m$. We then have $a_k > \alpha^k a_0$ for $1 \leq k \leq m+1$. Thus

$$\left| \frac{a_{m+2}}{a_{m+1}} - \frac{a_1}{a_0} \right| \leq \sum_{k=1}^{m+1} \left| \frac{a_{k+1}}{a_k} - \frac{a_k}{a_{k-1}} \right| \leq \sum_{k=1}^{m+1} \frac{1}{a_k},$$

which is less than

$$\frac{1}{a_0} \sum_{k=1}^{m+1} \alpha^{-k} < \frac{1}{a_0(\alpha-1)} = \frac{1}{\sqrt{a_0}}.$$

Therefore

$$\frac{a_{m+2}}{a_{m+1}} > \frac{a_1}{a_0} - \frac{1}{\sqrt{a_0}} > 1 + \frac{1}{\sqrt{a_0}};$$

thus (1.1) holds also for $n = m + 1$.

Let $p > q > 0$ be any integers. In the same way,

$$\left| \frac{a_{p+1}}{a_p} - \frac{a_{q+1}}{a_q} \right| \leq \sum_{k=q+1}^{p} \left| \frac{a_{k+1}}{a_k} - \frac{a_k}{a_{k-1}} \right| \leq \sum_{k=q+1}^{p} \frac{1}{a_k},$$

which is less than

$$\frac{1}{a_q} \sum_{k=1}^{p-q} \frac{1}{\alpha^k} < \frac{\sqrt{a_0}}{a_q}.$$

This means that the sequence $\{a_{n+1}/a_n\}$ satisfies the Cauchy criterion, because a_q diverges to $+\infty$ as $q \to \infty$. Letting $p \to \infty$ in the above two inequalities, we get

$$\left| \theta - \frac{a_{q+1}}{a_q} \right| \le \frac{\sqrt{a_0}}{a_q}.$$

Multiplying both sides by a_q/θ^{q+1}, we have

$$\left| \frac{a_{q+1}}{\theta^{q+1}} - \frac{a_q}{\theta^q} \right| \le \frac{\sqrt{a_0}}{\theta^{q+1}},$$

which shows that the sequence $\{a_n/\theta^n\}$ also satisfies the Cauchy criterion. □

SOLUTION 1.10

Suppose that $|V(x_1, \ldots, x_{n+1})|$ attains its maximum M_{n+1} at $\xi_1, \ldots, \xi_{n+1}$. Since

$$\frac{V(\xi_1, \ldots, \xi_{n+1})}{V(\xi_1, \ldots, \xi_n)} = (\xi_1 - \xi_{n+1}) \cdots (\xi_n - \xi_{n+1}),$$

we have

$$\frac{M_{n+1}}{M_n} \le |\xi_1 - \xi_{n+1}| \cdots |\xi_n - \xi_{n+1}|.$$

Applying the same argument to each point $\xi_1, \ldots, \xi_n$, we get $n + 1$ similar inequalities, whose product gives

$$\left(\frac{M_{n+1}}{M_n} \right)^{n+1} \le \prod_{i \ne j} |\xi_i - \xi_j| = M_{n+1}^2.$$

Hence the sequence $M_n^{2/(n(n-1))}$ is monotonously decreasing. □

SOLUTION 1.11

The complex logarithm whose imaginary part lies in the interval $[-\pi, \pi]$ is called the principal value and denoted by $\operatorname{Log} z$. Consider the function

$$f(z) = \frac{1}{z} + \frac{1}{\operatorname{Log}(1 - z)}.$$

Since $z = 0$ is a removable singularity of $f(z)$, it can be expanded into the Taylor series as

$$f(z) = \sum_{n=1}^{\infty} b_n z^{n-1},$$

whose radius of convergence is clearly 1. Comparing the coefficients of z^{n+1} on both sides of

$$\sum_{n=1}^{\infty} \frac{z^n}{n} \sum_{n=1}^{\infty} b_n z^n = \sum_{n=2}^{\infty} \frac{z^n}{n},$$

we get

$$b_n + \frac{b_{n-1}}{2} + \frac{b_{n-2}}{3} + \cdots + \frac{b_1}{n} = \frac{1}{n+1}$$

for $n \geq 1$. Thus the sequence $\{b_n\}$ coincides with $\{a_n\}$; so,

$$a_n = \frac{1}{2\pi i} \int_C \frac{f(z)}{z^n} dz = \frac{1}{2\pi i} \int_C \frac{dz}{z^n \operatorname{Log}(1-z)},$$

where C is a small circle centered at $z = 0$ and oriented counterclockwise. By Cauchy's theorem the path C can be replaced with a bigger circle centered at $z = 1$ with a detour, along which one moves backward and forward, as illustrated in Fig. 1.1.

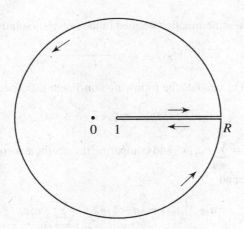

Fig. 1.1 The contour changed from C.

On the circle $z = 1 + Re^{i\theta}$ ($0 \leq \theta \leq 2\pi$) we have $\operatorname{Log}(1-z) = \log R + i(\theta - \pi)$; so,

$$\frac{1}{|z^n \operatorname{Log}(1-z)|} \leq \frac{1}{(R-1)^n \log R} = O\left(\frac{1}{R \log R}\right),$$

because $n \geq 1$. On the upper edge we have $\operatorname{Log}(1 - x) = \log(x - 1) - \pi i$ and on the lower edge $\text{Log}(1 - x) = \log(x - 1) + \pi i$ for $x > 1$. Thus, letting $R \to \infty$, we obtain

$$a_n = \frac{1}{2\pi i} \int_1^\infty \left(\frac{1}{\log(x - 1) - \pi i} - \frac{1}{\log(x - 1) + \pi i} \right) \frac{dx}{x^n}$$

$$= \int_1^\infty \frac{1}{\log^2(x - 1) + \pi^2} \cdot \frac{dx}{x^n}.$$

Substituting $x = 1/s$, we finally get

$$a_n = \int_0^1 s^{n-2} \mu(s)\, ds \quad \text{where} \quad \mu(s) = \frac{1}{\log^2\left(\dfrac{1 - s}{s}\right) + \pi^2}.$$

Note that $\mu(s)/s(1 - s)$ is an integrable function on $(0, 1)$. The sequence $\{a_n\}$ is thus monotonously decreasing and converges to 0 as $n \to \infty$. $\qquad\square$

SOLUTION 1.12

Let $f(z)$ be the same function defined in the previous solution. Differentiating

$$\log(1 - x) = \frac{x}{xf(x) - 1},$$

we see that $y = f(x)$ satisfies the following non-linear differential equation

$$x(1 - x)y' - xy^2 + 2y - 1 = 0.$$

Substituting $f(x) = \sum_{n=0}^\infty a_{n+1} x^n$ and comparing the coefficients of x^n on both sides, we have $a_1 = 1/2$ and

$$(n + 2)a_{n+1} = (n - 1)a_n + \sum_{j+k=n+1} a_j a_k$$

for $n \geq 1$. Hence the sequence $\{c_n\}$ in the problem coincides with $\{a_n\}$. $\qquad\square$

SOLUTION 1.13

Without loss of generality we can assume that $a_1 = 1$. Suppose, contrary to the conclusion, that there is an integer $N \geq 1$ satisfying

$$\left(\frac{1 + a_{n+1}}{a_n} \right)^n \leq e$$

for all $n \geq N$. Put

$$s(j,k) = \exp\left(\frac{1}{j} + \cdots + \frac{1}{k}\right)$$

for any positive integer $j \leq k$. Since $0 < a_{n+1} \leq e^{1/n}a_n - 1$, we get successively

$$0 < a_{n+1} \leq s(n,n)a_n - 1,$$
$$0 < a_{n+2} \leq s(n,n+1)a_n - s(n+1,n+1) - 1,$$

$$\vdots$$

$$0 < a_{n+k+1} \leq s(n,n+k)a_n - s(n+1,n+k) - \cdots - s(n+k,n+k) - 1$$

for any integer $k \geq 0$. Hence it follows that

$$a_n > \frac{1}{s(n,n)} + \frac{1}{s(n,n+1)} + \cdots + \frac{1}{s(n,n+k)}.$$

Then, using

$$\frac{1}{s(n,n+j)} > \exp\left(-\int_{n-1}^{n+j} \frac{dx}{x}\right) = \frac{n-1}{n+j},$$

we thus get

$$a_n > \sum_{j=0}^{k} \frac{n-1}{n+j}.$$

This is a contradiction, since the right-hand side diverges to ∞ as $k \to \infty$.

To see that the bound e cannot be replaced by any larger number, consider the case $a_n = n \log n$ for $n \geq 2$. Then

$$\left(\frac{a_1 + (n+1)\log(n+1)}{n \log n}\right)^n = \exp\left(n \log\left(1 + \frac{1}{n} + O\left(\frac{1}{n \log n}\right)\right)\right)$$
$$= \exp\left(1 + O\left(\frac{1}{\log n}\right)\right)$$

converges to e as $n \to \infty$. $\qquad\square$

SOLUTION 1.14

Since $0 < f(x) < x$, the sequence $\{x_n\}$ is strictly monotone decreasing and converges to 0 as $n \to \infty$. For any $0 < \epsilon < 1$ and $n \geq 0$ we put

$$y_n = \frac{\alpha\beta}{1 + \epsilon} x_n^\beta,$$

which is also a strictly monotone decreasing sequence converging to 0. We have

$$y_{n+1} = \frac{\alpha\beta}{1+\epsilon}\left(x_n - \alpha x_n^{\beta+1} + o\left(x_n^{\beta+1}\right)\right)^{\beta}$$

$$= y_n\left(1 - \frac{1+\epsilon}{\beta}y_n + o(y_n)\right)^{\beta};$$

hence there exists some integer $n_0 = n_0(\epsilon)$ such that

$$y_{n+1} \le \frac{y_n}{1+y_n}$$

for all $n \ge n_0$. We next define $z_n = ny_n > 0$ so that

$$\frac{z_{n+1}}{z_n} \le \frac{n+1}{n+z_n}. \tag{1.2}$$

We put $n_1 = \max(n_0, 1/\epsilon)$. If $z_n \ge 1 + \epsilon$ for all $n \ge n_1$, then $z_{n+1} < z_n$ by (1.2) and $\{z_n\}$ converges to some $\delta \ge 1 + \epsilon$ as $n \to \infty$. However, for all $n \ge n_1$ we have

$$\frac{z_{n+1}}{z_{n_1}} \le \prod_{k=n_1}^{n} \frac{k+1}{k+z_k} < \prod_{k=n_1}^{n}\left(1 - \frac{\epsilon}{k+1+\epsilon}\right).$$

This is a contradiction, because the left-hand side converges to δ/z_{n_1}, while the right-hand side converges to 0 as $n \to \infty$. Thus there exists an integer $N \ge n_1$ with $z_N < 1 + \epsilon$. Suppose next that $z_n \ge 1 + \epsilon$ for some $n \ge N$. Then we can take an integer $M \ge N$ satisfying $z_M < 1 + \epsilon \le z_{M+1}$. However it follows from (1.2) that $z_M < 1$ and

$$1 + \frac{1}{M} \ge \frac{z_{M+1}}{z_M} > \frac{1+\epsilon}{1} = 1 + \epsilon.$$

This is a contradiction, because $M \ge N \ge n_1 \ge 1/\epsilon$. Therefore $z_n < 1 + \epsilon$ for all $n \ge N$ and

$$\limsup_{n\to\infty} nx_n^{\beta} = \frac{1+\epsilon}{\alpha\beta}\limsup_{n\to\infty} z_n \le \frac{(1+\epsilon)^2}{\alpha\beta}.$$

To show the opposite inequality we put

$$u_n = \frac{\alpha\beta}{1-\epsilon}x_n^{\beta}$$

for all $n \ge 0$, which is a strictly monotone decreasing sequence converging to 0 as

$n \to \infty$. We have

$$u_{n+1} = \frac{\alpha\beta}{1-\epsilon}\left(x_n - \alpha x_n^{\beta+1} + o\left(x_n^{\beta+1}\right)\right)^{\beta}$$

$$= u_n\left(1 - \frac{1-\epsilon}{\beta}u_n + o(u_n)\right)^{\beta};$$

hence there exists an integer $n_2 = n_2(\epsilon)$ such that

$$u_{n+1} \geq \frac{u_n}{1+u_n}$$

for all $n \geq n_2$. We define $v_n = nu_n > 0$ so that

$$\frac{v_{n+1}}{v_n} \geq \frac{n+1}{n+v_n}. \tag{1.3}$$

Since the sequence $\{z_n\}$ is bounded, there exists a constant $K > 0$ with $z_n < K$ for all $n \geq 0$. Then

$$v_n = \frac{1+\epsilon}{1-\epsilon}z_n < \frac{1+\epsilon}{1-\epsilon}K.$$

We also put

$$n_3 = \max\left(n_2, K + \frac{K}{\epsilon}\right).$$

If $v_n \leq 1 - \epsilon$ for all $n \geq n_3$, then $v_{n+1} > v_n$ by (1.3) and $\{v_n\}$ converges to some $\delta' \leq 1 - \epsilon$ as $n \to \infty$. However, for all $n \geq n_3$ we have

$$\frac{v_{n+1}}{v_{n_3}} \geq \prod_{k=n_3}^{n}\frac{k+1}{k+v_k} > \prod_{k=n_3}^{n}\left(1 + \frac{\epsilon}{k+1-\epsilon}\right),$$

which is a contradiction, because the left-hand side converges to δ'/v_{n_3}, while the right-hand side diverges to $+\infty$ as $n \to \infty$. Thus there exists an integer $N' \geq n_3$ with $v_{N'} > 1 - \epsilon$. Suppose next that $v_n \leq 1 - \epsilon$ for some $n \geq N'$. Then we can take an integer $M' \geq N'$ satisfying $v_{M'} > 1 - \epsilon \geq v_{M'+1}$. However it follows from (1.3) that $v_{M'} > 1$ and

$$\frac{M'+1}{M'+(1+\epsilon)K/(1-\epsilon)} \leq \frac{v_{M'+1}}{v_{M'}} < \frac{1-\epsilon}{1} = 1 - \epsilon.$$

This is a contradiction, because $M' \geq N' \geq n_3 \geq K + K/\epsilon$. Therefore $v_n > 1 - \epsilon$ for all $n \geq N'$ and

$$\liminf_{n\to\infty} nx_n^{\beta} = \frac{1-\epsilon}{\alpha\beta}\liminf_{n\to\infty} v_n \geq \frac{(1-\epsilon)^2}{\alpha\beta}.$$

This completes the proof. $\qquad\square$

Chapter 2

Infinite Series

[Summary of Basic Points]

1. An infinite series $\displaystyle\sum_{n=1}^{\infty} a_n$ converges if and only if for any $\epsilon > 0$ there exists an integer $N > 0$ satisfying $|a_q + \cdots + a_p| < \epsilon$ for all integers $p \geq q \geq N$.

2. An infinite series $\displaystyle\sum_{n=1}^{\infty} a_n$ is said to *converge absolutely* if $\displaystyle\sum_{n=1}^{\infty} |a_n|$ converges.

3. If $\displaystyle\sum_{n=1}^{\infty} a_n$ converges but $\displaystyle\sum_{n=1}^{\infty} |a_n|$ diverges, then $\displaystyle\sum_{n=1}^{\infty} a_n$ is said to *converge conditionally*.

4. An absolutely convergent series converges to the same sum in whatever order the terms are taken. Therefore, for a non-negative sequence $\{a_n\}$, it is allowed to be written as $\displaystyle\sum_{n \in E} a_n$ for any subset E of $\mathbb{N}$, whether it converges or not.

5. Any conditionally convergent series can always be rearranged to yield a series which converges to any sum prescribed whatever, or diverges to ∞ or to $-\infty$.

6. Given two series $\displaystyle\sum_{n=0}^{\infty} a_n$ and $\displaystyle\sum_{n=0}^{\infty} b_n$, the series

$$\sum_{n=0}^{\infty} (a_0 b_n + a_1 b_{n-1} + \cdots + a_n b_0)$$

is called the Cauchy product of $\displaystyle\sum_{n=0}^{\infty} a_n$ and $\displaystyle\sum_{n=0}^{\infty} b_n$.

PROBLEM 2.1

As is well-known, the harmonic series

$$\frac{1}{1} + \frac{1}{2} + \frac{1}{3} + \cdots + \frac{1}{n} + \cdots$$

diverges to $+\infty$. A subset E of $\mathbb{N}$ is said to be 'dominant' if $\displaystyle\sum_{n \notin E} \frac{1}{n} < \infty$. Let E be the set of all positive integers whose decimal notation contains all the digits from 0 to 9. Show that E is dominant.

Kempner (1914) showed that the set consisting of all positive integers whose decimal notation contains one prescribed figure is dominant, although he did not discuss the case for the figure 0.

PROBLEM 2.2

(i) *Suppose that $\displaystyle\sum_{n=0}^{\infty} a_n$ and $\displaystyle\sum_{n=0}^{\infty} b_n$ converge to α and β respectively and that the Cauchy product converges to δ. Show then that $\delta = \alpha\beta$.*

(ii) *Suppose that $\displaystyle\sum_{n=0}^{\infty} a_n$ converges absolutely to α and that $\displaystyle\sum_{n=0}^{\infty} b_n$ converges to β. Show that the Cauchy product converges to $\alpha\beta$.*

(iii) *Suppose that $\displaystyle\sum_{n=0}^{\infty} a_n$ and $\displaystyle\sum_{n=0}^{\infty} b_n$ converge absolutely to α and β respectively. Show that the Cauchy product also converges absolutely to $\alpha\beta$.*

(iv) *Give an example of two convergent series whose Cauchy product is divergent.*

The assertion (ii) is due to Mertens (1875). From a viewpoint of digital signal processing the Cauchy product is the infinite sum of the discrete convolution $\{a * b(n)\}_{n \in \mathbb{Z}}$ of two infinite sequences $\{a(n)\}_{n \in \mathbb{Z}}$ and $\{b(n)\}_{n \in \mathbb{Z}}$, which is defined by

$$a * b(n) = \sum_{k \in \mathbb{Z}} a(k) b(n - k).$$

PROBLEM 2.3

Suppose that $\displaystyle\sum_{n=1}^{\infty} a_n b_n$ converges for any sequence $\{b_n\}$ such that $\{b_n\}$ converges to 0. Show that $\displaystyle\sum_{n=1}^{\infty} a_n$ converges absolutely.

PROBLEM 2.4

The limit

$$\gamma = \lim_{n \to \infty} \left(1 + \frac{1}{2} + \cdots + \frac{1}{n} - \log n \right)$$

$$= 0.57721\,56649\,01532\,86060\,65120...$$

is called Euler's constant or sometimes the Euler-Mascheroni constant. Show that the following series converges to γ:

$$\frac{1}{2} - \frac{1}{3} + 2\left(\frac{1}{4} - \frac{1}{5} + \frac{1}{6} - \frac{1}{7} \right) + 3\left(\frac{1}{8} - \frac{1}{9} + \cdots - \frac{1}{15} \right) + \cdots .$$

Vacca (1910) proved this formula and stated that it is simple and has its natural place near to the Gregory-Leibniz series

$$\frac{\pi}{4} = 1 - \frac{1}{3} + \frac{1}{5} - \frac{1}{7} + \cdots$$

and Mercator's series

$$\log 2 = 1 - \frac{1}{2} + \frac{1}{3} - \frac{1}{4} + \cdots .$$

It is even not known whether γ is rational or irrational, though conjectured to be transcendental. Hence it is desirable to approximate γ by rational numbers. Hilbert mentioned that the irrationality of γ is an unsolved problem that seems unapproachable. Nowadays the numerical value of γ is computed to more than 100 million decimal places. Papanikolaou pointed out that, if γ were rational, then the denominator would have at least 242080 digits. Improbable!

PROBLEM 2.5

Making use of the formula

$$\frac{\sin(2n+1)\theta}{(2n+1)\sin\theta} = \prod_{k=1}^{n} \left(1 - \frac{\sin^2\theta}{\sin^2 k\pi/(2n+1)} \right),$$

show that

$$\frac{\sin \pi x}{\pi x} = \prod_{n=1}^{\infty} \left(1 - \frac{x^2}{n^2} \right)$$

holds for all real x.

The proof was given in §408 of Fichtenholz (1964). Kortram (1996) gave almost the same proof. This product representation of $\sin x$ is usually proved in Complex Analysis as an application of the canonical product of an entire function of order 1.

PROBLEM 2.6

Show that the convergence of $\displaystyle\sum_{n=1}^{\infty} n a_n$ implies that of $\displaystyle\sum_{n=1}^{\infty} a_n$.

PROBLEM 2.7

Show that

$$\lim_{n\to\infty} \frac{a_1 + a_2 + \cdots + a_n}{n} = 0$$

when $\displaystyle\sum_{n=1}^{\infty} \frac{a_n}{n}$ converges.

PROBLEM 2.8

Suppose that $a_n > 0$ for all $n \geq 1$. Show that $\displaystyle\sum_{n=1}^{\infty} a_n$ converges if and only if

$$\sum_{n=1}^{\infty} \frac{a_n}{a_1 + a_2 + \cdots + a_n}$$

converges.

PROBLEM 2.9

Let a_n be the sequence defined in **PROBLEM 1.11**. Show that

$$\sum_{n=1}^{\infty} a_n = 1 \quad and \quad \sum_{n=1}^{\infty} \frac{a_n}{n} = \gamma$$

where γ is Euler's constant.

PROBLEM 2.10

Let $\alpha, \beta > 1$ be any constant satisfying

$$\frac{1}{\alpha} + \frac{1}{\beta} = 1.$$

Suppose that $\displaystyle\sum_{n=1}^{\infty} |a_n b_n|$ converges for any sequence $\{b_n\}$ such that $\displaystyle\sum_{n=1}^{\infty} |b_n|^{\beta}$ converges. Then show that $\displaystyle\sum_{n=1}^{\infty} |a_n|^{\alpha}$ converges.

PROBLEM 2.11

For any positive sequence $\{a_n\}_{n\geq 1}$ show the inequality

$$\sum_{n=1}^{\infty} (a_1 a_2 \cdots a_n)^{1/n} < e \sum_{n=1}^{\infty} a_n.$$

Prove further that the constant e on the right-hand side cannot be replaced by any smaller number in general.

Carleman(1922) showed this inequality with equality sign for non-negative sequences using Lagrange multipliers. At least four other proofs are known. Pólya(1926) proved that the equality cannot occur unless all a_n's vanish. Knopp(1928) gave a simpler but technical proof using the arithmetic-geometric mean inequality. Carleson(1954) proved it as an application of the integral inequality

$$\int_0^{\infty} \exp\left(-\frac{f(x)}{x}\right) dx \leq e \int_0^{\infty} \exp(-f'(x))\, dx.$$

See **PROBLEM 9.11** for details. Redheffer(1967) also gave another proof by introducing further parameters into the problem. See a nice survey of Duncan and McGregor(2003) for details, in which they claimed that, in many proofs, the elementary inequality

$$\left(1 + \frac{1}{n}\right)^n < e$$

holding for all positive integers n, plays a significant role.

PROBLEM 2.12

For any positive sequence $\{a_n\}_{n\geq 1}$ show the inequality

$$\left(\sum_{n=1}^{\infty} a_n\right)^4 < \pi^2 \left(\sum_{n=1}^{\infty} a_n^2\right)\left(\sum_{n=1}^{\infty} n^2 a_n^2\right).$$

Prove further that the constant π^2 on the right-hand side cannot be replaced by any smaller number in general.

This is due to Carlson(1935), who also derived the integral version:

$$\left(\int_0^{\infty} f(x)\, dx\right)^4 \leq \pi^2 \left(\int_0^{\infty} f^2(x)\, dx\right)\left(\int_0^{\infty} x^2 f^2(x)\, dx\right). \qquad (2.1)$$

Note that the equality holds when

$$f(x) = \frac{1}{1 + x^2}.$$

To my surprise Carlson improved the inequality in the above problem using (2.1). For details see the remark after **SOLUTION 2.12**.

Solutions for Chapter 2

SOLUTION 2.1

Let E_i be the set of all positive integers whose decimal notation does not contain the digit 'i'. The decimal notation of any integer in the interval $[10^{k-1}, 10^k)$ has k digits. Among these the number of integers which lack the digit 'i' in their decimal notations is exactly $8 \cdot 9^{k-1}$ for $1 \le i \le 9$ and 9^k for $i = 0$. Thus we have

$$\sum_{n \in E_i} \frac{1}{n} < \sum_{k=1}^{\infty} \frac{9^k}{10^{k-1}} = 90$$

for every i. Hence the set

$$\mathbb{N} \setminus (E_0 \cup E_1 \cup \cdots \cup E_9)$$

which is the set of all positive integers whose decimal notation contains all the digits from 0 to 9, is dominant. $\qquad \square$

SOLUTION 2.2

(i) Since $a_n \to 0$ and $b_n \to 0$ as $n \to \infty$, both power series

$$f(x) = \sum_{n=0}^{\infty} a_n x^n \quad \text{and} \quad g(x) = \sum_{n=0}^{\infty} b_n x^n$$

converge absolutely for $|x| < 1$ and the product

$$f(x)g(x) = \sum_{n=0}^{\infty} (a_0 b_n + a_1 b_{n-1} + \cdots + a_n b_0) \, x^n$$

also converges for $|x| < 1$. Hence it follows from Abel's continuity theorem (see **PROBLEM 7.7**) that $f(x)$, $g(x)$ and $f(x)g(x)$ converge to α, β and $\alpha\beta$ as $x \to 1-$ respectively. Thus we have $\delta = \alpha\beta$.

(ii) Let

$$M = \sum_{n=0}^{\infty} |a_n| \quad \text{and} \quad s_n = b_0 + b_1 + \cdots + b_n$$

with $|s_n| \leq K$ for some constant $K > 0$. By (i) it suffices to show the convergence of the Cauchy product. To see this put

$$c_n = \sum_{k=0}^{n} (a_0 b_k + a_1 b_{k-1} + \cdots + a_k b_0)$$

$$= a_0 s_n + a_1 s_{n-1} + \cdots + a_n s_0.$$

For any $\epsilon > 0$ there exists an integer N satisfying $|s_p - s_q| < \epsilon$ and

$$|a_q| + \cdots + |a_p| < \epsilon$$

for any integers p and q with $p > q \geq N$. Then for all $p > q > 2N$ we get

$$|c_p - c_q| = \left| \sum_{k=0}^{N} a_k(s_{p-k} - s_{q-k}) + \sum_{k=N+1}^{q} a_k(s_{p-k} - s_{q-k}) + \sum_{k=q+1}^{p} a_k s_{p-k} \right|.$$

Clearly the first sum on the right-hand side is estimated above by

$$M \max_{0 \leq k \leq N} |s_{p-k} - s_{q-k}| < M\epsilon.$$

Similarly the second and third sums are estimated above by

$$2K \sum_{k=N+1}^{p} |a_k| < 2K\epsilon$$

respectively. Thus the sequence $\{c_n\}$ satisfies the Cauchy criterion.

(iii) The Cauchy product of $\sum_{n=0}^{\infty} |a_n|$ and $\sum_{n=0}^{\infty} |b_n|$ converges by (ii).

(iv) For example, put

$$a_n = b_n = \frac{(-1)^n}{\sqrt{n+1}}.$$

Obviously $\sum_{n=0}^{\infty} a_n$ and $\sum_{n=0}^{\infty} b_n$ converge. However,

$$|a_0 b_n + a_1 b_{n-1} + \cdots + a_n b_0| = \sum_{k=1}^{n+1} \frac{1}{\sqrt{k(n+2-k)}}$$

$$\geq \sum_{k=1}^{n+1} \frac{2}{k+n+2-k},$$

which shows the divergence of the Cauchy product, because the last sum is greater than 1. $\qquad\square$

SOLUTION 2.3

Without loss of generality we can assume that $\{a_n\}$ is a positive sequence, because we can take $-b_n$ instead of b_n if necessary and remove a zero term if it exists. Suppose, on the contrary, that $\sum_{n=1}^{\infty} a_n = \infty$. Then it follows from **PROBLEM 2.8** that

$$\sum_{n=1}^{\infty} \frac{a_n}{a_1 + a_2 + \cdots + a_n} = \infty.$$

However this is contrary to the assumption, because

$$b_n = \frac{1}{a_1 + a_2 + \cdots + a_n}$$

converges to 0 as $n \to \infty$.   □

SOLUTION 2.4

For brevity put

$$\sigma_n = 1 - \frac{1}{2} + \frac{1}{3} - \cdots + \frac{1}{2^n - 1} - \log 2$$

for any positive integer n. It is easily seen that

$$\sigma_1 + \sigma_2 + \cdots + \sigma_n = 1 + \frac{1}{2} + \cdots + \frac{1}{2^n - 1} - n \log 2;$$

therefore

$$\gamma = \lim_{n \to \infty} (\sigma_1 + \sigma_2 + \cdots + \sigma_n).$$

We have Vacca's formula by noticing that $\sigma_n = \tau_n + \tau_{n+1} + \cdots$, where

$$\tau_n = \frac{1}{2^n} - \frac{1}{2^n + 1} + \frac{1}{2^n + 2} - \cdots - \frac{1}{2^{n+1} - 1}.$$ □

REMARK. Hardy (1912) pointed out that Vacca's formula may be deduced from the formula

$$\gamma = 1 - \int_0^1 \frac{F(x)}{1 + x} \, dx$$

found by Catalan (1875), where $F(x) = \sum\limits_{n=1}^{\infty} x^{2^n}$, because Vacca's formula can be written in the form

$$\gamma = \int_0^1 \left(\frac{x - x^3}{1 + x} + 2\frac{x^3 - x^7}{1 + x} + 3\frac{x^7 - x^{15}}{1 + x} + \cdots \right) dx$$

$$= \int_0^1 \frac{F(x)}{x(1 + x)} dx$$

and

$$\int_0^1 \frac{F(x)}{x} dx = \sum_{n=1}^{\infty} \frac{1}{2^n} = 1.$$

The power series $F(x)$ satisfies a simple functional equation

$$F(x) = x^2 + F(x^2),$$

which is a typical example of Mahler's functional equations in the theory of transcendental numbers.

SOLUTION 2.5

Since it is easily verified that $\sin(2n+1)\theta$ is a polynomial of $\sin\theta$ by induction, we can write

$$p_n(\sin^2\theta) = \frac{\sin(2n + 1)\theta}{\sin\theta},$$

where $p_n(x)$ is a polynomial of degree n satisfying $p_n(0) = 2n + 1$. The zeros of $p_n(x)$ can be obtained by solving the equation $\sin(2n + 1)\theta = 0$ with $\sin\theta \neq 0$; so we get the following n zeros of p_n:

$$\sin^2\xi_{1,n} < \sin^2\xi_{2,n} < \cdots < \sin^2\xi_{n,n}$$

in the interval $(0, 1)$ where $\xi_{k,n} = k\pi/(2n + 1) \in (0, \pi/2)$. Hence we have

$$\sin(2n + 1)\theta = (2n + 1)\sin\theta \prod_{k=1}^{n} \left(1 - \frac{\sin^2\theta}{\sin^2\xi_{k,n}} \right).$$

Substituting $x = (2n + 1)\theta/\pi$, we obtain

$$\frac{\sin\pi x}{\pi x} \cdot \frac{x_n}{\sin x_n} = \prod_{k=1}^{n} \left(1 - \frac{\sin^2 x_n}{\sin^2\xi_{k,n}} \right)$$

where $x_n = \pi x/(2n + 1)$.

We can assume that x is not an integer; otherwise the expression clearly holds.

Take any positive integers n and m satisfying $n > m > |x|$ so that $|x_n| < \xi_{m,n}$. Putting

$$\eta_{m,n} = \prod_{k=m+1}^{n} \left(1 - \frac{\sin^2 x_n}{\sin^2 \xi_{k,n}} \right),$$

we get

$$\lim_{n \to \infty} \frac{1}{\eta_{m,n}} = \frac{\pi x}{\sin \pi x} \prod_{k=1}^{m} \left(1 - \frac{x^2}{k^2} \right).$$

On the other hand, we have

$$1 > \eta_{m,n} \geq 1 - \sum_{k=m+1}^{n} \frac{\sin^2 x_n}{\sin^2 \xi_{k,n}},$$

because

$$(1 - \alpha_1) \cdots (1 - \alpha_n) \geq 1 - \alpha_1 - \cdots - \alpha_n$$

for any $0 \leq \alpha_k < 1$ and any positive integer n. Using now the inequalities

$$\frac{2\theta}{\pi} < \sin \theta < \theta$$

holding for $0 < \theta < \pi/2$, we have

$$\eta_{m,n} \geq 1 - \frac{\pi^2}{4} \sum_{k=m+1}^{n} \frac{x_n^2}{\xi_{k,n}^2} \geq 1 - \frac{\pi^2 x^2}{4m},$$

because

$$\sum_{k=m+1}^{\infty} \frac{1}{k^2} < \frac{1}{m}.$$

Hence, as a sequence in m, $\lim_{n \to \infty} \eta_{m,n}$ converges to 1 as $m \to \infty$. $\qquad \square$

SOLUTION 2.6

Put

$$b_n = a_1 + 2a_2 + \cdots + na_n$$

and let M be the least upper bound of $|b_n|$. For any $\epsilon > 0$ and any integers p, q with $p > q > 2M/\epsilon$ we have

$$
\left| \sum_{n=q}^{p} a_n \right| = \left| \frac{b_q - b_{q-1}}{q} + \frac{b_{q+1} - b_q}{q+1} + \cdots + \frac{b_p - b_{p-1}}{p} \right|
$$

$$
= \left| -\frac{b_{q-1}}{q} + \left(\frac{1}{q} - \frac{1}{q+1} \right) b_q + \cdots + \left(\frac{1}{p-1} - \frac{1}{p} \right) b_{p-1} + \frac{b_p}{p} \right|
$$

$$
\leq \frac{2M}{q},
$$

which is less than ϵ. This is nothing but the Cauchy criterion for the convergence of the series $\sum_{n=1}^{\infty} a_n$. $\qquad\square$

SOLUTION 2.7

Put

$$
\frac{a_1}{1} + \frac{a_2}{2} + \cdots + \frac{a_n}{n} = \alpha + \epsilon_n \quad \text{and} \quad \sigma_n = \frac{a_1 + a_2 + \cdots + a_n}{n},
$$

where ϵ_n is a sequence converging to 0 as $n \to \infty$.

We now show that

$$
\sigma_n = \frac{\alpha}{n} - \frac{\epsilon_1 + \epsilon_2 + \cdots + \epsilon_{n-1}}{n} + \epsilon_n \tag{2.2}
$$

by induction on n. The case $n = 1$ is clear, because $\sigma_1 = a_1 = \alpha + \epsilon_1$. Suppose next that (2.2) holds for $n = m$. We then have

$$
\sigma_{m+1} = \frac{m}{m+1} \sigma_m + \frac{a_{m+1}}{m+1}
$$

$$
= \frac{m}{m+1} \left(\frac{\alpha}{m} - \frac{\epsilon_1 + \cdots + \epsilon_{m-1}}{m} + \epsilon_m \right) + \epsilon_{m+1} - \epsilon_m
$$

$$
= \frac{\alpha}{m+1} - \frac{\epsilon_1 + \cdots + \epsilon_m}{m+1} + \epsilon_{m+1},
$$

as required. Obviously (2.2) implies that σ_n converges to 0 as $n \to \infty$. $\qquad\square$

SOLUTION 2.8

Put $s_n = a_1 + a_2 + \cdots + a_n$. If s_n converges to $\alpha > 0$, then there exists an integer N such that $s_n > \alpha/2$ for all $n \geq N$; so,

$$
\sum_{n \geq N} \frac{a_n}{s_n} < \frac{2}{\alpha} \sum_{n \geq N} a_n < \infty.
$$

Suppose next that $s_n \to \infty$ as $n \to \infty$. If $s_{n-1} < a_n$ for infinitely many n's, then we have

$$\frac{a_n}{s_n} = \frac{a_n}{a_n + s_{n-1}} > \frac{1}{2}$$

for infinitely many n's, which implies $\sum a_n/s_n = \infty$. If $s_{n-1} \geq a_n$ for all n greater than some integer N, then

$$\frac{a_n}{s_{n-1}} = \frac{1}{s_{n-1}} \int_{s_{n-1}}^{s_n} 1\, dx > \int_{s_{n-1}}^{s_n} \frac{dx}{x} \, ;$$

hence, using $2s_{n-1} \geq s_n$ we have

$$\sum_{n>N} \frac{a_n}{s_n} \geq \frac{1}{2} \sum_{n>N} \frac{a_n}{s_{n-1}} \geq \frac{1}{2} \int_{s_N}^{\infty} \frac{dx}{x} = \infty. \qquad \square$$

SOLUTION 2.9

Let $f(x)$ and $\mu(s)$ be the same functions defined in **SOLUTION 1.11**. Since $\mu(s)/s(1-s)$ is an integrable function on $(0, 1)$, we have

$$\sum_{n=1}^{\infty} a_n = \int_0^1 \frac{\mu(s)}{s(1-s)}\, ds < \infty.$$

Therefore it follows from Abel's continuity theorem that

$$\sum_{n=1}^{\infty} a_n = \lim_{x \to 1-} f(x)$$

$$= 1 + \lim_{x \to 1-} \frac{1}{\log(1-x)} = 1.$$

We next have

$$\sum_{n=1}^{\infty} \frac{a_n}{n} = \int_0^1 f(x)\, dx$$

$$= \int_0^1 \left(\frac{1}{x} + \frac{1}{\log(1-x)} \right) dx,$$

which is equal to

$$1 + \int_0^1 \left(\frac{x}{1-x} + \frac{1}{\log x} \right) dx = 1 + \sum_{n=1}^{\infty} \int_0^1 x^{n-1} \left(x + \frac{1-x}{\log x} \right) dx$$

$$= 1 + \sum_{n=1}^{\infty} \left(\frac{1}{n+1} + \int_0^1 \frac{x^{n-1} - x^n}{\log x}\, dx \right).$$

Since

$$\int_0^1 \frac{x^{n-1} - x^n}{\log x} \, dx = -\int_0^1 \frac{e^{(n-1+y)\log x}}{\log x} \bigg|_{y=0}^{y=1} dx$$

$$= -\int_0^1\!\!\int_0^1 x^{n-1+y} \, dx\, dy$$

$$= -\int_0^1 \frac{dy}{n+y} = -\log\left(1 + \frac{1}{n}\right),$$

we get

$$\sum_{n=1}^{\infty} \frac{a_n}{n} = 1 + \lim_{N \to \infty}\left(\frac{1}{2} + \frac{1}{3} + \cdots + \frac{1}{N} - \log N\right) = \gamma,$$

as required. $\qquad\qquad\square$

SOLUTION 2.10

Put $\sigma_n = \sum_{k=1}^{n} |a_k|^{\alpha}$ for brevity. Suppose, on the contrary, that σ_n diverges to $+\infty$. We can assume that $a_n \neq 0$ for all n. It follows from **PROBLEM 2.8** that

$$\sum_{n=1}^{\infty} \frac{|a_n|^{\alpha}}{\sigma_n} = \infty.$$

Now put

$$b_n = \frac{|a_n|^{\alpha-1}}{\sigma_n} = \frac{|a_n|^{\alpha/\beta}}{\sigma_n}.$$

Then we have $\sum_{n=1}^{\infty} |b_n|^{\beta} < \infty$, because

$$|b_n|^{\beta} = \frac{|a_n|^{\alpha}}{\sigma_n^{\beta}} = \frac{1}{\sigma_n^{\beta}} \int_{\sigma_{n-1}}^{\sigma_n} 1 \, dx < \int_{\sigma_{n-1}}^{\sigma_n} \frac{dx}{x^{\beta}}.$$

However this is contrary to the assumption, because

$$\sum_{n=1}^{\infty} |a_n b_n| = \sum_{n=1}^{\infty} \frac{|a_n|^{\alpha}}{\sigma_n} = \infty. \qquad\qquad\square$$

SOLUTION 2.11

The proof is based on Pólya (1926). Let $\{b_n\}$ be an arbitrary positive sequence. First we write

$$\sum_{n=1}^{m} (a_1 a_2 \cdots a_n)^{1/n} = \sum_{n=1}^{m} \left(\frac{a_1 b_1 a_2 b_2 \cdots a_n b_n}{b_1 b_2 \cdots b_n}\right)^{1/n}.$$

Using the arithmetic-geometric mean inequality, the sum on the right-hand side is less than or equal to

$$\sum_{n=1}^{m} \frac{1}{(b_1 b_2 \cdots b_n)^{1/n}} \cdot \frac{a_1 b_1 + a_2 b_2 + \cdots + a_n b_n}{n}$$

$$= \sum_{k=1}^{m} a_k b_k \sum_{n=k}^{m} \frac{1}{n(b_1 b_2 \cdots b_n)^{1/n}} \cdot$$

We now take

$$b_n = n\left(1 + \frac{1}{n}\right)^n$$

so that $(b_1 b_2 \cdots b_n)^{1/n} = n + 1$. Therefore

$$\sum_{k=1}^{m} a_k b_k \sum_{n=k}^{m} \frac{1}{n(n+1)} = \sum_{k=1}^{m} a_k b_k \left(\frac{1}{k} - \frac{1}{m+1}\right)$$

$$< \sum_{k=1}^{m} a_k \left(1 + \frac{1}{k}\right)^k$$

and this is smaller than $e \sum_{k=1}^{m} a_k$. Note that the equality does not occur in the original inequality when $m \to \infty$.

To see that e cannot be replaced by any smaller number, we take, for example,

$$a_n = \begin{cases} 1/n & (1 \le n \le m) \\ 2^{-n} & (n > m) \end{cases}$$

where m is an integer parameter. It is not hard to see that

$$\sum_{n=1}^{\infty} (a_1 a_2 \cdots a_n)^{1/n} = \sum_{n=1}^{m} n!^{-1/n} + O(1)$$

$$= e \log m + O(1)$$

and

$$\sum_{n=1}^{\infty} a_n = 1 + \sum_{n=1}^{m} \frac{1}{n} = \log m + O(1).$$

Therefore the ratio of the above two sums converges to e as $m \to \infty$. $\qquad \square$

SOLUTION 2.12

Put $\alpha = \sum_{n=1}^{\infty} a_n^2$ and $\beta = \sum_{n=1}^{\infty} n^2 a_n^2$ for brevity. We of course assume that β is finite. Introducing two positive parameters σ and τ we can write

$$\left(\sum_{n=1}^{\infty} a_n\right)^2 = \left(\sum_{n=1}^{\infty} \frac{a_n\sqrt{\sigma + \tau n^2}}{\sqrt{\sigma + \tau n^2}}\right)^2.$$

By the Cauchy-Schwarz inequality, this is less than or equal to

$$\sum_{n=1}^{\infty} \frac{1}{\sigma + \tau n^2} \sum_{n=1}^{\infty} a_n^2(\sigma + \tau n^2) = (\alpha\sigma + \beta\tau) \sum_{n=1}^{\infty} \frac{1}{\sigma + \tau n^2}.$$

Since $1/(\sigma + \tau x^2)$ is a monotone decreasing function on $x \in [0, \infty)$, we obtain

$$\sum_{n=1}^{\infty} \frac{1}{\sigma + \tau n^2} < \int_0^{\infty} \frac{dx}{\sigma + \tau x^2} = \frac{\pi}{2\sqrt{\sigma\tau}},$$

which implies that

$$\left(\sum_{n=1}^{\infty} a_n\right)^2 < \frac{\pi}{2} \cdot \frac{\alpha\sigma + \beta\tau}{\sqrt{\sigma\tau}}.$$

The right-hand side, as a function in σ and τ, attains its minimum $\pi\sqrt{\alpha\beta}$ at $\sigma/\tau = \beta/\alpha$.

To see that π^2 cannot be replaced by any smaller number, we take, for example,

$$a_n = \frac{\sqrt{\rho}}{\rho + n^2}$$

where ρ is a positive parameter. It then follows from

$$\sqrt{\rho} \int_1^{\infty} \frac{dx}{\rho + x^2} < \sum_{n=1}^{\infty} a_n < \sqrt{\rho} \int_0^{\infty} \frac{dx}{\rho + x^2}$$

that

$$\sum_{n=1}^{\infty} a_n = \frac{\pi}{2} + O\left(\frac{1}{\sqrt{\rho}}\right)$$

as $\rho \to \infty$. Similarly we have

$$\sum_{n=1}^{\infty} a_n^2 = \frac{\pi}{4\sqrt{\rho}} + O\left(\frac{1}{\rho}\right)$$

and

$$\sum_{n=1}^{\infty} n^2 a_n^2 = \sqrt{\rho} \sum_{n=1}^{\infty} a_n - \rho \sum_{n=1}^{\infty} a_n^2 = \frac{\pi}{4} \sqrt{\rho} + O(1),$$

which imply that

$$\left(\sum_{n=1}^{\infty} a_n \right)^4 = \frac{\pi^4}{16} + O\left(\frac{1}{\sqrt{\rho}} \right),$$

and

$$\sum_{n=1}^{\infty} a_n^2 \sum_{n=1}^{\infty} n^2 a_n^2 = \frac{\pi^2}{16} + O\left(\frac{1}{\sqrt{\rho}} \right)$$

as $\rho \to \infty$. $\qquad\qquad \square$

REMARK. Carlson (1935) obtained the following sharper inequality:

$$\left(\sum_{n=1}^{\infty} a_n \right)^4 < \pi^2 \sum_{n=1}^{\infty} a_n^2 \sum_{n=1}^{\infty} \left(n^2 - n + \frac{7}{16} \right) a_n^2. \tag{2.3}$$

To see this consider the function

$$f_N(x) = e^{-x/2} \sum_{n=1}^{N} (-1)^{n-1} a_n L_{n-1}(x)$$

where

$$L_n(x) = \frac{e^x}{n!} \frac{d^n}{dx^n} (x^n e^{-x}) = \sum_{k=0}^{n} \binom{n}{k} \frac{(-x)^k}{k!}$$

is called the nth Laguerre polynomial, which forms a system of orthogonal polynomials over $(0, \infty)$ with weight function e^{-x}. We write the corresponding inner product, as follows:

$$\langle f, g \rangle = \int_0^{\infty} f(x) g(x) e^{-x} \, dx.$$

We then have

$$\int_0^{\infty} f_N(x) \, dx = 2 \sum_{n=1}^{N} a_n,$$

$$\int_0^{\infty} f_N^2(x) \, dx = \sum_{m,n=1}^{N} (-1)^{m+n} a_m a_n \langle L_{m-1}, L_{n-1} \rangle = \sum_{n=1}^{N} a_n^2$$

and

$$\int_0^\infty x^2 f^2(x)\, dx = \sum_{m,n=1}^N (-1)^{m+n} a_m a_n \langle x^2 L_{m-1}, L_{n-1} \rangle$$

$$= 2 \sum_{n=3}^N (n-1)(n-2) a_n a_{n-2}$$

$$+ 8 \sum_{n=2}^N (n-1)^2 a_n a_{n-1} + 2 \sum_{n=1}^N (3n^2 - 3n + 1) a_n^2.$$

Using the inequality (2.1) in the remark after **PROBLEM 2.12** and letting $N \to \infty$ we get a certain inequality with the sign of equality.

We next use the Cauchy-Schwarz inequality for each n so that

$$\sum_{n=3}^\infty (n-1)(n-2) a_n a_{n-2} < \sum_{n=1}^\infty \frac{(n-1)^2 + n^2}{2} a_n^2$$

and

$$\sum_{n=2}^\infty (n-1)^2 a_n a_{n-1} < \sum_{n=1}^\infty \frac{(n-1)^2 + n^2}{2} a_n^2,$$

which imply Carlson's improved inequality (2.3). Note that both inequalities exclude the sign of equality.

The constant $7/16$ in (2.3) may be replaced by $3/8$, because

$$\sum_{n=2}^\infty (n-1)^2 a_n a_{n-1} = \sum_{n=2}^\infty (n-1)\left(n - \frac{3}{2}\right) a_n a_{n-1} + \frac{1}{2} \sum_{n=2}^\infty (n-1) a_n a_{n-1}$$

$$< \sum_{n=1}^\infty \left(\frac{(n-1)^2 + (n-1/2)^2}{2} + \frac{1}{2} \cdot \frac{n-1+n}{2} \right) a_n^2$$

$$= \sum_{n=1}^\infty \left(n^2 - n + \frac{3}{8} \right) a_n^2.$$

Chapter 3

Continuous Functions

[Summary of Basic Points]

1. A function $f(x)$ defined on an interval $I \subset \mathbb{R}$ is said to be *uniformly continuous* on I provided that for any $\epsilon > 0$ there exists a positive number δ such that $|f(x) - f(x')| < \epsilon$ whenever $x, x' \in I$ and $|x - x'| < \delta$.

2. A sequence of functions $\{f_n(x)\}$ defined on an open interval I is said to converge to $f(x)$ *uniformly on compact sets* provided that it converges uniformly on any compact set K contained in I; that is,

$$\sup_{x \in K} |f_n(x) - f(x)|$$

converges to 0 as $n \to \infty$.

3. If a sequence of continuous functions converges to $f(x)$ uniformly on compact sets in I, then $f(x)$ is continuous on I.

4. The pointwise convergence of a sequence of continuous functions does not imply the continuity of the limit function in general. For example, Dirichlet's function

$$\lim_{n \to \infty} \left(\lim_{m \to \infty} (\cos(n!\pi x))^m \right)$$

takes on the value 1 at every rational point and 0 at every irrational point.

5. Let $\{f_n(x)\}$ be a sequence of continuous functions defined on an interval I, converging pointwisely to the function $f(x)$. R. Baire(1874–1932) showed that the set of continuity points of $f(x)$ is dense in I.

6. The image of an interval under a continuous function is also an interval.

7. A point x_0 is said to be a *discontinuity point of the first kind* of $f(x)$ provided that both right and left limits exist but at least two among $f(x_0)$, $f(x_0+)$, $f(x_0-)$ are different from each other.

8. A function $f(x)$ defined on I is said to be *piecewise continuous* if the set of discontinuity points of f consists of finitely many points of the first kind. A piecewise continuous function is said to be *piecewise linear* if it is linear on each subinterval.

9. A function $f(x)$ defined on I is said to be a *Lipschitz function* with constant $L > 0$ provided that

$$|f(x) - f(y)| \le L|x - y|$$

for every x, y in I. Any Lipschitz function is clearly uniformly continuous on I. The least constant L with which f satisfies the above Lipschitz condition is said to be the Lipschitz constant of f.

10. For a function $f(x)$ on I the *support* of f is defined as the closure of the set $\{x \in I \mid f(x) \ne 0\}$ and is denoted by $\mathrm{supp}(f)$.

11. If a function $f(x)$ defined on $\mathbb{R}$ satisfies $f(x + c) = f(x)$ for all $x \in \mathbb{R}$ for some constant $c \ne 0$, then $f(x)$ is called a *periodic function* with period c. For example, Dirichlet's function is a periodic function whose period is any non-zero rational number.

＊ ∽ ＊ ∾ ＊

PROBLEM 3.1

Suppose that $f(x)$ defined on $\mathbb{R}$ is not a constant function and possesses an arbitrarily small period. Show that f is discontinuous everywhere.

PROBLEM 3.2

Suppose that $f \in \mathscr{C}(\mathbb{R})$ is periodic but not a constant function. Show that f has the least positive period τ and every period of f is an integral multiple of τ.

PROBLEM 3.3

Suppose that $f \in \mathscr{C}(\mathbb{R})$ satisfies

$$\lim_{x \to \infty} (f(x + 1) - f(x)) = 0.$$

Show that $\dfrac{f(x)}{x}$ also converges to 0 as $x \to \infty$.

PROBLEM 3.4

Suppose that $f \in \mathscr{C}(\mathbb{R})$ satisfies

$$\lim_{x \to \infty} (f(x+y) - f(x)) = 0$$

pointwisely in $y \in \mathbb{R}$. Show that the convergence is uniform on compact sets in $\mathbb{R}$.

PROBLEM 3.5

Let $c_1, c_2, \ldots, c_n$ and $\lambda_1, \lambda_2, \ldots, \lambda_n$ be real numbers with $\lambda_j \neq \lambda_k$ for all $j \neq k$. Show that $c_1 = c_2 = \cdots = c_n = 0$ if

$$\sum_{k=1}^{n} c_k \exp(\lambda_k i x)$$

converges to 0 as $x \to \infty$.

PROBLEM 3.6

Suppose that $f \in \mathscr{C}[0, \infty)$ and $f(nx)$ converges to 0 as $n \to \infty$ for each $x > 0$. Prove or disprove that $f(x)$ converges to 0 as $x \to \infty$.

PROBLEM 3.7

Let $f_n \in \mathscr{C}[a, b]$ be a monotone increasing sequence

$$f_1(x) \leq f_2(x) \leq \cdots,$$

which converges pointwisely to $f(x) \in \mathscr{C}[a, b]$. Show that the convergence is uniform on $[a, b]$.

This is known as Dini's theorem, which can be seen in Dini (1892). It will be generalized to a monotone increasing sequence of continuous functions defined on a compact topological space.

PROBLEM 3.8

Let $P(x)$ be a non-constant monic polynomial and let I be any closed interval of length ≥ 4. Show that there exists at least one point x in I satisfying

$$|P(x)| \geq 2.$$

PROBLEM 3.9

Suppose that three functions f, g defined on $(0, \infty)$ and h defined on $\mathbb{R}$ satisfy

$$h(f(x) + g(y)) = xy$$

for all $x, y > 0$. For example, $f(x) = g(x) = c \log x$ and $h(x) = e^{x/c}$ satisfy the relation for any constant $c > 0$. Show, however, that if we assume in addition that

$$f(x) = o\left(\log \frac{1}{x}\right) \quad as \quad x \to 0+,$$

then the set of discontinuity points of g is dense in $(0, \infty)$.

---

12. First put $E_1 = \{0, 1\}$ and suppose that a finite sequence $E_n \subset [0, 1]$ is given. Define E_{n+1} by inserting new fractions $(a + c)/(b + d)$ between every two consecutive fractions a/b and c/d in E_n. Of course we understand that $0 = 0/1$ and $1 = 1/1$. The sequence E_n consists of $2^{n-1} + 1$ terms. For example,

$$E_2 = \left\{0, \frac{1}{2}, 1\right\}, \quad E_3 = \left\{0, \frac{1}{3}, \frac{1}{2}, \frac{2}{3}, 1\right\}, \quad \dots$$

In this context, the sequence of all reduced fractions with denominators not exceeding n, listed in order of their size, is called the *Farey sequence* or *Farey series* of order n. See, for example, Niven and Zuckerman (1960) for details.

13. Let a/b and c/d be any two consecutive fractions in E_n. We call the interval $J = [a/b, c/d]$ a *Farey interval* of order n, although we do not here impose any restriction on the size of denominators like the Farey series. We also define

$$J_L = \left[\frac{a}{b}, \frac{a+c}{b+d}\right] \quad and \quad J_R = \left[\frac{a+c}{b+d}, \frac{c}{d}\right].$$

The set of all Farey intervals of order n is denoted by $\mathcal{I}_n$. For each Farey interval $J = [a/b, c/d]$ we define

$$\phi_J(x) = \frac{b+d}{2}(|a - bx| + |c - dx| - |a + c - (b+d)x|),$$

being a piecewise linear continuous function satisfying $\text{supp}(\phi_J) = J$ and

$$\phi_J\left(\frac{a+c}{b+d}\right) = 1.$$

Clearly $\phi_J(x)$ is linear on J_L and on J_R, as illustrated in Fig. 3.1.

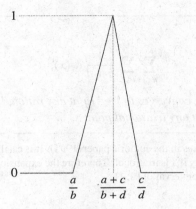

Fig. 3.1 The graph of ϕ_J.

14. For any Farey interval $J = [a/b, c/d]$ of order n we have $bc - ad = 1$ and $b + d \geq n + 1$, which can be easily shown by induction on the order.

PROBLEM 3.10

For any $f \in \mathscr{C}[0,1]$ and Farey interval $J = [a/b, c/d]$ we define

$$c_J(f) = f\left(\frac{a+c}{b+d}\right) - \frac{b}{b+d} f\left(\frac{a}{b}\right) - \frac{d}{b+d} f\left(\frac{c}{d}\right)$$

and put

$$\Phi_n(x) = \sum_{J \in I_n} c_J(f)\phi_J(x).$$

Then show that the series

$$f(0) + (f(1) - f(0))x + \sum_{n=1}^{\infty} \Phi_n(x)$$

converges to $f(x)$ uniformly on $[0,1]$.

This means that the system of piecewise linear functions

$$1, \ x, \ \{\phi_J(x)\}_{J \in I_1}, \ \{\phi_J(x)\}_{J \in I_2}, \ \cdots$$

forms a *Schauder basis* of $\mathscr{C}[0,1]$. See Schauder (1928) for details. Another system of piecewise linear functions, known as Faber-Schauder system, associates with dyadic rationals in $[0,1]$, instead of all rationals. See Faber (1910).

PROBLEM 3.11 ────────────────────────────────────

Show that the series

$$\sum_{n=1}^{\infty} \sum_{J \in I_n} \frac{1}{b+d}\, \phi_J(x),$$

where $J = [a/b, c/d]$, converges to $1 - 1/q$ at any rational point $x = p/q$ with $(p, q) = 1$, and to 1 at any irrational point x.

This expansion arises in the author's paper (1995). It is easily seen that $f(x)$ is continuous if and only if x is irrational. Therefore the expansion never converges uniformly on any subinterval of $[0, 1]$.

PROBLEM 3.12 ────────────────────────────────────

Put

$$\Psi_n(x) = \sum_{J \in I_n} \phi_J(x).$$

Let α be a quadratic irrational in the interval $(0, 1)$ and $Ax^2 + Bx + C$ be its minimal polynomial. Show that

$$\liminf_{n \to \infty} \Psi_n(\alpha) \geq \frac{1}{\sqrt{B^2 - 4AC}}.$$

✦   ❊   ✦

Solutions for Chapter 3

SOLUTION 3.1

Suppose, on the contrary, that $f(x)$ is continuous at some x_0. Since f is not a constant function, there exists a point x_1 with $f(x_0) \neq f(x_1)$. By the continuity of f at $x = x_0$, there exists a $\delta > 0$ such that

$$|f(x) - f(x_0)| < |f(x_1) - f(x_0)|$$

for any x in the interval $(x_0 - \delta, x_0 + \delta)$. This means that any period c of f must satisfy $|c| > 2\delta$, contrary to the assumption. $\square$

SOLUTION 3.2

Let E be the set of all periods of f. It follows from the previous solution that

$$\tau = \inf_{c \in E} |c| > 0.$$

We first show that E is closed, which will imply that τ is the least positive period of f. Let $\{c_n\}$ be any sequence in E converging to τ. Since $f \in \mathscr{C}(\mathbb{R})$, we have

$$f(x + \tau) = \lim_{n \to \infty} f(x + c_n) = f(x)$$

for any x; so $\tau \in E$.

For any $c \in E$, take an integer n satisfying $n\tau \leq c < (n + 1)\tau$. If $n\tau < c$, then $c' = c - n\tau \in E \cap (0, \tau)$, a contradiction. Thus we have $c = n\tau$ for some n. $\square$

SOLUTION 3.3

For any $\epsilon > 0$ there exists an integer N satisfying

$$-\epsilon < f(x + 1) - f(x) < \epsilon$$

for any $x > N$. Summing the following $\ell = [x] - N$ inequalities

$$-\epsilon < f(x - j + 1) - f(x - j) < \epsilon$$

for $j = 1, \ldots, \ell$ and for $x \geq N + 1$, we get

$$-\epsilon([x] - N) < f(x) - f(x - \ell) < \epsilon([x] - N).$$

Since $N \leq x - \ell < N + 1$, it follows that

$$-M_N - \epsilon(x - N + 1) < f(x) < M_N + \epsilon(x - N),$$

where M_N is the maximum of $|f(x)|$ on the closed interval $[N, N + 1]$. Therefore

$$\left| \frac{f(x)}{x} \right| < \frac{M_N + \epsilon(x - N + 1)}{x} < \epsilon + \frac{M_N}{x},$$

which implies that $|f(x)/x| < 2\epsilon$ for any $x > \max(N + 1, M_N/\epsilon)$. $\square$

SOLUTION 3.4

For any positive integer n we put

$$g_n(y) = \sup_{x \geq n} |f(x + y) - f(x)|.$$

Clearly $\{g_n(y)\}$ is a monotonously decreasing sequence converging to 0 as $n \to \infty$ pointwisely in $y \in \mathbb{R}$. The difficulty of this problem may lie in a situation that we do not know whether $g_n(y)$ is continuous or not.

If $g_n(y_0) > s$ for some y_0, then there exists $x_0 \geq n$ satisfying

$$|f(x_0 + y_0) - f(x_0)| > s.$$

By the continuity of f we have $|f(x + y) - f(x)| > s$ for any (x, y) sufficiently close to (x_0, y_0). This means that $g_n(y) > s$ for any y sufficiently close to y_0; in other words, the set

$$\{y \in \mathbb{R} \mid g_n(y) > s\}$$

is open. Hence $\{g_n(y)\}$ is a sequence of Borel measurable functions converging pointwisely to 0. By Egoroff's theorem we can find a measurable set F in $[-1, 1]$, whose measure is greater than $3/2$, such that $g_n(y)$ converges to 0 uniformly on F; therefore for any $\epsilon > 0$ there exists an integer N satisfying $g_n(y) < \epsilon$ for all $n > N$ and any $y \in F$. Since $E = F \cap (-F)$ has a positive measure, it follows from the theorem due to Steinhaus (1920) that 0 is an interior point of $E - E$. Thus an interval $I = [-r, r]$ with some $r > 0$ is contained in $E - E$ and any point $y \in I$ can be expressed as $y = u - v$ with $u, v \in E$.

Note that for any $y \geq 0$ and $y' \in \mathbb{R}$ we have

$$g_n(y + y') = \sup_{x \geq n} \left| f(x + y + y') - f(x) \right|$$

$$\leq \sup_{x \geq n} |f(x + y) - f(x)| + \sup_{x \geq n} \left| f(x + y + y') - f(x + y) \right|$$

$$\leq g_n(y) + g_n(y').$$

Applying the above inequality to $y = u - v \in [0, r]$ we get

$$g_n(y) = g_n(u - v) \le g_n(u) + g_n(-v) < 2\epsilon$$

for all $n > N$, because either $u \in F$ or $-v \in F$ is non-negative. Thus $\{g_n(y)\}$ converges to 0 uniformly on $[0, r]$. This is also true on any interval $[c, c + r]$. $\square$

<hr>

| SOLUTION 3.5 |

Put

$$f(x) = \sum_{k=1}^{n} c_k \exp(\lambda_k i x).$$

For any $\epsilon > 0$ we can find a sufficiently large integer N satisfying $|f(x)| < \epsilon$ for all x greater than N. For each $1 \le k \le n$ we have

$$\frac{1}{T} \int_T^{2T} f(x) \exp(-\lambda_k i x) \, dx$$

$$= c_k + \frac{1}{T} \sum_{\ell \ne k} c_\ell \int_T^{2T} \exp((\lambda_\ell - \lambda_k) i x) \, dx$$

$$= c_k + \frac{1}{T} \sum_{\ell \ne k} c_\ell \frac{\exp(2(\lambda_\ell - \lambda_k) i T) - \exp((\lambda_\ell - \lambda_k) i T)}{(\lambda_\ell - \lambda_k) i}.$$

Therefore

$$|c_k| \le \frac{1}{T} \int_T^{2T} |f(x)| \, dx + \frac{2}{T} \sum_{\ell \ne k} \frac{|c_\ell|}{|\lambda_\ell - \lambda_k|}$$

$$< \epsilon + O\left(\frac{1}{T}\right)$$

for any $T > N$, which implies $c_k = 0$, because ϵ is arbitrary. $\square$

<hr>

| SOLUTION 3.6 |

We prove that the assertion is true. Suppose, on the contrary, that $f(x)$ does not converge to 0 as $x \to \infty$. We then find a strictly monotone increasing sequence $1 < x_1 < x_2 < \cdots$ diverging to ∞ and a positive constant δ satisfying

$$|f(x_k)| > 2\delta$$

for all k. By the continuity of f there exists a sufficiently small $\epsilon_k > 0$ for each k such that $|f(x)| \ge \delta$ holds on the interval $[x_k - \epsilon_k, x_k + \epsilon_k]$. For all positive integers n put

$$E_n = \bigcup_{k=n}^{\infty} \bigcup_{m=-\infty}^{\infty} \left(\frac{m - \epsilon_k}{x_k}, \frac{m + \epsilon_k}{x_k} \right),$$

which is an open dense set, because x_k diverges to ∞ as $k \to \infty$. Since $\mathbb{R}$ is a Baire space, the intersection

$$\bigcap_{n=1}^{\infty} E_n$$

is also dense; so we can choose a point $x^* > 1$ which belongs to all the sets E_n. Namely there exist two integers $k_n \geq n$ and m_n satisfying

$$\left| x^* - \frac{m_n}{x_{k_n}} \right| < \frac{\epsilon_{k_n}}{x_{k_n}}$$

for every n. Since m_n diverges to ∞ as $n \to \infty$, we obtain

$$\left| x_{k_n} - \frac{m_n}{x^*} \right| < \frac{\epsilon_{k_n}}{x^*} < \epsilon_{k_n},$$

which implies that $|f(m_n/x^*)| \geq \delta$, contrary to the assumption that $f(nx)$ converges to 0 as $n \to \infty$ at $x = 1/x^*$. $\qquad \square$

SOLUTION 3.7

For any $\epsilon > 0$ define the set

$$E_n = \{ x \in [a, b] \mid f(x) - f_n(x) \geq \epsilon \},$$

which is a sequence of monotone decreasing compact sets in view of the continuity of f_n and f. Suppose that E_n is not empty for any positive integer n. It then follows that

$$\bigcap_{n=1}^{\infty} E_n \neq \emptyset.$$

Let x_0 be a point belonging to all the sets E_n. This means that $\{f_n(x_0)\}$ does not converge to $f(x_0)$, contrary to the assumption. Thus E_n is empty for all sufficiently large n; in other words, $|f(x) - f_n(x)| < \epsilon$ for every $x \in [a, b]$. $\qquad \square$

SOLUTION 3.8

Let I be any closed interval of length 4. Put

$$P(x) = x^n + a_{n-1} x^{n-1} + \cdots + a_0$$

with $a_{n-1}, \ldots, a_0 \in \mathbb{R}$ and $n \geq 1$. Since the leading coefficient is invariant under parallel translation, we can assume that $I = [-2, 2]$. Let M be the difference of the maximum and the minimum of $P(x)$ on the interval I. It may be convenient to introduce a new notation for the following special sum:

$$\overset{n}{\underset{k=0}{\bigtimes}} b_k = \sum_{k=0}^{n-1} (-1)^k (b_k - b_{k+1})$$

$$= b_0 - 2b_1 + 2b_2 - \cdots + 2(-1)^{n-1} b_{n-1} + (-1)^n b_n,$$

being a linear function in $b_0, b_1, \ldots, b_n$. For any integer $0 \leq s < n$ we put $\omega = e^{s\pi i/n} \neq -1$. Then we have

$$\sum_{k=0}^{n} \omega^k = (1 - \omega) \frac{1 - (-\omega)^n}{1 + \omega} = (1 - (-1)^{s+n}) \frac{1 - \omega}{1 + \omega}.$$

Since $\overline{\omega} = \omega^{-1}$, it can be seen that the real part of the above expression vanishes; in other words,

$$\sum_{k=0}^{n} \cos \frac{ks\pi}{n} = 0.$$

On the other hand, it is clear that $2 \cos s\theta = e^{si\theta} + e^{-si\theta}$ is a monic polynomial in $2 \cos \theta = e^{i\theta} + e^{-i\theta}$ with integer coefficients of degree s. So we can write

$$2 \cos s\theta = \tau_s(2 \cos \theta)$$

for some polynomial $\tau_s(x)$, and therefore

$$\overset{n}{\underset{k=0}{\bigtimes}} \alpha_k^s = 0 \quad \text{with} \quad \alpha_k = 2 \cos \frac{k\pi}{n}$$

for all $0 \leq s < n$. Moreover, for $s = n$, we have

$$\overset{n}{\underset{k=0}{\bigtimes}} \alpha_k^n = \overset{n}{\underset{k=0}{\bigtimes}} \tau_n(\alpha_k) = 2 \overset{n}{\underset{k=0}{\bigtimes}} \cos k\pi = 4n.$$

Hence

$$4n = \left| \overset{n}{\underset{k=0}{\bigtimes}} P(\alpha_k) \right| \leq \sum_{k=0}^{n-1} |P(\alpha_k) - P(\alpha_{k+1})| \leq nM,$$

which implies that $M \geq 4$. We thus have $\max_{x \in I} |P(x)| \geq 2$. $\qquad \square$

REMARK. The maximum of $|\tau_n(x)|$ on the interval $[-2, 2]$ is clearly equal to 2, which means that we cannot replace 2 by any larger constant in general.

$$T_n(x) = \frac{1}{2}\tau_n(2x)$$

is a polynomial with integer coefficients of degree n and these form a system of orthogonal polynomials over the interval $[-1, 1]$. $T_n(x)$ is called the nth Chebyshev polynomial of the first kind and satisfies $T_n(\cos\theta) = \cos n\theta$. See Chapter 15 for various properties of the Chebyshev polynomials.

SOLUTION 3.9

Suppose, on the contrary, that $g(x)$ is continuous on some interval $[a, b]$. Since $h(f(1) + g(y)) = y$ holds for any $y > 0$, the function $g(y)$ is one-to-one on $(0, \infty)$ and therefore g is a strictly monotone continuous function on $[a, b]$. We can assume that g is monotone increasing, because $H(F(x) + G(y)) = xy$ if we put $F(x) = -f(x)$, $G(y) = -g(y)$ and $H(x) = h(-x)$.

We put $g(a) = \alpha < g(b) = \beta$. The inverse function $g^{-1} : [\alpha, \beta] \to [a, b]$ is also strictly monotone increasing and continuous. Put $\lambda = b/a > 1$ for brevity. For any integer n we define $U_n = [\alpha + f(\lambda^{-n}), \beta + f(\lambda^{-n})]$. Since

$$h(f(\lambda^{-n}) + s) = \lambda^{-n}g^{-1}(s)$$

for $\alpha \le s \le \beta$, the function $h(x)$ is also strictly monotone increasing and continuous on each U_n. Since

$$h(U_n) = [\lambda^{-n}a, \lambda^{-n}b] = \left[\frac{a^{n+1}}{b^n}, \frac{a^n}{b^{n-1}} \right],$$

the sequence of intervals $\{h(U_n)\}_{n\in\mathbb{Z}}$ covers $(0, \infty)$ adjacently. Therefore $U_i \cap U_j$ is a single point or empty if $i \ne j$. For any positive integer N let k_N be the number of U_n's intersecting with the interval $[(\alpha - \beta)N, (\beta - \alpha)N]$. Since the length of U_n is $\beta - \alpha$ for all n, we have

$$(\beta - \alpha)(k_N - 2) \le 2(\beta - \alpha)N;$$

that is, $k_N \le 2N + 2$. This means that at least one among $U_1, U_2, ..., U_{2N+3}$ does not intersect with $[(\alpha - \beta)N, (\beta - \alpha)N]$. Therefore there exists an integer $m_N \in [1, 2N + 3]$ such that

$$\left| f(\lambda^{-m_N}) \right| > (\beta - \alpha)N - \max(|\alpha|, |\beta|);$$

in particular, m_N tends to ∞ as $N \to \infty$. Then we have

$$\frac{|f(\lambda^{-m_N})|}{\log \lambda^{m_N}} > \frac{(\beta - \alpha)N - \max(|\alpha|, |\beta|)}{m_N \log \lambda}$$

$$\geq \frac{(\beta - \alpha)N - \max(|\alpha|, |\beta|)}{(2N + 3)\log \lambda},$$

and hence

$$\limsup_{x \to 0+} \left| \frac{f(x)}{\log x} \right| \geq \frac{\beta - \alpha}{2 \log \lambda} > 0,$$

contrary to the assumption. □

SOLUTION 3.10

Without loss of generality, we can assume that $f(0) = f(1) = 0$, because $c_J(f) = c_J(\tilde{f})$ for every Farey interval J, where

$$\tilde{f}(x) = f(x) - f(0)(1 - x) - f(1)x.$$

We first show by induction on n that $\sigma_n(x) = f(x)$ for every $x \in E_{n+1}$, where

$$\sigma_n(x) = \sum_{k=1}^{n} \Phi_k(x).$$

By the assumption $\sigma_1(0) = \Phi_1(0) = 0 = f(0)$, $\sigma_1(1) = \Phi_1(1) = 0 = f(1)$ and

$$\sigma_1\left(\frac{1}{2}\right) = \Phi_1\left(\frac{1}{2}\right) = c_{[0,1]} = f\left(\frac{1}{2}\right);$$

so the case $n = 1$ holds. We next suppose that $\sigma_n(x) = f(x)$ for every $x \in E_{n+1}$ for some $n \geq 1$. Since $\sigma_{n+1}(x) = \sigma_n(x) + \Phi_{n+1}(x)$, we get $\sigma_{n+1}(x) = f(x)$ for every $x \in E_{n+1}$. Let $J = [a/b, c/d]$ be any Farey interval of order $n + 1$. Since $c_J(g) = 0$ for any g whose restriction to J is linear, we obtain $c_J(\sigma_n) = 0$ and therefore

$$c_J(\sigma_{n+1}) = c_J(\sigma_n) + c_J(\Phi_{n+1})$$
$$= c_J(c_J(f)\phi_J) = c_J(f).$$

This implies that

$$\sigma_{n+1}\left(\frac{a + c}{b + d}\right) = f\left(\frac{a + c}{b + d}\right),$$

because $\sigma_{n+1}(x) = f(x)$ for $x = a/b, c/d$. So $\sigma_{n+1}(x) = f(x)$ for any $x \in E_{n+2}$. Thus $\sigma_n(x)$ is the zigzag function drawing a line between each two adjacent points $(x, f(x))$, $x \in E_{n+1}$ successively.

Since f is uniformly continuous on $[0, 1]$, for any $\epsilon > 0$ there exists a $\delta > 0$

such that $|f(x) - f(y)| < \epsilon$ for any $x, y \in [0, 1]$ satisfying $|x - y| < \delta$. Take a sufficiently large integer n such that all the lengths of Farey intervals of order n is less than δ. For any $x \in [0, 1]$ let $J_x = [a/b, c/d]$ be a Farey interval of order $n + 1$ containing x. We then have

$$\left| f(x) - \sigma_n(x) \right| \le \left| f(x) - f\left(\frac{a}{b}\right) \right| + \left| \sigma_n(x) - \sigma_n\left(\frac{a}{b}\right) \right|$$

$$< \epsilon + \left| \sigma_n(x) - \sigma_n\left(\frac{a}{b}\right) \right|.$$

Since σ_n is linear on J_x, it follows that

$$\left| \sigma_n(x) - \sigma_n\left(\frac{a}{b}\right) \right| \le \left| \sigma_n\left(\frac{c}{d}\right) - \sigma_n\left(\frac{a}{b}\right) \right|$$

$$= \left| f\left(\frac{c}{d}\right) - f\left(\frac{a}{b}\right) \right| < \epsilon.$$

Therefore $|f(x) - \sigma_n(x)| < 2\epsilon$ for any $x \in [0, 1]$. $\qquad\qquad\square$

SOLUTION 3.11

Put

$$\Phi_n(x) = \sum_{\substack{J \in I_n \\ J = [a/b, c/d]}} \frac{1}{b + d}\, \phi_J(x)$$

for $n \ge 1$. We first show by induction on n that

$$\sum_{k=1}^{\infty} \Phi_k\left(\frac{p}{q}\right) = 1 - \frac{1}{q} \tag{3.1}$$

for any fraction p/q belonging to E_n. Note that the sum on the left-hand side is only finite. Clearly (3.1) holds for $p/q = 0/1$ and $1/1$, because $\Phi_k(0) = \Phi_k(1) = 0$ for all k. Suppose next that (3.1) holds for every p/q in E_n for some $n \ge 1$. So we have

$$\Phi_1\left(\frac{p}{q}\right) + \Phi_2\left(\frac{p}{q}\right) + \cdots + \Phi_{n-1}\left(\frac{p}{q}\right) = 1 - \frac{1}{q},$$

because $p/q \in E_k$ for all $k \ge n$.

Let a/b and c/d be any successive fractions in E_n and consider the Farey interval $J = [a/b, c/d]$. Since

$$\tau_{n-1}(x) = \Phi_1(x) + \cdots + \Phi_{n-1}(x)$$

is linear on J, we get

$$\tau_{n-1}\left(\frac{a+c}{b+d}\right) = \frac{\frac{c}{d} - \frac{a+c}{b+d}}{\frac{c}{d} - \frac{a}{b}} \tau_{n-1}\left(\frac{a}{b}\right) + \frac{\frac{a+c}{b+d} - \frac{a}{b}}{\frac{c}{d} - \frac{a}{b}} \tau_{n-1}\left(\frac{c}{d}\right)$$

$$= \frac{b}{b+d}\left(1 - \frac{1}{b}\right) + \frac{d}{b+d}\left(1 - \frac{1}{d}\right)$$

$$= 1 - \frac{2}{b+d}.$$

Hence we have

$$\sum_{k=1}^{\infty} \Phi_k\left(\frac{a+c}{b+d}\right) = \tau_{n-1}\left(\frac{a+c}{b+d}\right) + \Phi_n\left(\frac{a+c}{b+d}\right) = 1 - \frac{1}{b+d}$$

and (3.1) holds for any fraction p/q belonging to E_{n+1}.

Let m be any positive integer. Since $\tau_m(x)$ is piecewisely linear continuous function and $\tau_m(x) < 1$ at any point x belonging to E_{m+1}, it follows obviously that $\tau_m(x) < 1$ for all $x \in [0, 1]$. Thus the series

$$\sum_{n=1}^{\infty} \Phi_n(x)$$

converges at every irrational point x. Let a/b and c/d be any adjacent fractions in E_{m+1} satisfying $a/b < x < c/d$. We have

$$\tau_m(x) \geq \min\left(\tau_m\left(\frac{a}{b}\right), \tau_m\left(\frac{c}{d}\right)\right)$$

$$= \min\left(1 - \frac{1}{b}, 1 - \frac{1}{d}\right).$$

Since x is irrational, both b and d must diverge to $+\infty$ as $m \to \infty$; so, we have $f(x) = 1$, as required. $\qquad\square$

SOLUTION 3.12

Put $P(x) = Ax^2 + Bx + C$. For any rational number p/q it follows from the mean value theorem that

$$\left|P\left(\frac{p}{q}\right)\right| = \left|P(\alpha) - P\left(\frac{p}{q}\right)\right| = |P'(\theta)| \cdot \left|\alpha - \frac{p}{q}\right|$$

for some θ between α and p/q. For any $\epsilon > 0$ there exists a $\delta > 0$ such that $|P'(x) - P'(\alpha)| < \epsilon$ for any $|x - \alpha| < \delta$. Then,

$$\left| \alpha - \frac{p}{q} \right| = \left| \frac{P(p/q)}{P'(\theta)} \right| \geq \frac{1}{q^2 |P'(\theta)|}$$

$$> \frac{1}{q^2 (|P'(\alpha)| + \epsilon)}$$

for all p/q with $|\alpha - p/q| < \delta$, because $P(p/q) \neq 0$ and $q^2 P(p/q) \in \mathbb{Z}$.

Take a sufficiently large integer N such that all the lengths of Farey intervals of order N is less than δ. Let $J = [a/b, c/d]$ be a Farey interval of order $n \geq N$ containing α. If $\alpha \in J_L'$, then we have

$$\Psi_n(\alpha) = \frac{\alpha - \dfrac{a}{b}}{\dfrac{a+c}{b+d} - \dfrac{a}{b}} = b(b+d)\left(\alpha - \frac{a}{b} \right) > \frac{1}{|P'(\alpha)| + \epsilon}.$$

Similarly, if $\alpha \in J_R$, then

$$\Psi_n(\alpha) = \frac{\dfrac{c}{d} - \alpha}{\dfrac{c}{d} - \dfrac{a+c}{b+d}} = d(b+d)\left(\frac{c}{d} - \alpha \right) > \frac{1}{|P'(\alpha)| + \epsilon}.$$

Since ϵ is arbitrary, we conclude that

$$\liminf_{n \to \infty} \Psi_n(\alpha) \geq \frac{1}{|P'(\alpha)|} = \frac{1}{\sqrt{B^2 - 4AC}}. \qquad \square$$

✦ ❈ ✦

Chapter 4

Differentiation

1. A function $f(x)$ defined on an open interval I is differentiable at $a \in I$ if and only if there exists a function $\varphi_a(x)$ on I, which is continuous at a, satisfying

$$f(x) - f(a) = \varphi_a(x)(x - a).$$

Obviously the function $\varphi_a(x)$ is uniquely determined on $I \setminus \{a\}$ and $f'(a)$ is given by the value $\varphi_a(a)$. This definition is due to Carathéodory (1954). See also a survey by Kuhn (1991).

2. If $f(x)$ is continuous on a closed interval $[a, b]$ and differentiable on the open interval (a, b), then there exists a point c in (a, b) satisfying

$$f'(c) = 0$$

whenever $f(a) = f(b)$. This is known as *Rolle's theorem*, which is equivalent to the following *mean value theorem*: there exists a point c in (a, b) satisfying

$$\frac{f(b) - f(a)}{b - a} = f'(c),$$

whatever $f(a)$ and $f(b)$ are.

3. If $f(x)$ and $g(x)$ are continuous on a closed interval $[a, b]$ and differentiable on the open interval (a, b) with $g(a) \neq g(b)$, and if $f'(x)$ and $g'(x)$ never vanish simultaneously, then there exists a point $c \in (a, b)$ satisfying

$$\frac{f(b) - f(a)}{g(b) - g(a)} = \frac{f'(c)}{g'(c)}.$$

This is known as *Cauchy's mean value theorem*.

4. For any differentiable function $f(x)$ on an interval I, the image of the derivative $f'(I)$ is always an interval. In other words, if $f(x)$ is differentiable on $[a, b]$,

$f'(a) = \alpha$, $f'(b) = \beta$, and if η lies between α and β, then there is a ξ in (a, b) for which $f'(\xi) = \eta$. This theorem is due to J. G. Darboux (1842–1917).

5. If $f(x)$ is n times differentiable on an open interval I, then for a fixed a in I,

$$f(x) = f(a) + \frac{f'(a)}{1!}(x - a) + \cdots + \frac{f^{(n-1)}(a)}{(n-1)!}(x-a)^{n-1} + R_n.$$

This is called *Taylor's formula* of order n and R_n is called the remainder term.

6. J. L. Lagrange (1736–1813) showed that

$$R_n = \frac{f^{(n)}(c)}{n!}(x - a)^n$$

for some c between a and x, known as the Lagrange form of the remainder term or simply as the Lagrange remainder.

7. Moreover if $f^{(n)}(x)$ is continuous on I, then the remainder term can be expressed in the integral form

$$R_n = \frac{1}{(n-1)!}\int_a^x (x-t)^{n-1} f^{(n)}(t)\, dt.$$

PROBLEM 4.1

Suppose that a function $f(x)$ defined on an open interval I is differentiable at $a \in I$. Show that

$$\lim_{\substack{(h,h')\to(0,0)\\ h>0, h'>0}} \frac{f(a+h) - f(a-h')}{h + h'} = f'(a).$$

PROBLEM 4.2

Suppose that a function $f(x)$ defined on an open interval I is continuously differentiable on I. Show that

$$\lim_{\substack{(h,h')\to(0,0)\\ h+h'\neq 0}} \frac{f(a+h) - f(a-h')}{h + h'} = f'(a).$$

PROBLEM 4.3

Suppose that $f \in \mathscr{C}^{n+1}[0, 1]$ satisfies

$$f(0) = f'(0) = \cdots = f^{(n)}(0) = f'(1) = \cdots = f^{(n)}(1) = 0$$

and $f(1) = 1$. Show then that

$$\max_{0 \leq x \leq 1} \left| f^{(n+1)}(x) \right| \geq 4^n n!.$$

PROBLEM 4.4

Show that any $f \in \mathscr{C}^2(\mathbb{R})$ satisfies the inequality

$$\left(\sup_{x \in \mathbb{R}} |f'(x)| \right)^2 \leq 2 \sup_{x \in \mathbb{R}} |f(x)| \cdot \sup_{x \in \mathbb{R}} |f''(x)|.$$

Prove moreover that the constant 2 on the right-hand side cannot in general be replaced by any smaller number.

This was proved by Hadamard (1914). Kolmogorov (1939) generalized to the following inequality for $f \in \mathscr{C}^n(\mathbb{R})$:

$$\left(\sup_{x \in \mathbb{R}} |f^{(k)}(x)| \right)^n \leq C_{k,n} \left(\sup_{x \in \mathbb{R}} |f(x)| \right)^{n-k} \left(\sup_{x \in \mathbb{R}} |f^{(n)}(x)| \right)^k$$

for $0 < k < n$ with the best possible constant $C_{k,n}$, which is a rational number expressible in terms of the Euler numbers. de Boor and Schoenberg (1976) gave a proof using spline functions. We present first several values of $C_{k,n}$ as follows:

$n \backslash k$	1	2	3	4
2	2			
3	$\dfrac{9}{8}$	3		
4	$\dfrac{512}{375}$	$\dfrac{36}{25}$	$\dfrac{24}{5}$	
5	$\dfrac{1953125}{1572864}$	$\dfrac{125}{72}$	$\dfrac{225}{128}$	$\dfrac{15}{2}$

Landau (1913) showed that any $f \in \mathscr{C}^2(0, \infty)$ satisfies

$$\left(\sup_{x > 0} |f'(x)| \right)^2 \leq 4 \sup_{x > 0} |f(x)| \cdot \sup_{x > 0} |f''(x)|$$

with the best possible constant 4. An explicit formula for general $C_{k,n}$ in this case seems to be not known.

PROBLEM 4.5

Let $Q_n(x)$ be a polynomial with real coefficients of degree n and M be the maximum of $|Q_n(x)|$ on the interval $[-1, 1]$. Show that

$$\sqrt{1 - x^2} \; |Q_n'(x)| \le nM$$

for any $-1 \le x \le 1$. Show next that

$$|Q_n'(x)| \le n^2 M$$

for any $-1 \le x \le 1$.

The latter is called Markov's inequality, which first appeared in A. A. Markov (1889). He was famous for his study of Markov chains. The equality occurs for Chebyshev polynomial $T_n(x)$ of the first kind.

Concerning higher derivatives V. A. Markov (1892), a younger brother of A. A. Markov, showed that

$$\max_{-1 \le x \le 1} \left| Q_n^{(k)}(x) \right| \le \frac{n^2(n^2 - 1^2) \cdots (n^2 - (k-1)^2)}{1 \cdot 3 \cdot 5 \cdots (2k - 1)} \max_{-1 \le x \le 1} |Q_n(x)|.$$

Note that the coefficient of $\|Q_n\|$ on the right-hand side is equal to $T_n^{(k)}(1)$ (see **PROBLEM 15.5**). V. A. Markov's paper was published when he was 21 years old, a student of St. Petersburg University, who died at the age of 25. Duffin and Schaeffer (1941) gave another proof for this. Rogosinski (1955) discussed this problem using only the classical Lagrange interpolation polynomials.

PROBLEM 4.6

Suppose that $f \in \mathscr{C}^\infty(\mathbb{R})$ satisfies $f(0)f'(0) \ge 0$ and that $f(x)$ converges to 0 as $x \to \infty$. Show then that there exists an increasing sequence $0 \le x_1 < x_2 < x_3 < \cdots$ satisfying

$$f^{(n)}(x_n) = 0.$$

PROBLEM 4.7

Show that the maximum of $\left| Q^{(n)}(x) \right|$ over $[-1, 1]$ is equal to $2^n n!$ where

$$Q(x) = (1 - x^2)^n.$$

Multiplying $Q^{(n)}(x)$ by

$$\frac{(-1)^n}{2^n n!}$$

one gets the nth Legendre polynomial $P_n(x)$. See Chapter 14 for various properties of the Legendre polynomials.

PROBLEM 4.8

Define a piecewise linear continuous function

$$\psi(x) = \begin{cases} x & (0 \leq x < 1/2) \\ 1 - x & (1/2 \leq x < 1) \end{cases}$$

and extend it to $\mathbb{R}$ *periodically. Show that*

$$T(x) = \sum_{n=0}^{\infty} \frac{1}{2^n} \psi(2^n x)$$

is continuous but nowhere differentiable.

This function was derived by Takagi (1903) as a simpler example of a nowhere differentiable function than Weierstrass'

$$W(x) = \sum_{n=0}^{\infty} a^n \cos b^n \pi x.$$

Many types of nowhere differentiable continuous functions were reported after Weierstrass' discovery in 1874. Lerch (1888) examined various trigonometric series like $W(x)$. However the Takagi function did not seem to be well-known in European mathematical circles of those days. Takagi constructed his function using dyadic expansions of x in the interval $[0, 1)$, as stated in the next problem. Cesàro (1906) also used them to define many such functions.

The piecewise linear function $\psi(x) = \text{dist}(\{x\}, \mathbb{Z})$ in this problem was used by Faber (1907) to define

$$f(x) = \sum_{n=1}^{\infty} \frac{1}{10^n} \psi(2^{n!} x).$$

Landsberg (1908) also used the function $\psi(x)$ and dyadic expansions of x. Later van der Waerden (1930) found the same kind of function

$$f(x) = \sum_{n=1}^{\infty} \frac{1}{10^n} \psi(10^n x)$$

and eventually the Takagi function itself was rediscovered by de Rham (1957).

PROBLEM 4.9

Let $x = \sum_{n=1}^{\infty} \dfrac{a_n}{2^n}$ *be a dyadic expansion of* $x \in [0, 1)$ *and put* $v_n = a_1 + a_2 + \cdots + a_n$, *where* $a_n \in \{0, 1\}$. *Show that*

$$T(x) = \sum_{n=1}^{\infty} \frac{(1 - a_n)v_n + a_n(n - v_n)}{2^n}.$$

This is the original definition of $T(x)$ by Takagi (1903).

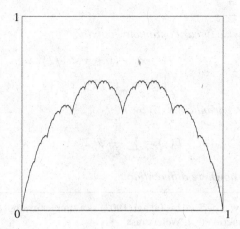

Fig. 4.1   The graph of Takagi function $T(x)$ on $[0, 1]$.

PROBLEM 4.10

Show that the maximum of $T(x)$ is equal to $2/3$. Let E be the set of all points in $[0, 1]$ at which the maximum of T is attained. Show then that

$$E = \left\{ x \in [0, 1] \;\middle|\; x = \sum_{n=1}^{\infty} \frac{c_n}{4^n}, \; c_n \in \{1, 2\} \right\}.$$

Consequently the set E is homeomorphic to the so-called Cantor's ternary set.

PROBLEM 4.11

Let $f, g \in \mathscr{C}^1(0, \infty)$. Suppose that $f(x) > 0$ and $g(x)$ is bounded above. Show that

$$\liminf_{x \to \infty} g'(f(x))f'(x) \le 0.$$

PROBLEM 4.12

Suppose that $f \in \mathscr{C}^2(0, \infty)$ converges to α as $x \to \infty$ and that

$$f''(x) + \lambda f'(x)$$

is bounded above, where λ is some constant. Then show that $f'(x)$ converges to 0 as $x \to \infty$.

PROBLEM 4.13

Suppose that all roots of an algebraic equation $x^n + a_{n-1}x^{n-1} + \cdots + a_0 = 0$ have negative real parts and that $f \in \mathscr{C}^n(0, \infty)$. Show that if

$$f^{(n)}(x) + a_{n-1}f^{(n-1)}(x) + \cdots + a_0 f(x)$$

converges to 0 as $x \to \infty$, then $f^{(k)}(x)$ also converges to 0 as $x \to \infty$ for all $0 \le k \le n$.

This does not hold if the algebraic equation has a root ξ with non-negative real part, because $e^{\xi x}$ is a solution of the differential equation

$$f^{(n)}(x) + a_{n-1}f^{(n-1)}(x) + \cdots + a_0 f(x) = 0$$

and does not converge to 0 as $x \to \infty$.

◆ ✳ ◆

Solutions for Chapter 4

SOLUTION 4.1

By Carathéodory's definition of the derivative there exists a function $\varphi(x)$ on I satisfying

$$f(x) - f(a) = \varphi_a(x)(x - a)$$

with $f'(a) = \varphi_a(a)$. Hence, for any sufficiently small $h, h' > 0$ we have

$$\left| \frac{f(a+h) - f(a-h')}{h + h'} - f'(a) \right|$$

$$= \left| \frac{h}{h+h'} (\varphi_a(a+h) - \varphi_a(a)) + \frac{h'}{h+h'} (\varphi_a(a-h') - \varphi_a(a)) \right|$$

$$\leq |\varphi_a(a+h) - \varphi_a(a)| + |\varphi_a(a-h') - \varphi_a(a)|.$$

The right-hand side converges to 0 as $h \to 0+$ and $h' \to 0+$.   □

SOLUTION 4.2

For any sufficiently small h, h' with $h + h' \neq 0$ it follows from the mean value theorem that

$$f(a+h) - f(a-h') = f'(\xi_{h,h'})(h + h'),$$

where $\xi_{h,h'}$ is some point between two points $a + h$ and $a - h'$. Therefore

$$\left| \frac{f(a+h) - f(a-h')}{h+h'} - f'(a) \right| = \left| f'(\xi_{h,h'}) - f'(a) \right|$$

and the right-hand side converges to 0 as $h, h' \to 0$, because

$$|\xi_{h,h'} - a| \leq \max(|h|, |h'|). \qquad \square$$

SOLUTION 4.3

Let

$$P(x) = x^n + a_{n-1}x^{n-1} + \cdots + a_0$$

be a polynomial with real coefficients. Integrating by parts repeatedly we have

$$\int_0^1 P(x)f^{(n+1)}(x)\,dx = -\int_0^1 P'(x)f^{(n)}(x)\,dx$$
$$\vdots$$
$$= (-1)^n \int_0^1 P^{(n)}(x)f'(x)\,dx = (-1)^n n!$$

by using $f(1) = 1$. Now, taking P as the polynomial attaining the minimum in **PROBLEM 15.7**, we get

$$n! = \left| \int_0^1 P(x)f^{(n+1)}(x)\,dx \right| \le \max_{0\le x\le 1} \left| f^{(n+1)}(x) \right| \int_0^1 |P(x)|\,dx$$
$$= \frac{1}{4^n} \max_{0\le x\le 1} \left| f^{(n+1)}(x) \right|. \qquad \square$$

SOLUTION 4.4

Put

$$\alpha = \sup_{x\in\mathbb{R}} |f(x)| \quad \text{and} \quad \beta = \sup_{x\in\mathbb{R}} |f''(x)|.$$

We can of course assume that both α and β are finite. If $\beta = 0$, then α is finite if and only if $f(x)$ vanishes everywhere; therefore we can also assume that β is positive. For every $x \in \mathbb{R}$ and $y > 0$ it follows from Taylor's formula that there exists a $\xi_{x,y}$ satisfying

$$f(x+y) = f(x) + f'(x)y + f''(\xi_{x,y})\frac{y^2}{2}.$$

Therefore

$$f(x+y) - f(x-y) = 2f'(x)y + (f''(\xi_{x,y}) - f''(\xi_{x,-y}))\frac{y^2}{2},$$

which implies that

$$2|f'(x)|y = \left| f(x+y) - f(x-y) + (f''(\xi_{x,-y}) - f''(\xi_{x,y}))\frac{y^2}{2} \right|$$
$$\le 2\alpha + \beta y^2.$$

Thus we have

$$\sup_{x\in\mathbb{R}} |f'(x)| \le \frac{\alpha}{y} + \frac{\beta y}{2}$$

and the right-hand side attains its minimum $\sqrt{2\alpha\beta}$ at $y = \sqrt{2\alpha/\beta}$.

To see that 2 is the best possible constant we first consider an even step

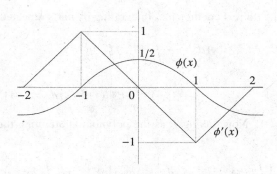

Fig. 4.2   The graphs of $\phi(x)$ and $\phi'(x)$.

function

$$\phi''(x) = \begin{cases} 0 & (|x| > 2) \\ 1 & (1 \leq |x| \leq 2) \\ -1 & (|x| < 1) \end{cases}.$$

Then

$$\phi'(x) = \int_{-2}^{x} \phi''(t)\,dt$$

is an odd piecewise linear continuous function and in turn

$$\phi(x) = \int_{-2}^{x} \phi'(t)\,dt - \frac{1}{2}$$

is an even $\mathscr{C}^1$ function. Obviously the maxima of $|\phi(x)|$, $|\phi'(x)|$ and $|\phi''(x)|$ are $1/2$, 1 and 1 respectively and $\phi(x)$ certainly satisfies the equality of the problem, although it does not belong to $\mathscr{C}^2(\mathbb{R})$. To tide over this difficulty it suffices to transform ϕ'' slightly to a continuous one in the neighborhood of the discontinuity points of ϕ'' so that its influence on ϕ and ϕ' becomes arbitrarily small.   □

SOLUTION 4.5

This proof is based on Cheney (1966), pp. 89−91. We first show that any polynomial $Q(x)$ with complex coefficients of degree $n - 1$ satisfies the inequality

$$\max_{-1 \leq x \leq 1} |Q(x)| \leq n \max_{-1 \leq x \leq 1} \sqrt{1 - x^2}\,|Q(x)|.$$

Let M denote the right-hand side. For any x satisfying $x^2 \leq 1 - 1/n^2$ the above inequality is obvious; so we can assume that

$$|x| > \sqrt{1 - 1/n^2} \quad \text{and hence} \quad |x| > \cos \frac{\pi}{2n},$$

because $\sqrt{1 - s^2} \geq \cos(\pi s/2)$ for any $|s| \leq 1$.

The nth Chebyshev polynomial $T_n(x)$ of the first kind (see Chapter 15) is factorized as

$$T_n(x) = (x - \xi_1) \cdots (x - \xi_n) \quad \text{where} \quad \xi_k = \cos \frac{2k-1}{2n} \pi \quad (1 \leq k \leq n).$$

The Lagrange interpolation polynomial for $Q(x)$ with nodes $\xi_1, \ldots, \xi_n$ is

$$\sum_{k=1}^{n} \frac{Q(\xi_k)}{T_n'(\xi_k)} \cdot \frac{T_n(x)}{x - \xi_k} = \frac{1}{n} \sum_{k=1}^{n} (-1)^{k-1} Q(\xi_k) \sqrt{1 - \xi_k^2} \frac{T_n(x)}{x - \xi_k},$$

which is a polynomial in x of degree less than n; hence it coincides with the polynomial $Q(x)$. Using the fact that $\mathrm{sgn}(x - \xi_k)$ is independent of k in view of $\xi_1 < |x| \leq 1$, we get

$$|Q(x)| \leq \frac{M}{n^2} \sum_{k=1}^{n} \left| \frac{T_n(x)}{x - \xi_k} \right| = \frac{M}{n^2} \left| \sum_{k=1}^{n} \frac{T_n(x)}{x - \xi_k} \right| = \frac{M}{n^2} \left| T_n'(x) \right|.$$

Since

$$\left| T_n'(\cos \theta) \right| = n \left| \frac{\sin n\theta}{\sin \theta} \right| \leq n^2,$$

we get $|Q(x)| \leq M$, as required.

By substituting $x = \cos \theta$ we can see that our inequality is equivalent to

$$\max_{\theta} |Q(\cos \theta)| \leq n \max_{\theta} |\sin \theta \, Q(\cos \theta)|,$$

which is valid for all polynomials Q with complex coefficients of degree $n - 1$. Let $S(\theta)$ be a linear combination over $\mathbb{C}$ of $1, \cos \theta, \cos 2\theta, \ldots, \cos n\theta$ and $\sin \theta$, $\sin 2\theta, \ldots, \sin n\theta$. For any ω and θ we put

$$S_0(\theta) = \frac{S(\omega + \theta) - S(\omega - \theta)}{2}.$$

Since $S_0(\theta)$ is an odd function, this is a linear combination of $1, \sin \theta, \ldots, \sin n\theta$ only. Thus $S_0(\theta)/\sin \theta$ is a polynomial in $\cos \theta$ of degree less than n, because $\sin k\theta / \sin \theta$ can be expressed as a polynomial in $\cos \theta$ of degree $k - 1$. Applying our inequality to this polynomial in $\cos \theta$, we have

$$\max_{\theta} \left| \frac{S_0(\theta)}{\sin \theta} \right| \leq n \max_{\theta} |S_0(\theta)| \leq n \max_{\theta} |S(\theta)|.$$

Therefore

$$\lim_{\theta \to 0} \frac{S_0(\theta)}{\sin\theta} = \lim_{\theta \to 0} \frac{S_0(\theta)}{\theta} = S_0'(0) = S'(\omega),$$

which implies that

$$\max_{\theta} |S'(\theta)| \leq n \max_{\theta} |S(\theta)|,$$

because ω is arbitrary. This is called Bernstein's inequality (1912b).

The two inequalities stated in the problem can be solved by using Bernstein's inequality. Let $P(x)$ be any polynomial with complex coefficients of degree n. Then $P(\cos\theta)$ is a linear combination of $1, \cos\theta, \ldots, \cos n\theta$ and it follows from Bernstein's inequality that

$$\max_{-1 \leq x \leq 1} \sqrt{1 - x^2} \, |P'(x)| = \max_{\theta} |\sin\theta P'(\cos\theta)|$$

$$\leq n \max_{\theta} |P(\cos\theta)| = nM.$$

Therefore, since $P'(x)$ is a polynomial of degree $n - 1$, we get

$$\max_{-1 \leq x \leq 1} |P'(x)| \leq n \max_{-1 \leq x \leq 1} \sqrt{1 - x^2} \, |P'(x)|$$

$$\leq n^2 M.$$

This completes the proof.   □

SOLUTION 4.6

First we suppose that $f'(x) > 0$ for all $x \geq 0$. Then $f(x)$ is strictly monotone increasing and $f(0) \geq 0$, because $f'(0) > 0$. This is a contradiction, because $f(x)$ converges to 0 as $x \to \infty$. Similarly we get a contradiction, if $f'(x) < 0$ for all $x \geq 0$. Hence there exists at least one point $x_1 \geq 0$ at which $f'(x)$ vanishes.

Suppose next that we could find n points $x_1 < \cdots < x_n$ satisfying $f^{(k)}(x_k) = 0$ for $1 \leq k \leq n$. If $f^{(n+1)}(x) > 0$ for all $x > x_n$, then clearly $f^{(n)}(x) \geq f^{(n)}(x_n + 1) > 0$ for all $x \geq x_n + 1$, because $f^{(n)}(x_n) = 0$. Thus we have

$$f(x) \geq \frac{1}{n!} f^{(n)}(x_n + 1) x^n$$

$$+ \text{(some polynomial of degree less than } n),$$

which is contrary to the assumption that $f(x)$ converges to 0 as $x \to \infty$. Similarly we would have a contradiction if $f^{(n+1)}(x) < 0$ for all $x > x_n$. Hence there exists at least one point x_{n+1} greater than x_n satisfying $f^{(n+1)}(x_{n+1}) = 0$.   □

SOLUTION 4.7

By Cauchy's integral formula we get

$$Q^{(n)}(x) = \frac{n!}{2\pi i} \int_C \frac{(1-z^2)^n}{(z-x)^{n+1}} \, dz$$

where C is a circle centered at $z = x$ with radius $r > 0$ and oriented counterclockwise. Putting

$$z = x + re^{i\theta}$$

for $0 \leq \theta < 2\pi$, we obtain

$$Q^{(n)}(x) = \frac{n!}{2\pi} \int_0^{2\pi} \frac{(1-(x+re^{i\theta})^2)^n}{r^{n+1}e^{(n+1)i\theta}} re^{i\theta} \, d\theta$$

$$= \frac{n!}{2\pi} \int_0^{2\pi} \left(\frac{1-(x+re^{i\theta})^2}{re^{i\theta}} \right)^n \, d\theta.$$

Observe that the expression in the large parentheses can be written as

$$\left(\frac{1-x^2}{r} - r \right) \cos \theta - 2x - i \left(\frac{1-x^2}{r} + r \right) \sin \theta.$$

We then take $r = \sqrt{1-x^2}$ for $|x| < 1$ so that $|P^{(n)}(x)| \leq 2^n n!$. This inequality clearly holds even for $x = \pm 1$. $\qquad \square$

SOLUTION 4.8

The continuity of $T(x)$ is obvious, because it is defined as the series of continuous functions converging uniformly. To show the non-differentiability it suffices to consider any point x in the interval $(0, 1]$, because $T(x)$ is a periodic function with period 1.

We first consider any point x which can be expressed in the form $k/2^m$ with some odd integer k and non-negative integer m. For any integer $n \geq m$ put $h_n = 1/2^n$ for brevity. Then for any integer ℓ in $[0, n)$ there are no integers nor half-integers in the interval $(2^\ell x, 2^\ell (x + h_n))$. For if $2^\ell x < p/2 < 2^\ell x + 2^{\ell-n}$ for some integer p, then we would have

$$2^n x = k2^{n-m} < 2^{n-\ell-1}p < k2^{n-m} + 1,$$

a contradiction. This implies that $g(x)$ is a linear function having the slope 1 or

-1 on this subinterval. Hence

$$
\begin{aligned}
\frac{T(x + h_n) - T(x)}{h_n} &= \sum_{\ell=0}^{\infty} \frac{g(2^{\ell}(x + h_n)) - g(2^{\ell}x)}{2^{\ell}h_n} \\
&= \sum_{\ell=0}^{n-1} \frac{g(2^{\ell}x + 2^{\ell-n}) - g(2^{\ell}x)}{2^{\ell-n}}
\end{aligned}
$$

is a finite sum of 1 or -1, which does not converge as $n \to \infty$.

Next consider any point x for which $2^n x$ is not an integer for all positive integers n. Since $2^n x$ is not an integer, we can find two positive numbers h_n and h'_n satisfying

$$
[2^n x] = 2^n(x - h'_n) \quad \text{and} \quad [2^n x] + 1 = 2^n(x + h_n).
$$

Note that $h_n + h'_n = 2^{-n}$. Then for any integer ℓ in $[0, n)$ there are no integers nor half-integers in the interval $(2^{\ell}(x - h'_n), 2^{\ell}(x + h_n))$. For if $p/2$ were contained in this interval for some integer p, then we have $[2^n x] < 2^{n-\ell-1}p < [2^n x] + 1$, a contradiction. Therefore

$$
\begin{aligned}
\frac{T(x + h_n) - T(x - h'_n)}{h_n + h'_n} &= \sum_{\ell=0}^{\infty} \frac{g(2^{\ell}(x + h_n)) - g(2^{\ell}(x - h'_n))}{2^{\ell}(h_n + h'_n)} \\
&= \sum_{\ell=0}^{n-1} \frac{g(2^{\ell}(x + h_n)) - g(2^{\ell}(x - h'_n))}{2^{\ell-n}}
\end{aligned}
$$

is a finite sum of 1 or -1, which does not converge as $n \to \infty$. Thus $T(x)$ is not differentiable at x. (See **PROBLEM 4.1**.) $\qquad\square$

SOLUTION 4.9

Put $\sigma_n(x) = a_n + (1 - 2a_n)x$ for brevity. Since

$$
2^m x = \sum_{n=1}^{\infty} \frac{a_n}{2^{n-m}} = (\text{some integer}) + \sum_{n=1}^{\infty} \frac{a_{m+n}}{2^n}
$$

for any $m \geq 0$, we have

$$
\psi(2^m x) = \psi\left(\sum_{n=1}^{\infty} \frac{a_{m+n}}{2^n} \right) = \begin{cases} \displaystyle\sum_{n=1}^{\infty} \frac{a_{m+n}}{2^n} & (a_{m+1} = 0) \\[4mm] \displaystyle\sum_{n=1}^{\infty} \frac{1 - a_{m+n}}{2^n} & (a_{m+1} = 1) \end{cases} ;
$$

so we can write

$$\psi(2^m x) = \sum_{n=1}^{\infty} \frac{\sigma_{m+1}(a_{m+n})}{2^n}$$

and therefore

$$T(x) = \sum_{m=0}^{\infty} \frac{1}{2^m} \psi(2^m x) = \sum_{m \geq 0} \sum_{n \geq 1} \frac{\sigma_{m+1}(a_{m+n})}{2^{m+n}}$$

$$= \sum_{\ell=1}^{\infty} \frac{1}{2^\ell} \sum_{k=1}^{\ell} \sigma_k(a_\ell),$$

where

$$\sum_{k=1}^{\ell} \sigma_k(a_\ell) = \sum_{k=1}^{\ell} (a_k + (1 - 2a_k)a_\ell) = v_\ell + (\ell - 2v_\ell)a_\ell. \qquad \square$$

SOLUTION 4.10

Let M be the maximum of $T(x)$ and Γ be the graph of $T(x)$ on the unit interval $[0, 1]$; that is, $\Gamma = \{(x, T(x)) \mid 0 \leq x \leq 1\}$. Since $T(x)$ satisfies the functional equations

$$T(x) = \begin{cases} 2x + T(4x)/4 & (0 \leq x < 1/4) \\ 1/2 + T(4x - 1)/4 & (1/4 \leq x < 1/2) \\ 1/2 + T(4x - 2)/4 & (1/2 \leq x < 3/4) \\ 2 - 2x + T(4x - 3)/4 & (3/4 \leq x \leq 1) \end{cases},$$

the set Γ is the self-similar set in a weak sense of the following four affine contractions on $\mathbb{R}^2$:

$$\Phi_0 \begin{pmatrix} x \\ y \end{pmatrix} = \begin{pmatrix} 1/4 & 0 \\ 1/2 & 1/4 \end{pmatrix} \begin{pmatrix} x \\ y \end{pmatrix},$$

$$\Phi_1 \begin{pmatrix} x \\ y \end{pmatrix} = \begin{pmatrix} 1/4 & 0 \\ 0 & 1/4 \end{pmatrix} \begin{pmatrix} x \\ y \end{pmatrix} + \begin{pmatrix} 1/4 \\ 1/2 \end{pmatrix},$$

$$\Phi_2 \begin{pmatrix} x \\ y \end{pmatrix} = \begin{pmatrix} 1/4 & 0 \\ 0 & 1/4 \end{pmatrix} \begin{pmatrix} x \\ y \end{pmatrix} + \begin{pmatrix} 1/2 \\ 1/2 \end{pmatrix},$$

$$\Phi_3 \begin{pmatrix} x \\ y \end{pmatrix} = \begin{pmatrix} 1/4 & 0 \\ -1/2 & 1/4 \end{pmatrix} \begin{pmatrix} x \\ y \end{pmatrix} + \begin{pmatrix} 3/4 \\ 1/2 \end{pmatrix}.$$

We denote by S_0 the unit square $[0, 1] \times [0, 1]$ and define

$$S_{n+1} = \bigcup_{k=0}^{3} \Phi_k(S_n)$$

for $n \geq 0$. Clearly $\{S_n\}$ is a monotone decreasing sequence of compact sets and we have

$$\Gamma = \bigcap_{n=0}^{\infty} S_n,$$

because a non-empty compact set X satisfying the set equation $X = \Phi_0(X) \cup \Phi_1(X) \cup \Phi_2(X) \cup \Phi_3(X)$ is unique. The sets S_1 and S_2 are illustrated in Fig. 4.3.

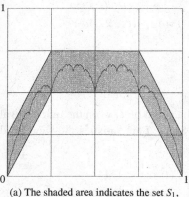

(a) The shaded area indicates the set S_1, which lies under the line $y = 3/4$.

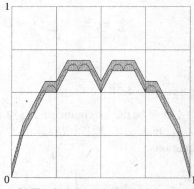

(b) The shaded area indicates the set S_2, which lies under the line $y = 11/16$.

Fig. 4.3

It is not hard to see that the set S_n lies under the horizontal line

$$y = 1 - \frac{1}{4} - \frac{1}{16} - \cdots - \frac{1}{4^n},$$

and therefore we have $M \leq 2/3$, which is the limit of the sum on the right-hand side as $n \to \infty$. Since

$$T\left(\frac{2}{3}\right) = \frac{1}{3} + \frac{1}{3 \cdot 2} + \frac{1}{3 \cdot 2^2} + \cdots = \frac{2}{3},$$

we get $M = 2/3$, as required. It is easy to see from Fig. 4.3(b) that the maximum of T is not attained over $[0, 1/4] \cup [3/4, 1]$. This holds true for every S_{2n}; so the maximum of T is attained at x if and only if every coefficient is either 1 or 2 in the quaternary expansion of x. Indeed the set $E \times \{2/3\}$ is self-similar for two contractions Φ_1 and Φ_2 and its Hausdorff dimension is $1/2$. □

SOLUTION 4.11

Suppose, contrary to the assertion, that there exist $\delta > 0$ and $N > 0$ satisfying

$$\delta < g'(f(x))f'(x)$$

for all $x \geq N$. Integrating from N to M we have

$$\delta(M - N) < \int_N^M g'(f(x))f'(x)\,dx$$

$$= g(f(M)) - g(f(N))$$

$$\leq K - g(f(N)),$$

where K is an upper bound of g. This is a contradiction, because the left-hand side tends to $+\infty$ as $M \to \infty$. $\qquad\square$

SOLUTION 4.12

The proof is substantially due to Hardy and Littlewood (1914). We use Taylor's formula with the integral remainder term

$$f(x + y) = f(x) + yf'(x) + y^2 \int_0^1 (1 - t)f''(x + yt)\,dt,$$

valid for any $x > 1$ and $|y| < 1$. Now let us consider the integral on the right-hand side replacing f'' by f'. By the mean value theorem there is a $\xi_{x,y}$ between 0 and y satisfying

$$y^2 \int_0^1 (1 - t)f'(x + yt)\,dt = \int_x^{x+y} f(s)\,ds - yf(x)$$

$$= yf(x + \xi_{x,y}) - yf(x).$$

Since there exists a positive constant K satisfying $f''(x) + \lambda f'(x) \leq K$, we have

$$f(x + y) - f(x) - yf'(x) + \lambda yf(x + \xi_{x,y}) - \lambda yf(x)$$

$$= y^2 \int_0^1 (1 - t)(f''(x + yt) + \lambda f'(x + yt))\,dt$$

$$\leq Ky^2 \int_0^1 (1 - t)\,dt = \frac{K}{2}y^2.$$

For the case in which $0 < y < 1$ we get therefore

$$f'(x) \geq \frac{f(x + y) - f(x)}{y} + \lambda f(x + \xi_{x,y}) - \lambda f(x) - \frac{K}{2}y;$$

we thus have

$$\liminf_{x\to\infty} f'(x) \ge -\frac{K}{2}y.$$

Hence the limit inferior of $f'(x)$ must be ≥ 0, because y is arbitrary.

Similarly, for the case in which $-1 < y < 0$, we have

$$f'(x) \le \frac{f(x) - f(x - |y|)}{|y|} + \lambda f(x + \xi_{x,y}) - \lambda f(x) + \frac{K}{2}|y|.$$

This implies that

$$\limsup_{x\to\infty} f'(x) \le \frac{K}{2}|y|\,;$$

so the limit superior of $f'(x)$ as $x \to \infty$ is ≤ 0, because y is arbitrary. Therefore $f'(x)$ converges to 0 as $x \to \infty$. □

SOLUTION 4.13

We first consider the case $n = 1$. Assume that $f(x) \in \mathscr{C}^1(0, \infty)$ is a complex-valued function and that $f'(x) + zf(x)$ converges to 0 as $x \to \infty$, where z is a complex number with $\lambda = \operatorname{Re} z > 0$. Differentiating $g(x) = e^{zx}f(x)$ we have

$$g'(x) = e^{zx}(f'(x) + zf(x))\,;$$

therefore $g'(x)e^{-zx}$ converges to 0 as $x \to \infty$. This means that for any $\epsilon > 0$ there exists an x_ϵ satisfying $|g'(x)|\,e^{-\lambda x} < \epsilon$ for all $x \ge x_\epsilon$. We have

$$e^{\lambda x}|f(x)| = |g(x)| \le |g(x_\epsilon)| + \int_{x_\epsilon}^x |g'(t)|\,dt$$

$$< e^{\lambda x_\epsilon}|f(x_\epsilon)| + \epsilon \int_{-\infty}^x e^{\lambda t}\,dt$$

$$= e^{\lambda x_\epsilon}|f(x_\epsilon)| + \frac{\epsilon}{\lambda}e^{\lambda x}$$

and hence

$$|f(x)| < e^{\lambda(x_\epsilon - x)}|f(x_\epsilon)| + \frac{\epsilon}{\lambda}.$$

Thus $|f(x)| < 2\epsilon/\lambda$ for all sufficiently large x; so $f(x)$ converges to 0 as $x \to \infty$.

We next prove the case $n + 1$ by assuming the case n. Let $-\xi$ be a root of

$$x^{n+1} + a_n x^n + \cdots + a_0 = 0$$

with Re $\xi > 0$. Since the polynomial in x on the left-hand side can be written as

$$(x + \xi)(x^n + b_{n-1}x^{n-1} + \cdots + b_0),$$

we can write

$$f^{(n+1)}(x) + a_n f^{(n)}(x) + \cdots + a_0 f(x) = \phi'(x) + \xi \phi(x)$$

where

$$\phi(x) = f^{(n)}(x) + b_{n-1}f^{(n-1)}(x) + \cdots + b_0 f(x).$$

By the assumption of induction $\phi(x)$ converges to 0 and hence so does $f^{(k)}(x)$ as $x \to \infty$ for each integer $0 \le k \le n$. Hence $f^{(n+1)}(x)$ converges to 0 as well. $\square$

✦ ✳ ✦

Chapter 5

Integration

1. For given points $a = x_0 < x_1 < \cdots < x_{n-1} < x_n = b$ we divide I into n subintervals $I_k = [x_{k-1}, x_k]$. This division is denoted by Δ and the maximum among $x_k - x_{k-1}$ is denoted by $|\Delta|$. Let $f(x)$ be a bounded function defined on the interval $I = [a, b]$. The finite sum

$$\sum_{k=1}^{n} f(\xi_k)(x_k - x_{k-1}),$$

where each $\xi_k \in I_k$ is taken arbitrarily, is called a *Riemann sum* associated with the division Δ. In particular, if $x_k - x_{k-1} = (b - a)/n$ for $1 \le k \le n$, the corresponding sum is called a Riemann sum divided equally.

2. A function $f(x)$ is said to be *integrable in the sense of Riemann* (or, simply R-*integrable*) on I provided that the Riemann sum associated with any division with arbitrarily chosen $\{\xi_k\}$ converges to a unique value as $|\Delta| \to 0$. This value is denoted by

$$\int_a^b f(x)\,dx.$$

3. A bounded function $f(x)$ is R-integrable on I if and only if the set of discontinuity points of f in I is a null set. A set $X \subset \mathbb{R}$ is said to be a null set provided that, for any $\epsilon > 0$, one can find a sequence of intervals $\{J_n\}$ satisfying

$$X \subset \bigcup_{n=1}^{\infty} J_n$$

and the total sum of the lengths of J_n is smaller than ϵ.

4. A function $f(x)$ defined on $I = [a, b]$ is said to be of bounded variation on I if

$$\sup_{\Delta} |f(x_{k+1}) - f(x_k)|$$

is finite, where the supremum is taken over all divisions Δ of I. Any function of bounded variation is R-integrable.

5. If $f(x)$ is R-integrable over $[0, 1]$, then of course any Riemann sum divided equally converges; that is,

$$\frac{1}{n} \sum_{k=1}^{n} f\left(\frac{k}{n}\right) \to \int_0^1 f(x)\, dx$$

as $n \to \infty$. This gives a simple definition of the Riemann integral for continuous functions. Then the following question naturally arises: Under what condition on a given sequence $\{a_n\} \subset [0, 1]$ can we assert that

$$\frac{1}{n} \sum_{k=1}^{n} f(a_k) \to \int_0^1 f(x)\, dx$$

as $n \to \infty$ for every continuous function $f(x)$ on $[0, 1]$? If you are interested in this question, you are certainly standing at the door of the theory of uniform distribution (see Chapter 12).

6. For any $f(x) \in \mathscr{C}[a, b]$ and a non-negative R-integrable function $g(x)$ on $[a, b]$, there exists a $c \in (a, b)$ satisfying

$$\int_a^b f(x)g(x)\, dx = f(c) \int_a^b g(x)\, dx.$$

This is called the first mean value theorem.

7. If $f(x)$ is a positive monotone decreasing function and $g(x)$ is an R-integrable function on $[a, b]$, then there exists a $c \in (a, b]$ satisfying

$$\int_a^b f(x)g(x)\, dx = f(a+) \int_a^c g(x)\, dx.$$

This is called the second mean value theorem.

8. For any R-integrable function $f(x)$ on $[a, b]$,

$$\frac{1}{b - a} \int_a^b f(x)\, dx$$

is called the mean of f over $[a, b]$ and denoted by $\mathscr{M}_{[a,b]}(f)$. Sometimes we write this simply as $\mathscr{M}(f)$ if there is no confusion.

9. Let $\alpha, \beta > 1$ be constants satisfying $1/\alpha + 1/\beta = 1$. For any non-negative functions $f, g \in \mathscr{C}[a, b]$ the inequality

$$\mathscr{M}(fg) \leq \mathscr{M}^{1/\alpha}(f^\alpha)\,\mathscr{M}^{1/\beta}(g^\beta)$$

holds and the equality occurs if and only if f^α and g^β are linearly dependent. This is known as *Hölder's inequality* and the case $\alpha = \beta = 2$ is called the *Cauchy-Schwarz inequality*.

10. The remainder terms of rational approximations to some transcendental numbers can be sometimes expressed in a comparatively simple integral. For example, one can find the following example in le Lionnais (1983):

$$\frac{22}{7} - \pi = \int_0^1 \frac{x^4(1-x)^4}{1+x^2}\,dx.$$

Historically Zu Chongzhi (429–501) found two good rational approximations to π: 22/7 and 355/113. It therefore may be interesting to look for an analogous integral formula for $355/113 - \pi$. The latter is correct to 6 decimal places and Delahaye (1997) stated that this represents a precision that Europe had to wait until the 16th century.

Compare with **PROBLEM 6.11** where a similar integral representation for the remainder term for Euler's constant is given, although we do not know whether it is transcendental or not.

◆ ⟿ ✳ ⟾ ◆

┌─ **PROBLEM 5.1** ─────────────────────────────

For any $f \in \mathscr{C}[a, b]$ and any division $\Delta : x_0 = a, x_1, \ldots, x_n = b$, show that there exists a point ξ_i in each subinterval (x_{i-1}, x_i) such that the corresponding Riemann sum of f is exactly equal to $\int_a^b f(x)\,dx$.

└──

┌─ **PROBLEM 5.2** ─────────────────────────────

Prove that π^2 is irrational by using the integral

$$\int_0^\pi x^{2n}(\pi - x)^{2n} \sin x\,dx.$$

└──

This is due to Niven (1947), giving a very simple irrationality proof of π.

PROBLEM 5.3

Suppose that $f \in \mathscr{C}[a, b]$ and $g \in \mathscr{C}(\mathbb{R})$ is a periodic function with period $r > 0$. Show that

$$\lim_{n \to \infty} \int_a^b f(x)g(nx)\,dx = \mathscr{M}_{[0,r]}(g) \int_a^b f(x)\,dx.$$

Taking $g(x) = \sin 2\pi x$, we have

$$\lim_{n \to \infty} \int_0^1 f(x) \sin 2\pi nx\,dx = 0.$$

Indeed, this holds for every integrable function $f(x)$ in the sense of Lebesgue and is known as the *Riemann-Lebesgue lemma*.

PROBLEM 5.4

Show that

$$\int_0^{\pi/2} \prod_{k=1}^n |\cos kx|\,dx = O\left(\frac{1}{n}\right)$$

as $n \to \infty$.

PROBLEM 5.5

For any integer $n \geq 1$ show that the inequality

$$\int_0^1 |f(x)|^n\,|f'(x)|\,dx \leq \frac{1}{n+1} \int_0^1 |f'(x)|^{n+1}\,dx$$

holds for any $f \in \mathscr{C}^1[0, 1]$ if $f(0) = 0$. Verify that the equality holds if and only if $f(x)$ is a linear function.

Opial (1960) showed this inequality for the case $n = 1$ and the general case was proved by Hua (1965).

PROBLEM 5.6

Suppose that both $f(x)$ and $g(x)$ are monotone increasing continuous functions defined on $[a, b]$. Show that

$$\mathscr{M}(fg) \geq \mathscr{M}(f)\mathscr{M}(g).$$

According to Franklin (1885), Hermite stated this theorem in his course, as communicated by Chebyshev.

PROBLEM 5.7

Let $\theta \in [0, 1]$ be a constant and $f \in \mathscr{C}^1[0, 1]$. Show that

$$\sum_{k=0}^{n-1} f\left(\frac{k+\theta}{n}\right) - n \int_0^1 f(x)\,dx$$

converges to $\left(\theta - \dfrac{1}{2}\right)(f(1) - f(0))$ as $n \to \infty$.

PROBLEM 5.8

Prove that

$$\sum_{n=1}^{m} \frac{\sin n\theta}{n} < \int_0^\pi \frac{\sin x}{x}\,dx = 1.8519\ldots$$

for all integers $m \geq 1$ and any $\theta \in [0, \pi]$. Show moreover that the right-hand side cannot be replaced by any smaller number.

This problem conjectured by Fejér(1910) are proved by Jackson(1911) and by Gronwall(1912) independently. See also **PROBLEM 1.7** and **PROBLEM 7.12**.

In general, if $f(x)$ is of bounded variation over $(-\pi, \pi)$, then the nth partial sum $s_n(x)$ of its Fourier series

$$\frac{a_0}{2} + \sum_{n=1}^{\infty} (a_n \cos nx + b_n \sin nx)$$

converges to

$$\frac{f(x+) + f(x-)}{2}$$

as $n \to \infty$ at every point x (see, for example, Zygmund(1979), p.57). Since the left-hand side of the inequality in the problem is equal to $s_n(\theta)$ for

$$f_0(x) = \begin{cases} (\pi - x)/2 & (0 < x < \pi) \\ 0 & (x = 0) \\ -(\pi + x)/2 & (-\pi < x < 0) \end{cases},$$

it follows that $s_n(\theta)$ converges to $f_0(\theta)$ pointwisely. Thus the convergence is not uniform [Why?]. The fact

$$\lim_{n \to \infty} s_n\left(\frac{\pi}{n}\right) = \int_0^\pi \frac{\sin x}{x}\,dx > \frac{\pi}{2} = f_0(0+)$$

is known as the Gibbs phenomenon; that is, the curves $y = s_n(x)$ are compressed into a longer interval than $[-\pi/2, \pi/2]$ on the y-axis as n increases. See also Fig. 13.1. Gibbs(1899) reported it in 'Nature' in response to Michelson(1898), known for his work on the measurement of the speed of light in 1887 and as the first American to receive a Nobel prize in Physics in 1907.

PROBLEM 5.9

Show that

$$\max_{0 \le x \le 1} |f'(x)| \ge 4 \int_0^1 |f(x)| \, dx$$

for any $f \in \mathscr{C}^1[0, 1]$ satisfying $f(0) = f(1) = 0$. Prove also that the coefficient 4 cannot be replaced by any larger number in general.

PROBLEM 5.10

Applying the expansion stated in **PROBLEM 3.10** *to*

$$f(x) = x \left\{ \frac{1}{x} \right\} \left(1 - \left\{ \frac{1}{x} \right\} \right),$$

show that Euler's constant γ can be expressed as

$$\gamma = \frac{1}{2} + \frac{1}{2} \sum \frac{1}{abcd(a+c)(b+d)}$$

where the sum extends over all Farey intervals $[a/b, c/d]$ satisfying $a \ge 1$. Here the symbol $\{x\}$ denotes the fractional part of x; that is $\{x\} = x - [x]$.

---

11. Put

$$A_n = \left\{ a_{j,n} = \frac{2j-1}{2 \cdot 3^n} \,\middle|\, 1 \le j \le 3^n \right\}$$

for $n \ge 0$. The set A_n forms a monotone increasing sequence of subsets in $[0, 1]$ and $A = \bigcup_{n=0}^{\infty} A_n$ is dense in $[0, 1]$. Then there exist two sequences $\{c_{j,n}\}$, $1 \le j \le 3^n$, $n \ge 0$ in $(0, 1)$ and $\{\lambda_n\}$, $n \ge 0$ with $\lambda_n \ge 3^{4n}$ such that

$$\sigma_n(x) = \sum_{k=0}^{n} \psi_k(x) \quad \text{where} \quad \psi_k(x) = \sum_{j=1}^{3^k} c_{j,k} \, \phi(\lambda_k(x - a_{j,k}))$$

satisfies the following two conditions:

$$\max_{0 \le x \le 1} \sigma_n(x) < 1 - \frac{1}{n+2} \tag{5.1}$$

and

$$\min_{1 \le j \le 3^n} \sigma_n(a_{j,n}) > 1 - \frac{1}{n+1}, \tag{5.2}$$

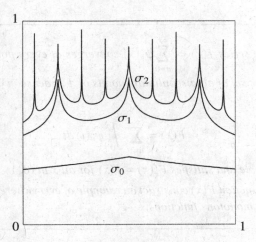

Fig. 5.1   The graphs of σ_0, σ_1 and σ_2 over $[0, 1]$.

where $\phi(x) = (1 + |x|)^{-1/2}$ is a function having a peak. So the graph of each σ_n has several sharp peaks, as illustrated in Fig. 5.1.

Indeed we can construct these sequences inductively, as follows. The case $n = 0$ holds if we define $c_{1,0} = 1/3$ and $\lambda_0 = 1$. Suppose next that (5.1) holds for $n \geq 0$. Then we can take a constant $c_{j,n+1}$ in $(0, 1)$ satisfying

$$1 - \frac{1}{n+2} < c_{j,n+1} + \sigma_n(a_{j,n+1}) < 1 - \frac{1}{n+3}$$

for each $1 \leq j \leq 3^{n+1}$. Since

$$\psi(x, \lambda) = \sum_{j=1}^{3^{n+1}} c_{j,n+1} \phi\big(\lambda(x - a_{j,n+1})\big)$$

converges to $c_{j,n+1}$ at $x = a_{j,n+1}$ and to 0 otherwise as $\lambda \to \infty$, we can take a sufficiently large $\lambda_{n+1} \geq 3^{4n+4}$ such that

$$1 - \frac{1}{n+2} < \sigma_{n+1}(a_{j,n+1}) < 1 - \frac{1}{n+3}$$

for $1 \leq j \leq 3^{n+1}$, and that

$$\max_{0 \leq x \leq 1} \sigma_{n+1}(x) < 1 - \frac{1}{n+3}$$

where $\sigma_{n+1}(x) = \sigma_n(x) + \psi(x, \lambda_{n+1})$. Thus (5.1) and (5.2) hold for $n + 1$.

PROBLEM 5.11

(i) *Clearly the series* $f(x) = \sum\limits_{n=0}^{\infty} \psi_n(x)$ *converges at every point* $x \in [0, 1]$.
Show that the set of all discontinuity points of f *is dense in* $[0, 1]$.

(ii) *Show that*

$$F(x) = \sum_{n=0}^{\infty} \int_0^x \psi_n(t)\, dt$$

is differentiable and satisfies $F'(x) = f(x)$ *for all* x *in* $(0, 1)$.

(iii) *Using the function* $F(x)$ *construct an example of everywhere differentiable but nowhere monotone functions.*

This is substantially due to Katznelson and Stromberg(1974). The first example of everywhere differentiable but nowhere monotone functions was constructed by Köpcke in a series of papers (1887, 89, 90) by graphical construction. Pereno(1897) gave a simpler example by a similar method, which is reproduced in the book of Hobson(1957), pp.412–421.

This kind of examples illustrates a peculiar aspect of 'derivatives'. Darboux showed that every derivative must satisfy the intermediate value theorem, irrespective of whether it is continuous or not. Moreover Baire's theorem implies that the derivative has so many continuity points, because $\varphi'(x)$ is the limit of continuous functions

$$n\left(\varphi\left(x + \frac{1}{n}\right) - \varphi(x)\right)$$

as $n \to \infty$. One may imagine that derivatives behave well. However the above example shows that the derivative may oscillate everywhere.

Some readers may notice that the example $g(x)$ constructed in (iii) is absolutely continuous. Then the derivative $g'(x)$ is not R-integrable on any subinterval. To see this, suppose, on the contrary, that $g'(x)$ is continuous almost everywhere on some subinterval. Then $g'(x)$ vanishes almost everywhere, because it vanishes at every continuity point. Since $g(x)$ is absolutely continuous, $g'(x)$ is integrable in the sense of Lebesgue and therefore

$$g(x) = g(c) + \int_c^x g'(t)\, dt = g(c);$$

so $g(x)$ is a constant function on this subinterval, a contradiction.

PROBLEM 5.12

For each integer $n \geq 1$ let μ_n be the largest positive constant satisfying

$$\int_0^1 P^2(x)\, dx \geq \mu_n \sum_{k=0}^n a_k^2$$

for any real polynomial $P(x) = a_0 + a_1 x + \cdots + a_n x^n$. Show that

$$\limsup_{n \to \infty} \frac{1}{n} \log \mu_n \leq -4 \log(\sqrt{2} + 1).$$

❖ ✳ ❖

Solutions for Chapter 5

SOLUTION 5.1

It follows from the first mean value theorem that there exists a ξ_i in (x_{i-1}, x_i) satisfying

$$\int_{x_{i-1}}^{x_i} f(x)\,dx = f(\xi_i)(x_i - x_{i-1}).$$

Adding these equalities from $i = 1$ to n, we see that the corresponding Riemann sum is equal to $\int_a^b f(x)\,dx$. $\qquad\square$

SOLUTION 5.2

Put

$$P(x) = \frac{1}{(2n)!}\,x^{2n}(\pi - x)^{2n}.$$

Integrating by parts repeatedly we have

$$I_n = \int_0^\pi P(x)e^{ix}dx = -iP(x)e^{ix}\Big|_0^\pi + i\int_0^\pi P'(x)e^{ix}\,dx$$
$$\vdots$$
$$= i(P(0) + P(\pi)) + i^2(P'(0) + P'(\pi))$$
$$+ \cdots + i^{4n+1}\big(P^{(4n)}(0) + P^{(4n)}(\pi)\big).$$

Since $P^{(k)}(0) = P^{(k)}(\pi) = 0$ for all $0 \le k < 2n$ and $P(x) = P(\pi - x)$, we obtain

$$\operatorname{Im} I_n = \sum_{k=n}^{2n}(-1)^k\big(P^{(2k)}(0) + P^{(2k)}(\pi)\big) = 2\sum_{k=n}^{2n}(-1)^k P^{(2k)}(0).$$

On the other hand, for $n \le k \le 2n$,

$$P^{(2k)}(x) = \frac{1}{(2n)!} \sum_{j=0}^{2n} (-1)^j \binom{2n}{j} \pi^{2n-j} (x^{2n+j})^{(2k)}$$

$$= \sum_{j=2k-2n}^{2n} (-1)^j \frac{(2n+j)!}{j!(2n-j)!(2n+j-2k)!} \pi^{2n-j} x^{2n+j-2k},$$

which implies that

$$P^{(2k)}(0) = \frac{(2k)!}{(2k-2n)!(4n-2k)!} \pi^{4n-2k};$$

therefore

$$\mathrm{Im}\, I_n = 2 \sum_{\ell=0}^{n} (-1)^\ell \frac{(4n-2\ell)!}{(2\ell)!(2n-2\ell)!} \pi^{2\ell}.$$

Suppose now that $\pi^2 = p/q$ for some integers $p, q \ge 1$. Since $\mathrm{Im}\, I_n$ is positive, the number $q^n I_n$ is a positive integer and less than

$$\frac{q^n}{(2n)!} \int_0^\pi x^{2n}(\pi - x)^{2n}\, dx < \frac{q^n \pi^{4n}}{2^{4n}(2n)!}.$$

Obviously the right-hand side converges to 0 as $n \to \infty$, a contradiction. $\qquad\square$

REMARK. The formula used in the above proof can be obtained by substituting $z = \pi i$ in a Padé approximation to e^z. One can prove the irrationality of $\log \alpha$ with $\alpha \in \mathbb{Q}, \alpha \ne 1$ in the same way.

$\boxed{\text{SOLUTION 5.3}}$

Without loss of generality we can replace $g(x)$ by $g(x) + c$ for any constant c; so we can assume that $g(x)$ is positive. Using the periodicity of g, we have

$$\int_a^b f(x)g(nx)\, dx = \frac{1}{n} \int_{na}^{nb} f\left(\frac{y}{n}\right) g(y)\, dy$$

$$= \frac{1}{n} \sum_{k=[na/r]+1}^{[nb/r]-1} \int_{kr}^{(k+1)r} f\left(\frac{y}{n}\right) g(y)\, dy + O\left(\frac{1}{n}\right)$$

$$= \frac{1}{n} \sum_{k=[na/r]+1}^{[nb/r]-1} \int_0^r f\left(\frac{kr+s}{n}\right) g(s)\, ds + O\left(\frac{1}{n}\right).$$

It follows from the first mean value theorem that each integral on the right-hand side can be written as

$$\mathcal{M}_{[0,r]}(g) \times r f\left(\frac{kr + s_k}{n}\right)$$

for some s_k in the interval $(0, r)$. Since

$$\frac{r}{n} \sum_{k=[na/r]+1}^{[nb/r]-1} f\left(\frac{kr + s_k}{n}\right)$$

is a Riemann sum of f over $[a, b]$, it converges to $\displaystyle\int_a^b f(x)\,dx$ as $n \to \infty$. $\square$

REMARK. It can be seen that this formula is valid also for any integrable function f in the sense of Lebesgue. To see this, for any epsilon $\epsilon > 0$, take a $\phi \in \mathscr{C}[a, b]$ satisfying $\displaystyle\int_a^b |f(x) - \phi(x)|\,dx < \epsilon$. Then

$$\limsup_{n \to \infty} \left| \int_a^b f(x)g(nx)\,dx - \mathcal{M}(g)\int_a^b f(x)\,dx \right| \le (M + |\mathcal{M}(g)|)\,\epsilon,$$

where M is the maximum of $|g|$ on the interval $[0, r]$.

SOLUTION 5.4

For $n \ge 3$ we define

$$\ell = \left\lceil \frac{\log(2n)}{\log 2} \right\rceil \quad \text{and} \quad m = \left\lceil \frac{\log(2n/3)}{\log 2} \right\rceil$$

so that $2^{\ell-1} \le n < 2^\ell$ and $3 \cdot 2^{m-1} \le n < 3 \cdot 2^m$ hold. Next define

$$f(x) = \prod_{k=0}^{\ell-1} \cos 2^k x = \frac{\sin 2^\ell x}{2^\ell \sin x} \quad \text{and} \quad g(x) = \prod_{k=0}^{m-1} \cos 3 \cdot 2^k x = \frac{\sin 3 \cdot 2^m x}{2^m \sin 3x}.$$

Obviously we have

$$I_n \le \int_0^{\pi/2} |f(x)g(x)|\,dx.$$

Let $\delta \in (0, \pi/6)$ be a constant to be determined later. On the interval $[0, \delta]$ it follows from a trivial estimate $|f(x)g(x)| \le 1$ that

$$\int_0^\delta |f(x)g(x)|\,dx \le \delta.$$

Next, on the interval $[\delta, \pi/6]$, using the fact that $\dfrac{\sin 3x}{3x}$ is monotone decreasing, we have

$$\frac{\sin x}{x} \cdot \frac{\sin 3x}{3x} \ge \frac{\sin \pi/6}{\pi/6} \cdot \frac{\sin \pi/2}{\pi/2} = \frac{6}{\pi^2};$$

that is, $\sin x \sin 3x \ge (18/\pi^2) x^2 > x^2$ holds on $[\delta, \pi/6]$. Hence

$$\int_\delta^{\pi/6} |f(x)g(x)|\, dx \le \frac{1}{2^{\ell+m}} \int_\delta^{\pi/6} \frac{dx}{\sin x \sin 3x}$$

$$\le \frac{1}{2^{\ell+m}} \int_\delta^{\pi/6} \frac{dx}{x^2}$$

$$< \frac{1}{2^{\ell+m}\delta} < \frac{3}{\delta n^2}.$$

Finally, on the interval $[\pi/6, \pi/2]$, using $|g(x)| \le 1$ and $\sin x \ge 1/2$, we get

$$\int_{\pi/6}^{\pi/2} |f(x)g(x)|\, dx \le \frac{1}{2^\ell} \int_{\pi/6}^{\pi/2} \frac{dx}{\sin x} < \frac{2\pi}{3 \cdot 2^\ell} < \frac{3}{n}.$$

Thus we have

$$\int_0^{\pi/2} \prod_{k=1}^n |\cos kx|\, dx \le \delta + \frac{3}{\delta n^2} + \frac{3}{n}.$$

The right-hand side attains its minimum $\dfrac{2\sqrt{3}+3}{n}$ at $\delta = \dfrac{\sqrt{3}}{n}$. $\qquad \square$

SOLUTION 5.5

We introduce the auxiliary function

$$\phi(x) = \frac{x^n}{n+1} \int_0^x |f'(s)|^{n+1}\, ds - \int_0^x |f(s)|^n\, |f'(s)|\, ds.$$

Obviously $\phi(0) = 0$ and

$$\phi'(x) = \frac{nx^{n-1}}{n+1} \int_0^x |f'(s)|^{n+1}\, ds + \frac{x^n}{n+1} |f'(x)|^{n+1} - |f(x)|^n\, |f'(x)|.$$

Applying Hölder's inequality to 1 and $|f'(x)|$ we get

$$|f(x)| = \left| \int_0^x f'(s)\, ds \right| \le \int_0^x 1 \cdot |f'(s)|\, ds$$

$$\le x^{n/(n+1)} \left(\int_0^x |f'(s)|^{n+1}\, ds \right)^{1/(n+1)}$$

and therefore

$$\int_0^x |f'(s)|^{n+1}\, ds \geq \frac{|f(x)|^{n+1}}{x^n},$$

whence

$$\phi'(x) \geq \frac{n}{n+1} \cdot \frac{|f(x)|^{n+1}}{x} + \frac{x^n}{n+1} |f'(x)|^{n+1} - |f(x)|^n\, |f'(x)|.$$

Multiplying the expression on the right-hand side by $(n + 1)x$, we see that it can be written as $\sigma(|f(x)|, x\,|f'(x)|)$, where

$$\sigma(u, v) = nu^{n+1} + v^{n+1} - (n + 1)u^n v.$$

Since $\sigma(u, 0) \geq 0$, we can assume $v > 0$. Put $t = a/b \geq 0$ for brevity. Then

$$\frac{\sigma(u, v)}{v^{n+1}} = nt^{n+1} + 1 - (n + 1)t^n$$

attains its minimum 0 at $t = 1$, which implies that $\phi(x)$ is monotone increasing. In particular we have $\phi(1) \geq 0$, as required.

The equality in Hölder's inequality occurs if and only if $f(x)$ is linear. Indeed the equality occurs actually in the inequality in question.   □

SOLUTION 5.6

It suffices to show that

$$\int_a^b f(x)\phi(x)\, dx \geq 0$$

where

$$\phi(x) = g(x) - \mathscr{M}_{[a,b]}(g).$$

By the mean value theorem there is a ξ in (a, b) satisfying $g(\xi) = \mathscr{M}_{[a,b]}(g)$. Since $\phi(x) \leq 0$ for $a \leq x \leq \xi$ and $\phi(x) \geq 0$ for $\xi \leq x \leq b$, we have

$$\int_a^b f(x)\phi(x)\, dx = \int_a^\xi f(x)\phi(x)\, dx + \int_\xi^b f(x)\phi(x)\, dx$$

$$\geq f(\xi)\int_a^\xi \phi(x)\, dx + f(\xi)\int_\xi^b \phi(x)\, dx$$

$$= f(\xi)\int_a^b \phi(x)\, dx = 0.$$ □

REMARK. Here is another proof due to Franklin(1885) using a double integral. Since

$$(f(x) - f(y))(g(x) - g(y))$$

is non-negative for any x, y in $[a, b]$, we have

$$\iint\limits_{S} (f(x) - f(y))(g(x) - g(y)) \, dx \, dy \geq 0$$

where S is the square $[a, b]^2$. Observe that the above double integral is equal to

$$2(b - a)^2(\mathcal{M}(fg) - \mathcal{M}(f)\mathcal{M}(g)).$$

Moreover it follows that the equality $\mathcal{M}(fg) = \mathcal{M}(f)\mathcal{M}(g)$ occurs if and only if either f or g is a constant function. To see this, suppose, on the contrary, that both f and g are not constant. Then there exist two points $x_0, x_1 \in [a, b]$ satisfying $f(x_0) \neq f(x_1)$. Since $(f(x) - f(y))(g(x) - g(y))$ vanishes identically, we get $g(x_0) = g(x_1)$. Take next the third point $x_2 \in [a, b]$ satisfying $g(x_0) \neq g(x_2)$. Then $f(x_2) = f(x_0)$ and $f(x_2) = f(x_1)$, a contradiction.

SOLUTION 5.7

We have

$$S_n = \sum_{k=0}^{n-1} f\left(\frac{k + \theta}{n}\right) - n \int_0^1 f(x) \, dx$$

$$= n \sum_{k=0}^{n-1} \int_{k/n}^{(k+1)/n} \left(f\left(\frac{k + \theta}{n}\right) - f(x) \right) dx$$

$$= n \sum_{k=0}^{n-1} \int_{k/n}^{(k+1)/n} \int_x^{(k+\theta)/n} f'(s) \, ds \, dx$$

$$= n \sum_{k=0}^{n-1} \left(\int_{k/n}^{(k+\theta)/n} \int_x^{(k+\theta)/n} - \int_{(k+\theta)/n}^{(k+1)/n} \int_{(k+\theta)/n}^x \right) f'(s) \, ds \, dx.$$

Changing the order of integration we get

$$\int_{k/n}^{(k+\theta)/n} \int_x^{(k+\theta)/n} f'(s) \, ds \, dx = \int_{k/n}^{(k+\theta)/n} \int_{k/n}^s f'(s) \, dx \, ds$$

$$= \int_{k/n}^{(k+\theta)/n} f'(s) \left(s - \frac{k}{n} \right) ds$$

and

$$\int_{(k+\theta)/n}^{(k+1)/n} \int_{(k+\theta)/n}^{x} f'(s) \, ds \, dx = \int_{(k+\theta)/n}^{(k+1)/n} \int_{s}^{(k+1)/n} f'(s) \, dx \, ds$$

$$= \int_{(k+\theta)/n}^{(k+1)/n} f'(s) \left(\frac{k+1}{n} - s \right) ds.$$

Therefore

$$S_n = \sum_{k=0}^{n-1} \int_{k/n}^{(k+1)/n} f'(s) \phi_k(s) \, ds$$

where

$$\phi_k(s) = ns - k - \begin{cases} 0 & (k \le ns < k + \theta) \\ 1 & (k + \theta \le ns < k + 1) \end{cases}.$$

Denoting by $\{x\}$ the fractional part of x and noticing that $k = [ns]$ for $k \le ns < k + 1$, we see that $\phi_k(s) = g(ns)$, where

$$g(x) = \{x\} - \begin{cases} 0 & (\{x\} < \theta) \\ 1 & (\{x\} \ge \theta) \end{cases}$$

$$= \{x - \theta\} + \theta - 1$$

is a piecewise linear periodic function of period 1. Since **Solution 5.3** is valid for any piecewise continuous periodic function $g(x)$, we conclude that

$$S_n = \int_0^1 f'(x) g(nx) \, dx \to \mathcal{M}_{[0,1]}(g) \times \int_0^1 f'(x) \, dx$$

$$= \left(\theta - \frac{1}{2} \right) (f(1) - f(0)),$$

as $n \to \infty$, because $f' \in \mathscr{C}[0, 1]$. □

Remark. This formula is valid also for every absolutely continuous function f, because f' is integrable in the sense of Lebesgue. See the remark stated after **Solution 5.3**.

However it does not hold for $f \in \mathscr{C}[0, 1]$ in general. For example, consider the Takagi function $T(x)$ defined in **Problem 4.8** with $\theta = 0$. For $0 \le k < 2^n$ we put

$$\frac{k}{2^n} = \frac{a_1^{(k)}}{2} + \frac{a_2^{(k)}}{2^2} + \cdots + \frac{a_n^{(k)}}{2^n}, \quad a_j^{(k)} \in \{0, 1\}.$$

Then it follows from **PROBLEM 4.9** that

$$T\left(\frac{k}{2^n}\right) = \sum_{m=1}^{\infty} \frac{v_m^{(k)} - 2a_m^{(k)} v_m^{(k)} + m a_m^{(k)}}{2^m}$$

where $v_m^{(k)} = a_1^{(k)} + \cdots + a_m^{(k)}$. Since

$$\sum_{0 \le k < 2^n} a_i^{(k)} = \begin{cases} 2^{n-1} & (i \le n) \\ 0 & (i > n) \end{cases} \quad \text{and} \quad \sum_{0 \le k < 2^n} a_i^{(k)} a_j^{(k)} = \begin{cases} 2^{n-2} & (i \ne j \le n) \\ 0 & (\min(i,j) > n) \end{cases},$$

one can see that

$$\sum_{0 \le k < 2^n} T\left(\frac{k}{2^n}\right) = \sum_{m=1}^{n} (m-1) 2^{n-m-1} + \sum_{m>n} n 2^{n-m-1} = 2^{n-1} - \frac{1}{2};$$

so,

$$S_{2^n} = 2^{n-1} - \frac{1}{2} - 2^n \int_0^1 T(x)\, dx = -\frac{1}{2},$$

while $T(1) = T(0) = 0$.

SOLUTION 5.8

As is already seen in **SOLUTION 1.7**, the maximal values in the interval $[0, \pi]$ of the function

$$s_m(\theta) = \sin\theta + \frac{\sin 2\theta}{2} + \cdots + \frac{\sin m\theta}{m}$$

are attained at $[(m+1)/2]$ points: $\pi/(m+1), 3\pi/(m+1), \ldots$, etc. Put $\vartheta = \theta/2$ for brevity. Since

$$s_m'(\theta) = \frac{1}{2} \sin 2(m+1)\vartheta \cot\vartheta - \cos^2(m+1)\vartheta,$$

we get

$$s_m(\beta) - s_m(\alpha) < \int_{\alpha/2}^{\beta/2} \sin 2(m+1)\vartheta \cot\vartheta\, d\vartheta$$

for any $0 < \alpha < \beta < \pi$. Substituting

$$t = 2(m+1)\vartheta - 2\ell\pi, \quad \alpha = \frac{2\ell - 1}{m+1}\pi \quad \text{and} \quad \beta = \frac{2\ell + 1}{m+1}\pi,$$

we see that the above integral on the right-hand side can be written as

$$\frac{1}{2(m+1)} \int_0^\pi \sin t \left(\cot \frac{2\ell\pi + t}{2(m+1)} - \cot \frac{2\ell\pi - t}{2(m+1)} \right) dt.$$

Since the function $\cot s$ is strictly monotone decreasing on the interval $(0, \pi/2)$, the above integral is clearly negative. This implies that the maximum of $s_m(\theta)$ on the interval $[0, \pi]$ is attained at $\theta = \pi/(m+1)$. Moreover, since

$$s_{m+1}\left(\frac{\pi}{m+2}\right) > s_{m+1}\left(\frac{\pi}{m+1}\right) = s_m\left(\frac{\pi}{m+1}\right),$$

the sequence $\{s_m(\pi/(m+1))\}$ is strictly monotone increasing and

$$s_m\left(\frac{\pi}{m+1}\right) = \frac{\pi}{m+1} \sum_{n=1}^{m+1} \frac{m+1}{n\pi} \sin \frac{n\pi}{m+1}$$

converges to the integral $\displaystyle\int_0^\pi \frac{\sin x}{x} \, dx$ as $m \to \infty$. $\qquad\square$

SOLUTION 5.9

Let $g(x)$ be one of the four functions $f(x)$, $-f(x)$, $f(1-x)$ and $-f(1-x)$. Let α be the maximum of $|g'(x)|$ over the interval $[0, 1]$. It also equals to the maximum of $|f'(x)|$ over $[0, 1]$. We can assume $\alpha > 0$; for otherwise, $f(x)$ would identically vanish. Suppose now that there is a point x_0 in $(0, 1)$ satisfying $g(x_0) > \alpha x_0$. By the mean value theorem there exists a point ξ in $(0, x_0)$ satisfying $g(x_0) = g'(\xi)x_0 > \alpha x_0$. However this implies $g'(\xi) > \alpha$, contrary to the definition of α. We thus have $g(x) \le \alpha x$ for all $0 < x < 1$ so that $|f(x)| \le \alpha \max\{x, 1-x\}$. Hence

$$\int_0^1 |f(x)| \, dx \le \alpha \int_0^1 \max\{x, 1-x\} \, dx = \frac{\alpha}{4}.$$

The equality does not occur, because the function $\max\{x, 1-x\}$ is not in $\mathscr{C}^1$ class. However we can modify it slightly in the neighborhood of $x = 1/2$ to become a continuously differentiable one so that the difference between $\displaystyle\int_0^1 |f(x)| \, dx$ and $\alpha/4$ becomes sufficiently small. $\qquad\square$

SOLUTION 5.10

Let $c_J(f)$ be the coefficient defined in **PROBLEM 3.10** for

$$f(x) = x\left\{\frac{1}{x}\right\}\left(1 - \left\{\frac{1}{x}\right\}\right) \in \mathscr{C}[0, 1].$$

Clearly $c_J(f) = 0$ for any Farey interval $J = [0/1, c/d]$, because such an interval must be $[0/1, 1/k]$ for some k. We next note that any Farey interval J with $a \ne 0$ does not contain $1/k$ as an interior point. For otherwise, $a/b < 1/k < c/d$ for some k and hence

$$\frac{1}{bd} = \frac{c}{d} - \frac{1}{k} + \frac{1}{k} - \frac{a}{b} \geq \frac{1}{dk} + \frac{1}{bk};$$

that is, $k \geq b + d$. This is a contradiction, because $k < b/a \leq b$. Therefore any Farey interval J with $a \neq 0$ is contained in some interval $[1/(k+1), 1/k]$ and we have

$$f(x) = 2k + 1 - k(k+1)x - \frac{1}{x}.$$

Since $c_J(g) = 0$ if g is linear on J, we have

$$c_J(f) = c_J\left(-\frac{1}{x}\right) = -\frac{b+d}{a+c} + \frac{b}{b+d} \cdot \frac{b}{a} + \frac{d}{b+d} \cdot \frac{d}{c}$$

$$= \frac{1}{ac(a+c)(b+d)}.$$

It follows from **PROBLEM 3.10** that

$$\sum_{n=1}^{\infty} \sum_{J \in I_n} c_J(f)\,\phi_J(x),$$

converges to $f(x)$ uniformly on $[0, 1]$. Hence

$$\int_0^1 f(x)\,dx = \sum_{n=1}^{\infty} \sum_{J \in I_n} c_J(f) \int_0^1 \phi_J(x)\,dx$$

$$= \frac{1}{2} \sum \frac{1}{abcd(a+c)(b+d)},$$

where the last sum extends over all Farey intervals $[a/b, c/d]$ satisfying $a \geq 1$. On the other hand, we get

$$\int_0^1 f(x)\,dx = \int_1^{\infty} \frac{\{x\}(1 - \{x\})}{x^3}\,dx$$

$$= \sum_{n=1}^{\infty} \int_n^{n+1} \frac{(x-n)(n+1-x)}{x^3}\,dx$$

$$= \sum_{n=1}^{\infty} \left(\frac{1}{2n} + \frac{1}{2n+2} - \log\left(1 + \frac{1}{n}\right)\right).$$

The last sum is equal to

$$\lim_{N \to \infty} \left(1 + \cdots + \frac{1}{N+1} - \log(N+1) - \frac{1}{2} - \frac{1}{2N+2}\right) = \gamma - \frac{1}{2}. \qquad \square$$

SOLUTION 5.11

(i) It follows from (5.1) and (5.2) that $0 < f(x) \leq 1$ and $f(x) = 1$ for any $x \in A$. To show the discontinuity of f we put

$$B_n = \left\{ b_{k,n} = \frac{k}{3^n} \,\middle|\, 0 \leq k \leq 3^n \right\} \quad \text{and} \quad B = \bigcup_{n=0}^{\infty} B_n.$$

The sequence $\{B_n\}$ is monotone increasing and B is a dense subset of $[0,1]$. Observe that $A \cap B = \emptyset$. Since

$$|a_{j,m} - b_{k,n}| = \left| \frac{2j-1}{2 \cdot 3^m} - \frac{k}{3^n} \right| \geq \frac{1}{2 \cdot 3^m}$$

holds if $m \geq n$, we have

$$\psi_m(b_{k,n}) = \sum_{j=1}^{3^m} c_{j,m} \phi(\lambda_m(b_{k,n} - a_{j,m}))$$

$$< 3^m \phi\left(\frac{\lambda_m}{2 \cdot 3^m} \right) < \frac{\sqrt{2}}{3^{m/2}}.$$

Therefore

$$f(b_{k,n}) = \sum_{\ell=0}^{n-1} \psi_\ell(b_{k,n}) + \sum_{m=n}^{\infty} \psi_m(b_{k,n})$$

$$< 1 - \frac{1}{n+1} + \sum_{m=n}^{\infty} \frac{\sqrt{2}}{3^{m/2}},$$

which is certainly less than 1 for sufficiently large n. Obviously f is discontinuous at every $b_{k,n}$.

(ii) We first show that the function $\phi(x) = (1 + |x|)^{-1/2}$ on $\mathbb{R}$ has the property

$$\mathcal{M}_{[a,b]}(\phi) < 4 \min(\phi(a), \phi(b)) \qquad (5.3)$$

for any $a < b$. To see this it suffices to consider only two cases: $0 \leq a < b$ and $a < 0 < b$, $|a| \leq b$, because ϕ is an even function. If $0 \leq a < b$, then

$$\mathcal{M}_{[a,b]}(\phi) = \frac{2}{\sqrt{1+a} + \sqrt{1+b}} < 2\phi(b).$$

Similarly, if $a < 0 < b$ and $|a| \leq b$, then

$$\mathcal{M}_{[a,b]}(\phi) = \frac{2}{|a|+b}\left(\sqrt{1+|a|} + \sqrt{1+b} - 2 \right),$$

which is less than

$$\frac{4}{b}\left(\sqrt{1+b}-1\right) = \frac{4}{\sqrt{1+b}+1} < 4\phi(b),$$

as required. For any $\lambda > 0$ the function $\tilde{\phi}(x) = \phi(\lambda x)$ also satisfies (5.3) in view of

$$\begin{aligned}
\mathcal{M}_{[a,b]}(\tilde{\phi}) &= \mathcal{M}_{[\lambda a, \lambda b]}(\phi) \\
&< 4\min(\phi(\lambda a), \phi(\lambda b)) \\
&= 4\min(\tilde{\phi}(a), \tilde{\phi}(b)).
\end{aligned}$$

Since $\mathcal{M}_I(f)$ is a linear functional and since

$$\min(a, b) + \min(c, d) \le \min(a+c, b+d)$$

for any a, b, c, d, it follows that $\alpha f + \beta g$ also satisfies (5.3) if both f and g satisfy (5.3) whenever $\alpha, \beta > 0$. Thus every $\psi_n(x)$ defined in p. 80 also satisfies (5.3). Hence, noting that

$$\left|\int_0^x \psi_n(t)\,dt\right| \le 4x\psi_n(0) \le 4\psi_n(0)$$

for $0 \le x \le 1$, we infer that the series

$$F(x) = \sum_{n=0}^{\infty} \int_0^x \psi_n(t)\,dt$$

converges absolutely and uniformly on $[0, 1]$. Therefore we have $F \in \mathscr{C}[0, 1]$. For an arbitrarily fixed $x \in (0, 1)$ and $\epsilon > 0$, we can take an integer $N = N(x, \epsilon)$ satisfying

$$\sum_{n=N}^{\infty} \psi_n(x) < \epsilon.$$

We next take a sufficiently small $\delta > 0$ such that

$$|\psi_k(t) - \psi_k(x)| < \frac{\epsilon}{N}$$

for any integer $0 \le k < N$ and all t with $|x - t| < \delta$. Then, for any $0 < |h| < \delta$, we obtain

$$\left| \frac{F(x+h) - F(x)}{h} - \sum_{n=0}^{\infty} \psi_n(x) \right| = \left| \sum_{n=0}^{\infty} \frac{1}{h} \int_x^{x+h} (\psi_n(t) - \psi_n(x))\, dt \right|$$

$$\leq \epsilon + \sum_{n=N}^{\infty} \Big(\mathcal{M}_{[x, x+h]}(\psi_n) + \psi_n(x) \Big)$$

$$\leq \epsilon + 5 \sum_{n=N}^{\infty} \psi_n(x) < 6\epsilon.$$

Therefore $F(x)$ is differentiable and

$$F'(x) = \sum_{n=0}^{\infty} \psi_n(x) = f(x)$$

for all $x \in (0, 1)$.

(iii) Finally we construct an example of everywhere differentiable but nowhere monotone functions. Put

$$g(x) = F(x) - F\left(x - \frac{1}{6} \right)$$

for $1/6 < x < 1$. Since $a_{j,n} - 1/6 = b_{k,n}$ if $k = j - (3^{n-1} + 1)/2 \geq 0$, we have

$$g'(a_{j,n}) = f(a_{j,n}) - f(b_{k,n}) > 0.$$

Similarly, since $b_{k,n} - 1/6 = a_{j,n}$ if $j = k - (3^{n-1} - 1)/2 \geq 1$, we get

$$g'(b_{k,n}) = f(b_{k,n}) - f(a_{j,n}) < 0.$$

Since both A and B are dense subsets of $[0, 1]$, $g(x)$ is nowhere monotone. $\square$

SOLUTION 5.12

The constant μ_n is the minimum of the function

$$f(a_0, a_1, \ldots, a_n) = \int_0^1 (a_0 + a_1 x + \cdots + a_n x^n)^2\, dx$$

under the condition

$$g(a_0, a_1, \ldots, a_n) = \sum_{k=0}^{n} a_k^2 - 1 = 0.$$

Since the constraint surface $g = 0$ is smooth and compact, the function f certainly attains its minimum μ_n at some point on the surface. Let λ be a Lagrange multiplier. Then

$$0 = \frac{\partial f}{\partial a_k} - \lambda \frac{\partial g}{\partial a_k}$$

$$= 2 \int_0^1 (a_0 + a_1 x + \cdots + a_n x^n) x^k \, dx - 2\lambda a_k;$$

therefore

$$\frac{a_0}{k+1} + \frac{a_1}{k+2} + \cdots + \frac{a_n}{k+n+1} = \lambda a_k \qquad (5.4)$$

for $0 \le k \le n$. Hence λ is an eigenvalue of the following symmetric matrix

$$A = \begin{pmatrix} 1 & \frac{1}{2} & \cdots & \frac{1}{n+1} \\ \frac{1}{2} & \frac{1}{3} & \cdots & \frac{1}{n+2} \\ \vdots & \vdots & \ddots & \vdots \\ \frac{1}{n+1} & \frac{1}{n+2} & \cdots & \frac{1}{2n+1} \end{pmatrix}.$$

Since every leading principal minors has a positive determinant, it follows from Sylvester's criterion that all the eigenvalues of A are positive. From (5.4) we have

$$f(a_0, a_1, \ldots, a_n) = \sum_{0 \le j,k \le n} \frac{a_j a_k}{j+k+1}$$

$$= \lambda \sum_{k=0}^{n} a_k^2 = \lambda,$$

which implies that μ_n is the least eigenvalue of A. However it seems to be difficult to obtain a good upper estimate of μ_n from the characteristic polynomial of A.

To obtain an upper estimate of μ_n we use the following polynomial

$$Q(x) = \frac{1}{n!} (x^n (1 - x^n))^{(n)}.$$

Since $Q(x) = (-1)^n P_n(2x - 1)$ where $P_n(x)$ is the nth Legendre polynomial, we have

$$\int_0^1 Q^2(x) \, dx = \frac{1}{2} \int_{-1}^1 (P_n(x))^2 \, dx = \frac{1}{2n+1}$$

by **Problem 14.1**. Therefore

$$\mu_n \le \frac{1}{(2n+1)A_n} \quad \text{where} \quad A_n = \sum_{k=0}^n \binom{n}{k}^2 \binom{n+k}{n}^2.$$

Now let M_n be the maximum of $\binom{n}{k}\binom{n+k}{n}$ for $0 \le k \le n$. Obviously we have

$$M_n \le B_n = \sum_{k=0}^{n} \binom{n}{k}\binom{n+k}{n} \le (n+1)M_n.$$

Since $B_n = Q(-1) = (-1)^n P_n(-3)$, it follows from (14.1) that

$$\lim_{n\to\infty} M_n^{1/n} = \lim_{n\to\infty} B_n^{1/n} = \lim_{n\to\infty} |P_n(-3)|^{1/n}$$
$$= \max_{0 \le \theta \le \pi} \left(3 + 2\sqrt{2}\cos\theta\right)$$
$$= \left(\sqrt{2}+1\right)^2.$$

Similarly, using the inequality $M_n^2 \le A_n \le (n+1)M_n^2$ we finally get

$$\lim_{n\to\infty} A_n^{1/n} = \lim_{n\to\infty} M_n^{2/n} = \left(\sqrt{2}+1\right)^4,$$

which implies that

$$\limsup_{n\to\infty} \frac{1}{n}\log\mu_n \le -4\log\left(\sqrt{2}+1\right). \qquad \square$$

❖ ❋ ❖

Chapter 6

Improper Integrals

In the previous chapter the integral was defined for bounded functions on closed bounded intervals. The notion of integrals can be extended to (unbounded) functions defined on open (unbounded) intervals.

[Summary of Basic Points]

1. Suppose that a function $f(x)$ defined on an interval $[a, b)$ is R-integrable on the interval $[a, b - \epsilon]$ for any ϵ in $(0, b - a)$. If the Riemann integral

$$\int_a^{b-\epsilon} f(x)\,dx$$

converges as $\epsilon \to 0+$, then this limit is denoted by $\int_a^b f(x)\,dx$ and called the *improper integral* of $f(x)$ over $[a, b)$. The improper integral can similarly be defined for bounded intervals like $(a, b]$ or (a, b).

2. If a function $f(x)$ defined on an interval $[a, \infty)$ is R-integrable on $[a, b]$ for any $b > a$, we define $\int_a^\infty f(x)\,dx$ as the limit of $\int_a^b f(x)\,dx$ as $b \to \infty$ if it exists. Similarly we can define the improper integral for unbounded intervals like $(-\infty, b]$ or $(-\infty, \infty)$.

3. If a continuous function $f(x)$ defined on $[1, \infty)$ is positive and monotone decreasing, then $\sum_{n=1}^\infty f(n)$ converges if and only if $\int_1^\infty f(x)\,dx$ converges.

4. The convergence of the improper integral of $f(x)$ does not imply that of $|f(x)|$

in general. For example, Fresnel's integral

$$\int_0^\infty \frac{\sin x}{\sqrt{x}}\, dx = \sqrt{\frac{\pi}{2}}$$

converges, while $\int_0^\infty \frac{|\sin x|}{\sqrt{x}}\, dx$ diverges. If $\int_a^b |f(x)|\, dx$ converges, then we say that $\int_a^b f(x)\, dx$ converges absolutely.

PROBLEM 6.1

Find an example of $f \in \mathscr{C}^\infty(0, 1]$ such that

$$\int_0^1 |f(x)|\, dx$$

converges, while a Riemann sum

$$\frac{1}{n} \sum_{k=1}^n f\!\left(\frac{k}{n}\right)$$

diverges as $n \to \infty$.

PROBLEM 6.2

Find the set A of $\alpha \in \mathbb{R}$ at which the improper integral

$$\int_0^\infty \sin x^\alpha\, dx$$

converges. Find also the set $B \subset A$ of α where the above integral converges absolutely.

PROBLEM 6.3

Let $f, g \in \mathscr{C}(\mathbb{R})$. Suppose that

$$\int_{-\infty}^\infty |f(x)|\, dx$$

converges and $g(x)$ is a periodic function with period $r > 0$. Show then that

$$\lim_{n \to \infty} \int_{-\infty}^\infty f(x)g(nx)\, dx = \mathscr{M}_{[0,r]}(g) \int_{-\infty}^\infty f(x)\, dx.$$

PROBLEM 6.4

Show that

$$\int_0^1 \frac{dx}{x^x} = \frac{1}{1^1} + \frac{1}{2^2} + \frac{1}{3^3} + \cdots + \frac{1}{n^n} + \cdots .$$

Although there is no trick used in the proof, the expression is interesting, because the relation

$$\int_0^1 f(x)\,dx = \sum_{n=1}^{\infty} f(n)$$

holds when $f(x) = x^{-x}$. As an another example, we have $f(x) = a^{-x}$ where

$$a = 2.536591355...$$

is a unique real root greater than 1 of the equation

$$a - 2 + \frac{1}{a} = \log a.$$

It is remarkable that

$$\int_0^1 x^x\,dx = \frac{1}{1^1} - \frac{1}{2^2} + \frac{1}{3^3} - \frac{1}{4^4} + \cdots$$

was already noticed by Johannis Bernoulli(1697). This appeared also in the William Lowell Putnam Mathematical Competition(1970) and in Elementary Problems and Solutions proposed by Klamkin(1970) in Amer. Math. Monthly. See also p. 308 of Ramanujan's Notebooks Part IV edited by Berndt(1994).

PROBLEM 6.5

Show that

$$\lim_{n \to \infty} \sqrt{n} \int_{-\infty}^{\infty} \frac{dx}{(1 + x^2)^n} = \sqrt{\pi}.$$

PROBLEM 6.6

Show that

$$\int_0^{\infty} \frac{e^{-x/s - 1/x}}{x}\,dx \sim \log s$$

as $s \to \infty$.

PROBLEM 6.7

Suppose that $g \in \mathscr{C}[0, \infty)$ is monotone decreasing and

$$\int_0^\infty g(x)\,dx$$

converges. (Note that $g(x)$ is non-negative.) Show then that

$$\lim_{h \to 0+} h \sum_{n=1}^\infty f(nh) = \int_0^\infty f(x)\,dx$$

for any $f \in \mathscr{C}[0, \infty)$ satisfying $|f(x)| \le g(x)$ for all $x \ge 0$.

PROBLEM 6.8

Evaluate

$$\int_0^\infty e^{-(x-s/x)^2}\,dx$$

for $s > 0$.

PROBLEM 6.9

Show that a rational function $R(x)$ satisfies

$$\int_{-\infty}^\infty f(R(x))\,dx = \int_{-\infty}^\infty f(x)\,dx$$

for all piecewise continuous functions $f(x)$ such that

$$\int_{-\infty}^\infty f(x)\,dx$$

exists, if and only if

$$R(x) = \pm\left(x - \alpha_0 - \sum_{k=1}^m \frac{c_k}{x - \alpha_k}\right)$$

for some integer $m \ge 0$, real constants $\alpha_0, \alpha_1, \ldots, \alpha_m$ with $\alpha_1 < \cdots < \alpha_m$ and positive constants $c_1, \ldots, c_m$.

This is the problem posed by Pólya (1931) and solved by Szegö (1934).

PROBLEM 6.10

Show that

$$\gamma = \int_0^1 \frac{1-\cos x}{x}\,dx - \int_1^\infty \frac{\cos x}{x}\,dx,$$

where γ is Euler's constant.

This formula will be used in the proof of Kummer's series for $\log \Gamma(s)$. See
PROBLEM 16.11. It can be easily verified by differentiation that

$$\log s + \gamma = \int_0^s \frac{1-\cos x}{x}\,dx - \int_s^\infty \frac{\cos x}{x}\,dx$$

for all $s > 0$.

PROBLEM 6.11

Show that

$$\frac{7}{12} - \gamma = \int_0^\infty \frac{\{x\}^2\,(1-\{x\})^2}{(1+x)^5}\,dx$$

where γ is Euler's constant and $\{x\}$ denotes the fractional part of x.

---

* ❋ *

Solutions for Chapter 6

SOLUTION 6.1

Let $\{a_n\}_{n\geq 2}$ be a positive sequence satisfying

$$a_n + a_{n+1} < \frac{1}{n(n+1)} \quad \text{and} \quad \sum_{n=2}^{\infty} n^2 a_n < \infty.$$

We also define

$$\phi_n(x) = \begin{cases} \exp\left(\dfrac{1}{a_n^2} + \dfrac{1}{x^2 - a_n^2}\right) & (-a_n < x < a_n) \\ 0 & (\text{otherwise}) \end{cases}$$

for all $n \geq 2$. Clearly $\phi_n \in \mathscr{C}^\infty(\mathbb{R})$, $\phi_n(x) \geq 0$, $\text{supp}(\phi_n) = [-a_n, a_n]$ and $\phi_n(0) = 1$ is the maximum of ϕ_n over $\mathbb{R}$. Consider the series

$$f(x) = \sum_{n=2}^{\infty} n^2 \phi_n\left(x - \frac{1}{n}\right).$$

Since $f(x)$ is the sum of $\mathscr{C}^\infty$-functions whose supports are disjoint mutually, we see that $f \in \mathscr{C}^\infty(0, 1]$, $f(x) \geq 0$ and $f(1/n) = n^2$. We also have

$$\int_0^1 f(x)\,dx = \sum_{n=2}^{\infty} n^2 \int_0^1 \phi_n\left(x - \frac{1}{n}\right) dx$$

$$< 2 \sum_{n=2}^{\infty} n^2 a_n < \infty.$$

However

$$\frac{1}{n} \sum_{k=1}^{n} f\left(\frac{k}{n}\right) \geq \frac{1}{n} f\left(\frac{1}{n}\right) = n.$$

$\square$

SOLUTION 6.2

Clearly A does not contain $\alpha = 0$. We first consider the case $\alpha > 0$.

Suppose that the improper integral converges. We have

$$\alpha \int_{(2n\pi)^{1/\alpha}}^{((2n+1)\pi)^{1/\alpha}} \sin x^\alpha \, dx = \int_{2n\pi}^{(2n+1)\pi} \frac{\sin t}{t^{1-1/\alpha}} \, dt$$

$$= \int_0^\pi \frac{\sin t}{(t + 2n\pi)^{1-1/\alpha}} \, dt$$

$$> \frac{2}{\pi^{1-1/\alpha}} (2n + \delta)^{1/\alpha - 1},$$

where $\delta = 1$ for $\alpha > 1$ and $\delta = 0$ for $0 < \alpha \le 1$. Since the left-hand side converges to 0 as $n \to \infty$, we must have $\alpha > 1$. Hence the improper integral diverges when $0 < \alpha \le 1$. If $\alpha > 1$, then it follows from integration by parts that

$$\int_1^a \sin x^\alpha \, dx = -\frac{\cos x^\alpha}{\alpha x^{\alpha-1}} \bigg|_1^a - \frac{\alpha - 1}{\alpha} \int_1^a \frac{\cos x^\alpha}{x^\alpha} \, dx,$$

which shows the convergence of $\int_1^\infty \sin x^\alpha \, dx$. However this does not converge absolutely for every $\alpha > 1$, because

$$\alpha \int_0^{(n\pi)^{1/\alpha}} |\sin x^\alpha| \, dx = \int_0^{n\pi} \frac{|\sin t|}{t^{1-1/\alpha}} \, dt$$

$$= \sum_{k=0}^{n-1} \int_0^\pi \frac{\sin t}{(t + k\pi)^{1-1/\alpha}} \, dt$$

$$> \frac{2}{\pi^{1-1/\alpha}} \sum_{k=0}^{n-1} \frac{1}{(k+1)^{1-1/\alpha}}$$

and the last sum diverges to $+\infty$.

When $\alpha < 0$, we have

$$\int_1^a \sin x^\alpha \, dx = \int_{1/a}^1 \frac{\sin t^{|\alpha|}}{t^2} \, dt.$$

Since $t^{-2} \sin t^{|\alpha|} \sim t^{|\alpha|-2}$ as $t \to 0+$, the above integral converges as $a \to \infty$ if and only if $|\alpha| > 1$. Moreover, in this case it converges absolutely. Similarly

$$\int_\epsilon^1 \sin x^\alpha \, dx = \int_1^{1/\epsilon} \frac{\sin t^{|\alpha|}}{t^2} \, dt$$

converges as $\epsilon \to 0+$ absolutely whatever α is. Therefore we have $A = \{|\alpha| > 1\}$

and $B = \{\alpha < -1\}$.   □

SOLUTION 6.3

This is an improper integral version of **PROBLEM 5.3**. Let M be the maximum of $|g(x)|$ over $[0, r]$. For any $\epsilon > 0$ we can take a sufficiently large number L satisfying

$$\int_{-\infty}^{-L} |f(x)| \, dx + \int_{L}^{\infty} |f(x)| \, dx < \epsilon.$$

On the other hand, applying the result in **PROBLEM 5.3**, we have

$$\lim_{n \to \infty} \int_{-L}^{L} f(x) g(nx) \, dx = \mathscr{M}_{[0,r]}(g) \int_{-L}^{L} f(x) \, dx.$$

Therefore

$$\limsup_{n \to \infty} \left| \int_{-\infty}^{\infty} f(x) g(nx) \, dx - \mathscr{M}_{[0,r]}(g) \int_{-\infty}^{\infty} f(t) \, dt \right| \le 2M\epsilon.$$

Since ϵ is arbitrary, this completes the proof.   □

SOLUTION 6.4

We have

$$\int_0^1 x^{-x} \, dx = \int_0^1 e^{-x \log x} \, dx = \sum_{n=0}^{\infty} \frac{(-1)^n}{n!} \int_0^1 x^n \log^n x \, dx,$$

where the termwise integration is allowed, because the series

$$\sum_{n=0}^{\infty} \frac{(-1)^n}{n!} x^n \log^n x$$

converges uniformly on the interval $(0, 1]$. Substituting $x = e^{-s}$ we get

$$\int_0^1 x^n \log^n x \, dx = (-1)^n \int_0^{\infty} e^{-(n+1)s} s^n \, ds = (-1)^n \frac{n!}{n^n}$$

by definition of the Gamma function (see Chapter 16).   □

SOLUTION 6.5

We divide $(-\infty, \infty)$ into three parts as

$$(-\infty, -n^{-1/3}) \cup [-n^{-1/3}, n^{-1/3}] \cup (n^{-1/3}, \infty),$$

which we denote by A_1, A_2, A_3 respectively. For $1 \le k \le 3$ we put

$$I_k = \sqrt{n} \int_{A_k} \frac{dx}{(1+x^2)^n}.$$

Substituting $t = \sqrt{n}\, x$ we have

$$I_2 = \sqrt{n} \int_{A_2} \exp(-n \log(1+x^2))\, dx$$

$$= \int_{-n^{1/6}}^{n^{1/6}} \exp\left(-n \log\left(1 + \frac{t^2}{n}\right)\right) dt.$$

Since

$$n \log\left(1 + \frac{t^2}{n}\right) = t^2 + O\left(\frac{1}{n^{1/3}}\right)$$

uniformly in $|t| \le n^{1/6}$, we obtain

$$I_2 = \left(1 + O\left(\frac{1}{n^{1/3}}\right)\right)\left(\sqrt{\pi} - \int_{-\infty}^{-n^{1/6}} e^{-t^2}\, dt - \int_{n^{1/6}}^{\infty} e^{-t^2}\, dt\right)$$

$$= \sqrt{\pi} + O\left(\frac{1}{n^{1/3}}\right)$$

as $n \to \infty$. On the other hand it follows that

$$0 < I_1 + I_3 < \frac{\sqrt{n}}{(1 + n^{-2/3})^{n-1}} \int_{-\infty}^{\infty} \frac{dx}{1 + x^2}$$

and the right-hand side converges to 0 as $n \to \infty$. $\qquad\square$

SOLUTION 6.6

Let $I(s)$ be the improper integral in the problem, which is invariant under the substitution $t = s/x$; hence we can write

$$I(s) = 2 \int_{\sqrt{s}}^{\infty} \frac{e^{-x/s - 1/x}}{x}\, dx.$$

Since $e^{-1/x} = 1 + O(s^{-1/2})$ uniformly in $x \ge \sqrt{s}$ as $s \to \infty$, we have

$$I(s) = 2\left(1 + O\left(\frac{1}{\sqrt{s}}\right)\right) \int_{\sqrt{s}}^{\infty} \frac{e^{-x/s}}{x}\, dx$$

$$= 2\left(1 + O\left(\frac{1}{\sqrt{s}}\right)\right) \int_{s^{-1/2}}^{\infty} \frac{e^{-t}}{t}\, dt.$$

Integrating the last integral by parts, we get

$$\int_{s^{-1/2}}^{\infty} \frac{e^{-t}}{t}\, dt = e^{-t} \log t \Big|_{s^{-1/2}}^{\infty} + \int_{s^{-1/2}}^{\infty} e^{-t} \log t \, dt$$

$$= \frac{1}{2} \log s + O(1);$$

hence $I(s) = \log s + O(1)$ as $s \to \infty$. □

SOLUTION 6.7

For any $\epsilon > 0$ we can take a sufficiently large number $L > 1$ satisfying

$$\int_{L-1}^{\infty} g(x)\, dx < \epsilon.$$

For any $h \in (0, 1)$ take a positive integer N satisfying

$$Nh \leq L < (N+1)h.$$

Then

$$nh \in \left(\frac{Ln}{N+1}, \frac{Ln}{N} \right] \subset \left(\frac{L(n-1)}{N}, \frac{Ln}{N} \right]$$

for all integers $1 \leq n \leq N$. Thus

$$\frac{L}{N} \sum_{n=1}^{N} f(nh)$$

is a Riemann sum for f over $[0, L]$ and it converges to the integral

$$\int_0^L f(x)\, dx$$

as $N \to \infty$. Since $N \to \infty$ as $h \to 0+$, there exists a sufficiently small $h_0 > 0$ such that

$$\left| \frac{L}{N} \sum_{n=1}^{N} f(nh) - \int_0^L f(x)\, dx \right| < \epsilon \qquad (6.1)$$

and

$$\frac{1}{N} \int_0^{\infty} g(x)\, dx < \epsilon$$

for any $0 < h < h_0$. On the other hand, we have

$$\left| \left(\frac{L}{N} - h \right) \sum_{n=1}^{N} f(nh) \right| \leq \left| \frac{L}{N} - h \right| \sum_{n=1}^{N} g(nh) \leq \frac{h}{N} \sum_{n=1}^{N} g(nh).$$

By the monotonicity of $g(x)$ the right-hand side is less than or equal to

$$\frac{1}{N} \int_0^L g(x)\,dx,$$

which is clearly less than ϵ, whence

$$\left| \left(\frac{L}{N} - h\right) \sum_{n=1}^N f(nh) \right| < \epsilon. \tag{6.2}$$

Moreover we have

$$h \left| \sum_{n>N} f(nh) \right| \leq h \sum_{n>N} g(nh) \leq \int_{L-1}^\infty g(x)\,dx < \epsilon; \tag{6.3}$$

therefore, by (6.1), (6.2) and (6.3),

$$\left| h \sum_{n=1}^\infty f(nh) - \int_0^\infty f(x)\,dx \right| < 3\epsilon + \int_L^\infty |f(x)|\,dx < 4\epsilon.$$

Since ϵ is arbitrary, this completes the proof. $\qquad\square$

REMARK. The special case $f(x) = g(x)$ was shown in Pólya and Szegö (1972).

SOLUTION 6.8

Let $f(s)$ be the integral in the problem. It is easily seen that

$$e^{-4s} f(s) = \int_0^\infty \exp(-(x + s/x)^2)\,dx.$$

Differentiating both sides we get

$$(f'(s) - 4f(s))\, e^{-4s} = -2 \int_0^\infty \left(1 + \frac{s}{x^2}\right) \exp(-(x + s/x)^2)\,dx,$$

where the differentiation under the integral sign is allowed (see Summary of Basic Points in Chapter 11). Substituting $t = x - s/x$, we thus have

$$f'(s) - 4f(s) = -2 \int_0^\infty \left(1 + \frac{s}{x^2}\right) \exp(-(x + s/x)^2)\,dx$$

$$= -2 \int_{-\infty}^\infty e^{-t^2}\,dt = -2\sqrt{\pi}.$$

Solving this linear differential equation we get

$$f(s) = ce^{4s} + \frac{\sqrt{\pi}}{2}$$

for some constant c. Since $0 < f(s) < (\sqrt{\pi}/2)\,e^{2s}$, we have $c = 0$; hence

$$f(s) = \frac{\sqrt{\pi}}{2}.$$ □

REMARK. This may be also solved by applying **PROBLEM 6.9**. Note that the function $f(s)$ can be defined for all $s \in \mathbb{R}$ and $f(s) = \sqrt{\pi}/2$ is valid also for $s = 0$. The above method is effective even for $s < 0$ and we get the differential equation

$$f'(s) = 4f(s),$$

from which we have

$$f(s) = \frac{\sqrt{\pi}}{2}e^{4s} \quad \text{for} \quad s < 0.$$

SOLUTION 6.9

The proof is based on Szegö (1934).

For any $0 < \epsilon < 1$ we can find a positive constant M satisfying

$$|R(x) - R(x_0)| \le |R'(x_0)|\,\epsilon + M\epsilon^2$$

for all $|x - x_0| \le \epsilon$ unless x_0 is a real pole of $R(x)$. Let $f_\epsilon(x)$ be the step function defined by

$$f_\epsilon(x) = \begin{cases} 1 & \left(|x - R(x_0)| \le |R'(x_0)|\,\epsilon + M\epsilon^2\right) \\ 0 & (\text{othewise}) \end{cases}.$$

Then we have

$$2\epsilon(|R'(x_0)| + M\epsilon) = \int_{-\infty}^{\infty} f_\epsilon(x)\,dx$$
$$\ge \int_{x_0-\epsilon}^{x_0+\epsilon} f_\epsilon(R(x))\,dx = 2\epsilon;$$

hence $|R'(x_0)| \ge 1$, because ϵ is arbitrary. This means that $R(x)$ maps each interval divided by real poles of $R(x)$, say $\alpha_1 < \alpha_2 < \cdots < \alpha_m$, bijectively on $\mathbb{R}$. Of course $m = 0$ if $R(x)$ has no real poles. Therefore, for any $s \in \mathbb{R}$, the equation $R(x) = s$ has a unique solution in each interval $(-\infty, \alpha_1), (\alpha_1, \alpha_2), \ldots, (\alpha_m, \infty)$, say $x_k(s)$ for $0 \le k \le m$.

Next, for an arbitrarily fixed s, let $g_\epsilon(x)$ be the step function defined by

$$g_\epsilon(x) = \begin{cases} 1 & (|x - s| \le \epsilon) \\ 0 & (\text{otherwise}) \end{cases}.$$

We then have

$$2\epsilon = \int_{-\infty}^{\infty} g_\epsilon(x)\,dx = \int_{-\infty}^{\infty} g_\epsilon(R(x))\,dx, \tag{6.4}$$

where the right-hand side is equal to the sum of lengths of $m+1$ intervals satisfying $|R(x) - s| \leq \epsilon$. Each of these intervals contains exactly one $x_k(s)$ and the length is

$$\frac{2\epsilon}{|R'(x_k(s))|} + O(\epsilon^2);$$

so letting ϵ to $0+$ in (6.4) we get

$$\sum_{k=0}^{m} \frac{1}{|R'(x_k(s))|} = 1. \tag{6.5}$$

In the case $m = 0$ the function $R(x)$ maps $\mathbb{R}$ homeomorphically onto $\mathbb{R}$ and $R'(x_0(s)) = \pm 1$ implies that

$$x_0'(s) = (R^{-1})'(s) = \frac{1}{R'(R^{-1}(s))} = \frac{1}{R'(x_0(s))} = \pm 1.$$

Therefore $R(x) = \pm(x - \alpha_0)$ for some constant α_0.

We hereafter assume that $m \geq 1$. Since $|R'(x_k(s))|$ diverges to ∞ as $|s| \to \infty$ for each $1 \leq k < m$, it follows from (6.5) that either

$$R'(x) \geq 1 \text{ on } (-\infty, \alpha_1) \cup (\alpha_m, \infty),$$

$$\lim_{s \to -\infty} x_0(s) = -\infty, \quad \lim_{s \to -\infty} x_m(s) = \alpha_m,$$

$$\lim_{s \to \infty} x_0(s) = \alpha_1, \quad \lim_{s \to \infty} x_m(s) = \infty$$

or

$$R'(x) \leq -1 \text{ on } (-\infty, \alpha_1) \cup (\alpha_m, \infty),$$

$$\lim_{s \to -\infty} x_0(s) = \alpha_1, \quad \lim_{s \to -\infty} x_m(s) = \infty,$$

$$\lim_{s \to \infty} x_0(s) = -\infty, \quad \lim_{s \to \infty} x_m(s) = \alpha_m$$

holds. We call these the first and the second cases respectively.

The rational function $R(x)$ can now be represented as

$$R(x) = P(x) - \sum_{k,\ell} \frac{c_{k,\ell}}{(x - \alpha_k)^{d_{k,\ell}}} + Q(x)$$

where $P(x)$ is a polynomial of degree r with some integer $r \geq 1$, $c_{k,\ell}$ are non-zero real constants, $d_{k,\ell}$ are positive integers, and $Q(x)$ is a rational function having no real poles and converging to 0 as $|x| \to \infty$. Let βx^r be the leading term of $P(x)$

where β is a non-zero constant. Since $R'(x) \sim \beta r x^{r-1}$ as $|x| \to \infty$, we must have $r = 1$ and hence $\beta = \pm 1$ by (6.5). Hence $P(x) = \pm(x - \alpha_0)$ for some constant α_0.

We first treat the first case. Since $R'(x) = 1 + O(x^{-2})$ as $|x| \to \infty$, we have

$$\sum_{k=0}^{m-1} \frac{1}{|R'(x_k(s))|} = O(s^{-2}) \quad \text{as} \quad s \to \infty,$$

$$\sum_{k=1}^{m} \frac{1}{|R'(x_k(s))|} = O(s^{-2}) \quad \text{as} \quad s \to -\infty. \tag{6.6}$$

Let d_k^* be the largest integer among $d_{k,\ell}$ and c_k^* be the corresponding coefficient $c_{k,\ell}$ for each $0 \le k \le m$. Clearly $x_k(s) \to \alpha_k$ either as $s \to \infty$ or $s \to -\infty$ for each $1 \le k \le m$. Then

$$R(x) \sim -\frac{c_k^*}{(x - \alpha_k)^{d_k^*}} \quad \text{as} \quad x \to \alpha_k,$$

and in particular,

$$s = R(x_k(s)) \sim -\frac{c_k^*}{(x_k(s) - \alpha_k)^{d_k^*}}.$$

Thus, solving in $x_k(s) - \alpha_k$ and substituting in

$$R'(x_k(s)) \sim \frac{c_k^* d_k^*}{(x_k(s) - \alpha_k)^{d_k^*+1}} \quad \text{as} \quad x \to \alpha_k,$$

we obtain

$$\frac{1}{|R'(x_k(s))|} \sim \frac{\left|c_k^*\right|^{1/d_k^*}}{d_k^*} \cdot \frac{1}{|s|^{1+1/d_k^*}}$$

either as $s \to \infty$ or $s \to -\infty$. Therefore, in view of (6.6), we have $d_k^* = 1$ and $R(x)$ can be written as

$$R(x) = x - \alpha_0 - \sum_{k=1}^{m} \frac{c_k}{x - \alpha_k} + Q(x) \tag{6.7}$$

for some real constants c_k. Since $R(x)$ is monotone increasing on (α_m, ∞) we have $c_m > 0$ and hence all coefficients c_k must be positive; in particular,

$$\sum_{k=0}^{m} \frac{1}{R'(x_k(s))} = 1. \tag{6.8}$$

Differentiating $s = R(x_k(s))$ in s and substituting in (6.8), we get

$$\sum_{k=0}^{m} x_k'(s) = 1,$$

from which we have

$$\sum_{k=0}^{m} x_k(s) = s + c'$$

for some constant c'. Since $x_m(s) = s + \alpha_0 + O(s^{-1})$ and $x_k(s) = \alpha_{k+1} + O(s^{-1})$ as $s \to \infty$ for each $0 \le k < m$, we have

$$\sum_{k=0}^{m} x_k(s) = s + \sum_{k=0}^{m} \alpha_k. \tag{6.9}$$

The rest of the proof is devoted to show that Q vanishes identically. (It may be interesting to find an easier real-analytic proof of this part.)

Suppose, on the contrary, that $Q(x)$ does not vanish identically. In view of (6.7), we can write $R(x) = V(x)/U(x)$ with

$$U(x) = A(x) \prod_{k=1}^{m} (x - \alpha_k)$$

where

$$A(x) = x^{2p} - c'' x^{2p-1} + O(x^{2p-2}),$$

$$V(x) = x^{m+2p+1} - \left(c'' + \sum_{k=0}^{m} \alpha_k\right) x^{m+2p} + O(x^{m+2p-1})$$

are polynomials with real coefficients for some integer $p \ge 1$ and some constant c''. We can assume that $U(x)$ and $V(x)$ are relatively prime; that is, they have no common factor except for constants. Let $w_1, \overline{w_1}, ..., w_p, \overline{w_p}$ be non-real zeros of $A(z)$. The algebraic equation

$$V(z) = sU(z)$$

has exactly $m + 1$ real simple roots $x_0(s), ..., x_m(s)$ and $2p$ non-real roots $z_1(s), \overline{z_1(s)}, ..., z_p(s), \overline{z_p(s)}$ counting with multiplicity for each $s \in \mathbb{R}$. From (6.9) we have

$$z_1(s) + \overline{z_1(s)} + \cdots + z_p(s) + \overline{z_p(s)} = c''. \tag{6.10}$$

Let C be a circle enclosing all the zeros of $U(z)$. Take a sufficiently large s such that

$$s \min_{z \in C} |U(z)| > \max_{z \in C} |V(z)|$$

and that $x_m(s)$ lies outside of C. Since $|V(z)| < s|U(z)|$ on C, it follows from Rouché's theorem that $V(z) = sU(z)$ has exactly $m + 2p$ roots inside of C, which are of course $x_1(s), ..., x_m(s)$ and $z_1(s), \overline{z_1(s)}, ..., z_p(s), \overline{z_p(s)}$. This implies that all the non-real roots are bounded as $s \to \infty$.

As a polynomial of two variables, $V(z) - sU(z)$ is irreducible; that is, it cannot be expressed as the product of two polynomials none of which is constant. We then consider $x_k(s)$ and $z_k(s)$ as function elements of the algebraic function uniquely determined by $V(z) - sU(z) = 0$. Since all solutions of $V(z) - sU(z) = 0$ are branches of the same algebraic function, it follows in particular that $x_m(s)$ can be continued to $z_1(s)$ along an arc on the Riemann surface of this algebraic function. This is a contradiction, because the relation (6.10) holds globally except for possible isolated singularities and because $x_m(s)$ is a unique solution which is unbounded as $s \to \infty$. Therefore $p = 0$ and hence $Q(x)$ must vanish identically, completing the proof of the first part. The similar argument can be applied to the second case.

Conversely, let

$$R(x) = \pm\left(x - \alpha_0 - \sum_{k=1}^{m} \frac{c_k}{x - \alpha_k}\right)$$

for some real numbers $\alpha_0, \ldots, \alpha_m$ with $\alpha_1 < \cdots < \alpha_m$ and some positive constants $c_1, \ldots, c_m$. Then it is clear that $\sum_{k=0}^{m} x_k'(s) = \pm 1$ and hence

$$\int_{-\infty}^{\infty} f(R(x))\, dx = \int_{-\infty}^{\alpha_1} f(R(x))\, dx + \cdots + \int_{\alpha_m}^{\infty} f(R(x))\, dx$$

$$= \pm \sum_{k=0}^{m} \int_{-\infty}^{\infty} f(s) x_k'(s)\, ds,$$

which is equal to $\int_{-\infty}^{\infty} f(x)\, dx$. $\qquad\square$

SOLUTION 6.10

The proof is essentially due to Gronwall (1918). Put

$$c_n = \int_0^{n\pi} \frac{1 - \cos x}{x}\, dx - \log(n\pi)$$

for any positive integer n. Then obviously

$$c_n = \int_0^1 \frac{1 - \cos x}{x}\, dx - \int_1^{n\pi} \frac{\cos x}{x}\, dx$$

and we are to show that c_n converges to Euler's constant γ as $n \to \infty$.

Substituting $x = 2n\pi s$ we have

$$\int_0^{n\pi} \frac{1 - \cos x}{x} \, dx = \int_0^{1/2} \frac{1 - \cos 2n\pi s}{s} \, ds,$$

which is equal to

$$\pi \int_0^{1/2} \frac{1 - \cos 2n\pi s}{\sin \pi s} \, ds + \int_0^{1/2} \phi(s) \, ds - \int_0^{1/2} \phi(s) \cos 2n\pi s \, ds, \qquad (6.11)$$

where

$$\phi(s) = \frac{1}{s} - \frac{\pi}{\sin \pi s}$$

is a continuous function on the interval $[0, 1/2]$ if we define $\phi(0) = 0$. Hence by the remark after **Problem 5.3** the third integral in (6.11) converges to 0 as $n \to \infty$. The second integral in (6.11) is equal to

$$\log \frac{s}{\tan(\pi s/2)} \bigg|_{0+}^{1/2} = \log \pi - 2 \log 2.$$

Finally it is easily verified that

$$\frac{1 - \cos 2n\pi s}{\sin \pi s} = 2 \sum_{k=1}^n \sin(2k - 1)\pi s;$$

so the first integral in (6.11) is equal to

$$2\pi \sum_{k=1}^n \int_0^{1/2} \sin(2k - 1)\pi s \, ds = \sum_{k=1}^n \frac{2}{2k - 1},$$

which is $\log n + 2 \log 2 + \gamma + o(1)$ as $n \to \infty$. Thus we obtain $c_n = \gamma + o(1)$, which completes the proof. $\qquad \square$

Solution 6.11

Let I be the integral on the right-hand side of the problem. We have

$$I = \sum_{k=1}^{\infty} \int_{k-1}^k \frac{\{x\}^2 (1 - \{x\})^2}{(1 + x)^5} \, dx$$

$$= \int_0^1 t^2 (1 - t)^2 H_5(t) \, dt,$$

where

$$H_m(x) = \sum_{k=1}^{\infty} \frac{1}{(x + k)^m}$$

for any integer $m > 1$. Then, by integration by parts, we get

$$I = \int_0^1 t(1-t)\left(\frac{1}{2} - t\right) H_4(t)\, dt$$

$$= \int_0^1 \left(\frac{1}{6} - t(1-t)\right) H_3(t)\, dt.$$

Since

$$\int_0^1 H_3(t)\, dt = -\frac{1}{2}\big(H_2(1) - H_2(0)\big) = \frac{1}{2},$$

it follows that

$$I = \frac{1}{12} + \int_0^1 \left(t - \frac{1}{2}\right) H_2(t)\, dt;$$

therefore

$$I = \frac{1}{12} + \lim_{n \to \infty} \sum_{k=1}^n \int_0^1 \frac{t - 1/2}{(t+k)^2}\, dt$$

$$= \frac{1}{12} - \lim_{n \to \infty} \sum_{k=1}^n \left(\frac{1}{2k} + \frac{1}{2(k+1)} - \log\frac{k+1}{k}\right),$$

which is equal to $7/12 - \gamma$ from the definition of Euler's constant. $\square$

◆ ❋ ◆

Chapter 7

Series of Functions

[Summary of Basic Points]

1. If each function $f_n(x)$ is continuous on a closed interval $[a, b]$ and if the series $\sum_{n=1}^{\infty} f_n(x)$ converges uniformly on $[a, b]$, then we have

$$\sum_{n=1}^{\infty} \int_a^b f_n(x)\, dx = \int_a^b \sum_{n=1}^{\infty} f_n(x)\, dx. \tag{7.1}$$

In other words, termwise integration is allowed.

2. A simple and useful test for uniform convergence, known as Dirichlet's test, is as follows: Suppose that the Nth partial sum of $\sum_{n=1}^{\infty} f_n(x)$ is uniformly bounded (with respect to both N and x) on an interval I and that $g_n(x)$ is a monotone decreasing sequence converging uniformly to 0. Then the series

$$\sum_{n=1}^{\infty} f_n(x) g_n(x)$$

converges uniformly on I.

3. If each function $f_n(x)$ is R-integrable over a closed interval $[a, b]$ and if the nth partial sum of the series $\sum_{n=1}^{\infty} f_n(x)$ is uniformly bounded on $[a, b]$ and converges pointwise to the limit function which is also R-integrable over $[a, b]$, then (7.1) holds true. This is known as Arzelà's theorem.

4. If each function $f_n(x)$ has the derivative $f_n'(x)$ at any point x in an open inter-

117

val (a, b), if the series $\sum_{n=1}^{\infty} f_n(x)$ converges at least one point c in (a, b) and if $\sum_{n=1}^{\infty} f_n'(x)$ converges uniformly on (a, b) to a function $g(x)$, then $\sum_{n=1}^{\infty} f_n(x)$ converges uniformly on (a, b) and is differentiable at any point x in (a, b), whose derivative is equal to $g(x)$. Namely,

$$\left(\sum_{n=1}^{\infty} f_n(x) \right)' = \sum_{n=1}^{\infty} f_n'(x);$$

that is, termwise differentiation is allowed.

5. An infinite series of the form

$$\sum_{n=0}^{\infty} a_n (z - z_0)^n$$

with a complex z_0, a complex variable z and a complex sequence $\{a_n\}$, is called a *power series* about $z = z_0$.

6. Given a power series, let

$$\frac{1}{\rho} = \limsup_{n \to \infty} |a_n|^{1/n}. \tag{7.2}$$

Of course, we adopt the rule $\rho = 0$ or $\rho = \infty$ according to the superior limit on the right-hand side is equal to ∞ or 0 respectively. This number ρ is called the *radius of convergence* of the power series and (7.2) is referred to as *Hadamard's formula*.

7. The circle $|z| = \rho$ is called the circle of convergence and has the following properties:

   (i) The series converges absolutely and uniformly on compact sets in $|z| < \rho$.

   (ii) The sum is an analytic function and the derivative is obtained by termwise differentiation in $|z| < \rho$. The derived series has the same radius of convergence.

   (iii) If $|z| > \rho$, then the terms of the series are unbounded and the series is divergent.

   Note that nothing is claimed for the convergence on the circle. Tauber's theorem (**PROBLEM 7.9**) describes the behavior on the circle of convergence.

8. If $f(x)$ has derivatives of every order at a point a in an open interval I, the

power series about $x = a$:

$$\sum_{n=0}^{\infty} \frac{f^{(n)}(a)}{n!} (x - a)^n$$

is called the *Taylor series* generated by f. It then follows from Taylor's formula that this series represents the given function $f(x)$ if and only if the remainder term $R_n(x, a)$ converges to 0 as $n \to \infty$. For the Lagrange remainder, see **Solution 7.6**. If the Taylor series of f represents $f(x)$ in some neighborhood, then f is called a *real analytic* function.

9. Let $f(x)$ be an R-integrable function defined on the interval $[0, 2\pi]$. We define

$$a_n = \frac{1}{\pi} \int_0^{2\pi} f(x) \cos nx \, dx \quad \text{and} \quad b_n = \frac{1}{\pi} \int_0^{2\pi} f(x) \sin nx \, dx$$

for all integers $n \geq 0$. The series

$$\frac{a_0}{2} + \sum_{n=1}^{\infty} (a_n \cos nx + b_n \sin nx)$$

is called the *Fourier series* of the function $f(x)$. If $f(x)$ is of bounded variation over $(0, 2\pi)$, then the Fourier series of f converges to

$$\frac{f(x+) + f(x-)}{2}$$

at every point x. This is known as the Dirichlet-Jordan test. In particular, it converges to $f(x)$ at every continuity point. See p. 57 in Zygmund (1979).

On the other hand, Kolmogorov (1926) showed that there exists a Lebesgue integrable function whose Fourier series diverges everywhere.

PROBLEM 7.1

    Let

$$\sum_{n=1}^{\infty} a_n x^n$$

be the Taylor series about $x = 0$ of the algebraic function

$$f(x) = \frac{1 - \sqrt{1 - 4x}}{2}.$$

Show that each a_n is a positive integer which is odd if and only if n is a power of 2.

PROBLEM 7.2

Show that

$$\lim_{x \to 1-} \sqrt{1 - x} \sum_{n=1}^{\infty} x^{n^2} = \frac{\sqrt{\pi}}{2}.$$

The behavior of the series

$$\sum_{n=1}^{\infty} z^{n^2}$$

when $z \to -1$ in a certain manner was used by Hardy (1914) to show that the Riemann zeta function $\zeta(z)$ has an infinitely many zeros on the critical line. This proof was materially simplified by Landau (1915), who showed that no property of this series was needed for the purpose of proof except for its upper estimate $O((1 - |z|)^{-1/2})$.

PROBLEM 7.3

Show that

$$\lim_{x \to 1-} (1 - x)^2 \sum_{n=1}^{\infty} \frac{nx^n}{1 - x^n} = \frac{\pi^2}{6}.$$

The series of the form

$$\sum_{n=1}^{\infty} \frac{a_n x^n}{1 - x^n}$$

is called the Lambert series and transformed formally to

$$\sum_{n=1}^{\infty} \left(\sum_{d|n} a_d \right) x^n,$$

where d runs over all divisors of n. The reader may be enticed to find any other example of simple power series $f(x)$ for which the limit of $(1 - x)^{\alpha} f(x)$ as $x \to 1-$ is equal to a rational multiple of π^{α}.

PROBLEM 7.4

Show that

$$1 + \sum_{n=1}^{\infty} \frac{1}{n!} \left(1 - \frac{1}{e} \right)^n x(x + 1) \cdots (x + n - 1)$$

converges to e^x uniformly on any compact set in $\mathbb{R}$.

The partial sum of this series is known as Newton's backward interpolation formula approximating e^x.

PROBLEM 7.5

Show that the series

$$f(x) = \sum_{n=0}^{\infty} e^{-n} \cos n^2 x$$

is infinitely differentiable everywhere, but the Taylor series about $x = 0$:

$$\sum_{n=0}^{\infty} \frac{f^{(n)}(0)}{n!} x^n$$

does not converge except for the origin.

This gives an example of a $\mathscr{C}^{\infty}$-function whose Taylor series has the radius of convergence $\rho = 0$.

PROBLEM 7.6

Suppose that $f \in \mathscr{C}^{\infty}(\mathbb{R})$ satisfies $f^{(n)}(x) \geq 0$ for all integer $n \geq 0$ and for any $x \in \mathbb{R}$. Show that the Taylor series about $x = 0$ generated by f converges for any x.

This is a special case of the result obtained by Bernstein(1928). Bernstein's theorem on a finite interval is explained in the book of Apostol(1957) on p. 418. For example, the function $f(x) = a^x$ for $a > 1$ satisfies the conditions stated in the problem.

PROBLEM 7.7

Suppose that a power series

$$f(x) = \sum_{n=0}^{\infty} a_n x^n$$

has the radius of convergence $\rho > 0$ and $\sum_{n=0}^{\infty} a_n \rho^n$ converges to α. Show that $f(x)$ converges to α as $x \to \rho-$.

This is known as Abel's continuity theorem(1826), which was published in the first volume of Crelle Journal. This is the first major mathematical journal, except for proceedings of academies, founded by August Leopold Crelle.

PROBLEM 7.8

Suppose that a power series

$$g(x) = \sum_{n=0}^{\infty} b_n x^n$$

has the radius of convergence $\rho > 0$, all b_n's are positive, and that

$$\sum_{n=0}^{\infty} b_n \rho^n = \infty.$$

Then show that $f(x)/g(x)$ converges to α as $x \to \rho-$ for any power series

$$f(x) = \sum_{n=0}^{\infty} a_n x^n$$

such that a_n/b_n converges to α as $n \to \infty$.

This generalization of Abel's continuity theorem is due to Cesàro (1893). Indeed, applying Cesàro's theorem to $a_n = c_0 + c_1 + \cdots + c_n$ and $b_n = 1$ for a convergent series $\sum_{n=0}^{\infty} c_n$, we get

$$\sum_{n=0}^{\infty} c_n x^n = \frac{a_0 + a_1 x + \cdots + a_n x^n + \cdots}{1 + x + \cdots + x^n + \cdots} \to \sum_{n=0}^{\infty} c_n$$

as $x \to 1-$.

Cesàro's theorem can be used to compute the limit in **PROBLEM 7.2** as follows: Put

$$f(x) = \sum_{n=1}^{\infty} x^{n^2} = (1 - x) \sum_{n=1}^{\infty} [\sqrt{n}] x^n = (1 - x) F(x)$$

and

$$G(x) = (1 - x)^{-3/2} = 1 + \sum_{n=1}^{\infty} b_n x^n,$$

where

$$b_n = \frac{1}{n!} \cdot \frac{3}{2} \cdots \left(n + \frac{1}{2}\right) \sim 2\sqrt{\frac{n}{\pi}}$$

as $n \to \infty$. This asymptotic formula follows from **PROBLEM 16.1** and $\Gamma(1/2) = \sqrt{\pi}$ where $\Gamma(s)$ is the Gamma function. We then have

$$\lim_{x \to 1-} \sqrt{1 - x} f(x) = \lim_{x \to 1-} \frac{F(x)}{G(x)} = \frac{\sqrt{\pi}}{2}$$

because $[\sqrt{n}]/b_n$ converges to $\sqrt{\pi}/2$ as $n \to \infty$.

PROBLEM 7.9

Suppose that

$$f(x) = \sum_{n=0}^{\infty} a_n x^n$$

has the radius of convergence $\rho > 0$, na_n converges to 0 as $n \to \infty$, and $f(x)$ converges to α as $x \to \rho-$. Show then that

$$\sum_{n=0}^{\infty} a_n \rho^n = \alpha.$$

This is known as Tauber's theorem (1897). Partial converses to Abel's continuity theorem are generally called Tauberian theorems. See the comment after **SOLUTION 7.9**.

PROBLEM 7.10

Suppose that

$$f(x) = \sum_{n=0}^{\infty} a_n x^n$$

has the radius of convergence 1, all a_n's are non-negative, and $(1 - x)f(x)$ converges to 1 as $x \to 1-$. Show then that

$$\lim_{n \to \infty} \frac{a_0 + a_1 + a_2 + \cdots + a_n}{n} = 1.$$

Karamata (1930) gave the elegant proof using Weierstrass' approximation theorem, which was a new proof to Littlewood's rather difficult theorem (1910) stated in **SOLUTION 7.9**. According to Nikolić (2002) Karamata's two-page paper created a sensation in mathematical circles. See also Wielandt (1952).

PROBLEM 7.11

Let $\{a_n\}$ be a monotone decreasing sequence converging to 0. Show that the trigonometric series

$$\sum_{n=1}^{\infty} a_n \sin n\theta$$

converges uniformly on $\mathbb{R}$ if and only if na_n converges to 0 as $n \to \infty$.

This is due to Zygmund (1979), p. 182. In particular, if a trigonometric series represents a discontinuous function, then na_n does not converge to 0. This will be illustrated by the following example.

PROBLEM 7.12

Show that the trigonometric series

$$\sum_{n=1}^{\infty} \frac{\sin n\theta}{n}$$

converges uniformly to $(\pi - \theta)/2$ on the interval $[\delta, 2\pi - \delta]$ for any $\delta > 0$.

This is the Fourier expansion of the first periodic Bernoulli polynomial defined by $\overline{B}_1(x) = x - [x] - 1/2$ $(x \notin \mathbb{Z})$. See also **PROBLEM 1.7** and **PROBLEM 5.8**.

PROBLEM 7.13

Suppose that $f \in \mathscr{C}^1(0, 2\pi)$ and $\displaystyle\int_0^{2\pi} f(x)\, dx$ converges absolutely. Show that the Fourier series converges to $f(x)$ in the interval $(0, 2\pi)$.

＊　＊　＊

Solutions for Chapter 7

SOLUTION 7.1

Since

$$\left(1 - 2\sum_{n=1}^{\infty} a_n x^n\right)^2 = 1 - 4x,$$

we have $a_1 = 1$ and the recursion formula

$$a_{n+1} = a_1 a_n + a_2 a_{n-1} + \cdots + a_n a_1$$

holds for all $n \geq 1$. This implies immediately that every a_n is a positive integer; therefore

$$a_{2k+1} = 2(a_1 a_{2k} + \cdots + a_k a_{k+1})$$

is an even integer for all $k \geq 1$.

Suppose now that there exists an integer $\ell \geq 0$ such that a_n is even for every $n = 2^\ell(2k + 1), k \geq 1$. Then

$$a_{2n} = 2(a_1 a_{2n-1} + a_2 a_{2n-2} + \cdots + a_{n-1} a_{n+1}) + a_n^2$$

implies that a_{2n} is also even for every $n = 2^\ell(2k + 1), k \geq 1$. Hence by induction every a_n is even except for the case in which n is a power of 2. However, if n is a power of 2, we can similarly show that a_n is odd because a_1 is odd. □

REMARK. By the Taylor expansion of $\sqrt{1 + x}$ it is easily seen that

$$a_n = \frac{1}{n}\binom{2n - 2}{n - 1}.$$

We adopt, of course, the convention that $0! = 1$. Thus the above result may give some information on the divisibility of the central binomial coefficients.

SOLUTION 7.2

Applying **PROBLEM 6.7** to $f(x) = e^{-x^2}$ we have

$$h\sum_{n=0}^{\infty} \exp(-n^2 h^2) \to \int_0^{\infty} e^{-x^2}\,dx = \frac{\sqrt{\pi}}{2}$$

as $h \to 0+$. Substituting $h = \sqrt{-\log x}$ and noting that h converges to 0+ if and only if x converges to 1−; hence, using $-\log x \sim 1 - x$ as $x \to 1-$, we obtain

$$\lim_{x \to 1-} \sqrt{1 - x} \sum_{n=0}^{\infty} x^{n^2} = \lim_{x \to 1-} \sqrt{-\log x} \sum_{n=0}^{\infty} x^{n^2}$$

$$= \lim_{h \to 0+} h \sum_{n=0}^{\infty} \exp(-n^2 h^2) = \frac{\sqrt{\pi}}{2}. \qquad \square$$

SOLUTION 7.3

Applying **PROBLEM 6.7** to $f(x) = x/(e^x - 1)$ we get

$$h \sum_{n=1}^{\infty} \frac{nh}{e^{nh} - 1} \to \int_0^{\infty} \frac{x\,dx}{e^x - 1} = \frac{\pi^2}{6}$$

as $h \to 0+$. Substituting $h = -\log x$ and noting that h converges to 0+ if and only if x converges to 1−; hence, using $-\log x \sim 1 - x$ as $x \to 1-$, we get

$$\lim_{x \to 1-} (1 - x)^2 \sum_{n=1}^{\infty} \frac{nx^n}{1 - x^n} = \lim_{x \to 1-} \log^2 x \sum_{n=1}^{\infty} \frac{nx^n}{1 - x^n}$$

$$= \lim_{h \to 0+} h \sum_{n=1}^{\infty} \frac{nh}{e^{nh} - 1} = \frac{\pi^2}{6}.$$

Note that the function $f(x) = x/(e^x - 1)$ is continuous on the interval $[0, \infty)$ if we define $f(0) = 1$. $\qquad \square$

SOLUTION 7.4

For an arbitrarily fixed $s \in \mathbb{R}$ the series

$$\sum_{n=0}^{\infty} \binom{s}{n} z^n$$

converges to $(1 + z)^s$ at least for $|z| < 1$, known as the 'binomial series'. Putting $z = \dfrac{1}{e} - 1$ and $s = -x$ we have

$$e^x = 1 + \sum_{n=1}^{\infty} \left(\frac{1}{e} - 1\right)^n \binom{-x}{n}$$

$$= 1 + \sum_{n=1}^{\infty} \frac{1}{n!} \left(1 - \frac{1}{e}\right)^n x(x + 1) \cdots (x + n - 1); \qquad (7.3)$$

that is, the series (7.3) converges pointwisely for all $x \in \mathbb{R}$. Moreover the convergence is uniform on $|x| \leq r$, because (7.3) has the majorant series

$$1 + \sum_{n=1}^{\infty} \frac{1}{n!}\left(1 - \frac{1}{e}\right)^n r(r+1)\cdots(r+n-1),$$

which converges to e^r by (7.3). $\qquad \square$

REMARK. The remainder term of this expansion is merely that of the binomial series. Thus, applying the integral form of the remainder term stated in the 7th basic point in Chapter 4 we have

$$R_{N+1} = e^x - 1 - \sum_{n=1}^{N} \frac{1}{n!}\left(1 - \frac{1}{e}\right)^n x(x+1)\cdots(x+n-1)$$

$$= \frac{x(x+1)\cdots(x+N)}{N!} e^x \int_0^1 (1 - e^{-s})^N e^{-sx}\, ds.$$

SOLUTION 7.5

By k times termwise differentiation of the given series we get

$$\sum_{n=1}^{\infty} n^{2k} e^{-n} \operatorname{Re}(i^k e^{in^2 x}),$$

which clearly converges uniformly on $\mathbb{R}$. Hence this series represents $f^{(k)}(x)$; in particular,

$$f^{(k)}(0) = \operatorname{Re}(i^k) \sum_{n=1}^{\infty} n^{2k} e^{-n}.$$

We thus have

$$\left|f^{(2\ell)}(0)\right| \geq \left(\frac{4\ell}{e}\right)^{4\ell}$$

for all integer $\ell \geq 1$ by looking at the 4ℓth term only. Let ρ be the radius of convergence of $f(x)$. It then follows from Hadamard's formula (7.2) that

$$\frac{1}{\rho} \geq \limsup_{\ell \to \infty} \left(\frac{(4\ell)^{4\ell}}{(2\ell)! e^{4\ell}}\right)^{1/(2\ell)} \geq \limsup_{\ell \to \infty} \frac{(4\ell)^2}{2e^2\ell} = \infty,$$

which means that $\rho = 0$. $\qquad \square$

SOLUTION 7.6

It follows from Taylor's formula about $x = a$ with Lagrange's remainder term that

$$f(x) = \sum_{k=0}^{n} \frac{f^{(k)}(a)}{k!} (x - a)^k + \frac{f^{(n+1)}(\xi)}{(n+1)!} (x - a)^{n+1}$$

for some ξ between x and a. If we take $x = 2a > 0$, then

$$f(2a) = \sum_{k=0}^{n} \frac{f^{(k)}(a)}{k!} a^k + \frac{f^{(n+1)}(\xi)}{(n+1)!} a^{n+1}$$

$$\geq \sum_{k=0}^{n} \frac{f^{(k)}(a)}{k!} a^k,$$

which implies the convergence of the series

$$\sum_{n=0}^{\infty} \frac{f^{(n)}(a)}{n!} a^n.$$

In particular, the sequence $f^{(n)}(a)a^n/n!$ converges to 0 as $n \to \infty$ for any $a > 0$. Hence

$$\left| f(x) - \sum_{k=0}^{n} \frac{f^{(k)}(0)}{k!} x^k \right| = \frac{f^{(n+1)}(\xi)}{(n+1)!} |x|^{n+1}$$

$$\leq \frac{f^{(n+1)}(|x|)}{(n+1)!} |x|^{n+1} \to 0$$

as $n \to \infty$, because all derivatives of $f(x)$ are monotone increasing on $\mathbb{R}$. $\qquad \square$

SOLUTION 7.7

Replacing x by ρx we can assume that $\rho = 1$ without loss of generality. Put

$$s_n = a_0 + a_1 + \cdots + a_n.$$

Since s_n converges to α as $n \to \infty$, for any $\epsilon > 0$ we can take a sufficiently large integer N satisfying $|s_n - \alpha| < \epsilon$ for all $n > N$. For $0 < x < 1$ we have

$$\frac{f(x)}{1 - x} = \sum_{n=0}^{\infty} s_n x^n = \frac{\alpha}{1 - x} + \sum_{n=0}^{\infty} (s_n - \alpha) x^n;$$

therefore

$$|f(x) - \alpha| \leq (1 - x) \sum_{n=0}^{N} |s_n - \alpha| x^n + (1 - x) \sum_{n>N} |s_n - \alpha| x^n$$

$$< (1 - x) \sum_{n=0}^{N} |s_n - \alpha| + (1 - x) \sum_{n=0}^{\infty} \epsilon x^n$$

$$= (1 - x) \sum_{n=0}^{N} |s_n - \alpha| + \epsilon.$$

The right-hand side can be less than 2ϵ by letting x be sufficiently close to $1-$. This means that $f(x)$ converges to α as $x \to 1-$. $\qquad\square$

SOLUTION 7.8

As in the previous problem we can assume that $\rho = 1$. For any $\epsilon > 0$ there is an integer $N \geq 1$ such that $|a_n - \alpha b_n| < \epsilon b_n$ for all $n > N$. Since

$$f(x) = \sum_{n=0}^{\infty} a_n x^n = \alpha g(x) + \sum_{n=0}^{\infty} (a_n - \alpha b_n) x^n,$$

we have

$$\left| \frac{f(x)}{g(x)} - \alpha \right| \leq \frac{1}{g(x)} \left(\sum_{n=0}^{N} |a_n - \alpha b_n| + \epsilon \sum_{n>N} b_n x^n \right)$$

$$< \frac{1}{g(x)} \sum_{n=0}^{N} |a_n - \alpha b_n| + \epsilon \qquad (7.4)$$

for any $0 < x < 1$. Now $g(x)$ diverges to ∞ as $x \to 1-$, because

$$\liminf_{x \to 1-} g(x) \geq \sum_{k=0}^{n} b_k$$

for all integers $n \geq 1$. Hence the right-hand side of (7.4) can be smaller than 2ϵ if we take x sufficiently close to 1. $\qquad\square$

SOLUTION 7.9

As in the previous problems we can assume that $\rho = 1$. Put

$$s_n = a_0 + a_1 + \cdots + a_n.$$

For any $\epsilon > 0$ there exists an integer $N \geq 1$ such that $n\,|a_n| < \epsilon$ for all $n > N$. For any $0 < x < 1$ and all $n > N$ we obtain

$$|s_n - f(x)| \leq \sum_{k=1}^{n} |a_k|\,(1 - x^k) + \sum_{k>n} k\,|a_k|\,\frac{x^k}{k}$$

$$\leq (1 - x) \sum_{k=1}^{n} k\,|a_k| + \frac{\epsilon}{n(1 - x)}.$$

Substituting $x = 1 - 1/n$ we infer that

$$\left| s_n - f\left(1 - \frac{1}{n}\right) \right| \leq \frac{|a_1| + 2\,|a_2| + \cdots + n\,|a_n|}{n} + \epsilon.$$

Therefore the right-hand side can be smaller than 2ϵ if we take n sufficiently large. This means that s_n converges to α as $n \to \infty$.   □

REMARK. Pringsheim (1900) weakened Tauber's condition $na_n = o(1)$ to

$$a_1 + 2a_2 + \cdots + na_n = o(n)$$

as $n \to \infty$. To see this, we put $\tau_0 = 0$ and $\tau_n = a_1 + 2a_2 + \cdots + na_n$. Since

$$a_n = \frac{\tau_n - \tau_{n-1}}{n}$$

converges to 0 as $n \to \infty$, the given series $f(x)$ converges in $|x| < 1$. Moreover for any $0 < x < 1$ we obtain

$$f(x) = a_0 + \sum_{n=1}^{\infty} \frac{\tau_n - \tau_{n-1}}{n} x^n$$

$$= a_0 + \sum_{n=1}^{\infty} \tau_n \left(\frac{x^n}{n} - \frac{x^{n+1}}{n+1} \right)$$

$$= a_0 + (1 - x) \sum_{n=1}^{\infty} \frac{\tau_n}{n+1} x^n + \sum_{n=1}^{\infty} \frac{\tau_n}{n(n+1)} x^n.$$

Since $\tau_n = o(n)$, the second term on the right-hand side clearly converges to 0 as $x \to 1-$. This implies that the third term converges to $\alpha - a_0$ as $x \to 1-$. Since $\tau_n/(n(n+1)) = o(1/n)$ as $n \to \infty$, we can apply Tauber's theorem to this power series so that

$$\alpha - a_0 = \lim_{n \to \infty} \sum_{k=1}^{n} \frac{\tau_k}{k(k+1)}$$

$$= \lim_{n \to \infty} \left(\tau_1 + \frac{\tau_2 - \tau_1}{2} + \cdots + \frac{\tau_n - \tau_{n-1}}{n} - \frac{\tau_n}{n+1} \right)$$

$$= \sum_{n=1}^{\infty} a_n,$$

as required.

Furthermore Littlewood (1910) has shown that Tauber's theorem holds true if the sequence na_n is bounded. Finally Hardy and Littlewood (1914) proved it even if na_n is either bounded above or bounded below. To see this we need the following result.

SOLUTION 7.10

It follows from the assumption that

$$(1 - x) \sum_{n=0}^{\infty} a_n x^{(k+1)n} = \frac{1}{1 + x + \cdots + x^k} (1 - x^{k+1}) \sum_{n=0}^{\infty} a_n (x^{k+1})^n$$

converges to

$$\frac{1}{k+1} = \int_0^1 t^k \, dt$$

as $x \to 1-$ for all integer $k \geq 0$. Therefore we have

$$\lim_{x \to 1-} (1 - x) \sum_{n=0}^{\infty} a_n x^n P(x^n) = \int_0^1 P(t) \, dt$$

for any polynomial $P(t)$.

We now introduce the discontinuous function $\phi(x)$ defined by

$$\phi(x) = \begin{cases} 0 & (0 \leq x < 1/e) \\ 1/x & (1/e \leq x \leq 1) \end{cases}.$$

For any $\epsilon > 0$ we can find two continuous functions $\phi_{\pm}(x)$ defined on the interval $[0, 1]$ such that $\phi_-(x) \leq \phi(x) \leq \phi_+(x)$ and $\phi_+(x) - \phi_-(x) < \epsilon$ for any x in $[0, 1]$. By Weierstrass' approximation theorem there are two polynomials $P_{\pm}(x)$ satisfying $|\phi_{\pm}(x) \pm \epsilon - P_{\pm}(x)| < \epsilon$ respectively. Hence it follows that $P_-(x) < \phi(x) < P_+(x)$

and $P_+(x) - P_-(x) < 5\epsilon$. Since a_n are non-negative, we obtain

$$\sum_{n=0}^{\infty} a_n x^n P_-(x^n) \le \sum_{n=0}^{\infty} a_n x^n \phi(x^n) \le \sum_{n=0}^{\infty} a_n x^n P_+(x^n)$$

for any $0 < x < 1$. If we put $x = e^{-1/N}$, then $x \to 1-$ if and only if $N \to \infty$; so, using $1 - e^{-1/N} \sim 1/N$, we have

$$\int_0^1 P_-(t)\,dt \le \liminf_{N\to\infty} \frac{1}{N} \sum_{n=0}^{N} a_n$$

and

$$\limsup_{N\to\infty} \frac{1}{N} \sum_{n=0}^{N} a_n \le \int_0^1 P_+(t)\,dt.$$

Therefore, since $P_+(x) - P_-(x) < 5\epsilon$ and

$$\int_0^1 \phi(x)\,dx = \int_{1/e}^1 \frac{dx}{x} = 1,$$

it follows that

$$1 < \int_0^1 P_+(t)\,dt < \int_0^1 P_-(t)\,dt + 5\epsilon < 1 + 5\epsilon.$$

Hence the sequence $\dfrac{1}{N} \sum_{n=0}^{N} a_n$ converges to 1 as $N \to \infty$. □

REMARK. Using this result we can now show that $\tau_n = o(n)$ as $n \to \infty$ even if $na_n < K$ for some positive constant K, where τ_n is defined in the remark after **SOLUTION 7.9.** This gives a simpler proof of the theorem due to Hardy and Littlewood(1914).

First of all we have

$$f''(x) = \sum_{n=2}^{\infty} n(n-1)a_n x^{n-2}$$

$$\le K \sum_{n=2}^{\infty} (n-1)x^{n-2} = \frac{K}{(1-x)^2}$$

for any $0 < x < 1$. Put

$$g(t) = f(1 - e^{-t}) \in \mathscr{C}^{\infty}(0, \infty)$$

for brevity. By the assumption $g(t)$ converges to α as $t \to \infty$. Moreover

$$g''(t) + g'(t) = e^{-2t}f''(1 - e^{-t}) \le K$$

holds for $t > 0$. It follows from **Problem 4.12** that $g'(t)$ converges to 0 as $t \to \infty$. This means that $(1 - x)f'(x)$ converges to 0 as $x \to 1-$. Then the power series

$$\sum_{n=1}^{\infty} \left(1 - \frac{na_n}{K}\right) x^{n-1} = \frac{1}{1 - x} - \frac{f'(x)}{K} \ .$$

satisfies all the conditions stated in **Problem 7.10**; therefore

$$\frac{1}{N} \sum_{n=1}^{N} \left(1 - \frac{na_n}{K}\right) = 1 - \frac{\tau_N}{KN}$$

converges to 1 as $N \to \infty$. Hence $\tau_n = o(n)$ as $n \to \infty$.

SOLUTION 7.11

Suppose first that the given series converges uniformly on $\mathbb{R}$. For any $\epsilon > 0$ we can take an integer $N \geq 1$ such that

$$\max_{\theta \in \mathbb{R}} \left| a_p \sin p\theta + a_{p+1} \sin(p + 1)\theta + \cdots + a_q \sin q\theta \right| < \epsilon$$

for all integers $q > p \geq N$. We choose $\theta = \pi/(4p)$ and $q = 2p$ so that $a_k \sin k\theta \geq 0$ for $p \leq k \leq 2p$. Then we have

$$\epsilon > \sum_{n=p}^{2p} a_n \sin \frac{n\pi}{4p} \geq \frac{1}{\sqrt{2}} \sum_{n=p}^{2p} a_n \geq \frac{p + 1}{\sqrt{2}} a_{2p}.$$

Hence

$$\max(2pa_{2p}, (2p + 1)a_{2p+1}) \leq 2(p + 1)a_{2p} < 2\sqrt{2}\epsilon,$$

and so $na_n \to 0$ as $n \to \infty$.

Conversely assume that na_n converges to 0 as $n \to \infty$. For any $\epsilon > 0$ we can take a positive integer N so that $na_n < \epsilon$ for any integer n greater than N. For any $\theta \in (0, \pi]$ let m_θ be a unique positive integer satisfying $\pi/(m + 1) < \theta \leq \pi/m$. For any integers $q > p \geq N$ let $S_0(\theta)$ and $S_1(\theta)$ be the sums of $a_n \sin n\theta$ from $n = p$ to r and from $n = r + 1$ to q respectively, where

$$r = \min(q, p + m_\theta - 1).$$

For the sum $S_0(\theta)$ we use an almost trivial estimate $\sin x < x$ for $x > 0$ to obtain

$$|S_0(\theta)| \leq \theta \sum_{n=p}^{r} na_n < \theta m_\theta \epsilon \leq \pi\epsilon.$$

Next for the sum $S_1(\theta)$ we can assume $q \geq p + m_\theta$; so $r = p + m_\theta - 1$. By partial summation we have

$$|S_1(\theta)| = \left| \sum_{n=r+1}^{q} a_n(\sigma_n - \sigma_{n-1}) \right|$$
$$\leq a_{r+1}|\sigma_r| + a_q|\sigma_q| + (a_{r+1} - a_q) \max_{r<k<q} |\sigma_k|$$

where

$$\sigma_n = \sin\theta + \sin 2\theta + \cdots + \sin n\theta.$$

Using Jordan's inequality $\pi \sin x \geq 2x$ valid on $[0, \pi/2]$, we get

$$|\sigma_n| = \left| \frac{\cos\vartheta - \cos(2n+1)\vartheta}{\sin\vartheta} \right| \leq \frac{2}{\sin\vartheta} \leq \frac{\pi}{\theta} < m_\theta + 1$$

where $\theta = 2\vartheta$; therefore

$$|S_1(\theta)| \leq (m_\theta + 1)(a_{r+1} + a_q + a_{r+1} - a_q)$$
$$\leq 2(r+1)a_{r+1} < 2\epsilon.$$

We then have

$$\left| \sum_{n=p}^{q} a_n \sin n\theta \right| \leq |S_0(\theta)| + |S_1(\theta)| < (\pi + 2)\epsilon,$$

which holds uniformly on the interval $[0, \pi]$; whence on $\mathbb{R}$ by the symmetry and periodicity. $\square$

SOLUTION 7.12

The uniform convergence on any interval $[\delta, 2\pi - \delta]$ with $\delta > 0$ can be easily derived from Dirichlet's test (see the second basic point on p. 117). Thus it suffices to show that the limit function is $(\pi - \theta)/2$.

Put $\theta = 2\vartheta$ for brevity. Integrating the formula

$$\frac{1}{2} + \sum_{n=1}^{m} \cos n\theta = \frac{\sin(2m+1)\vartheta}{2\sin\vartheta}$$

from 0 to $\omega \in [\delta, 2\pi - \delta]$, we obtain

$$\frac{\omega}{2} + \sum_{n=1}^{m} \frac{\sin n\omega}{n} = \int_0^\omega \frac{\sin(2m+1)\vartheta}{2\sin\vartheta} \, d\theta$$
$$= \int_0^\eta \frac{\sin(2m+1)\vartheta}{\sin\vartheta} \, d\vartheta$$

where $\omega = 2\eta$. We divide the last integral into two parts:

$$\int_0^\eta \frac{\sin(2m+1)t}{t}\,dt + \int_0^\eta \phi(t)\sin(2m+1)t\,dt,$$

and call them $I_m(\eta)$ and $J_m(\eta)$ respectively, where

$$\phi(t) = \frac{1}{\sin t} - \frac{1}{t}.$$

Note that $\phi \in \mathscr{C}^1(-\pi, \pi)$ if we define $\phi(0) = 0$.

Integrating by parts we will get

$$J_m(\eta) = -\frac{\phi(\eta)}{2m+1}\cos(2m+1)\eta$$
$$+ \frac{1}{2m+1}\int_0^\eta \phi'(t)\cos(2m+1)t\,dt,$$

from which it is easily seen that

$$|J_m(\eta)| < \frac{M_0 + \pi M_1}{2m+1}$$

where M_0 and M_1 are the maxima of $|\phi|$ and $|\phi'|$ on $[0, \pi - \delta/2]$ respectively. Thus $J_m(\eta)$ converges uniformly to 0 as $m \to \infty$ on $[0, \pi - \delta/2]$.

We next deal with the integral $I_m(\eta)$. Substituting $s = (2m+1)t$ we have

$$I_m(\eta) = \int_0^{(2m+1)\eta} \frac{\sin s}{s}\,ds.$$

Since $\pi/2 = I_m(\pi/2) + J_m(\pi/2)$, we have for any $x > \pi$

$$\left|\frac{\pi}{2} - \int_0^x \frac{\sin s}{s}\,ds\right| \leq \left|J_{m_x}\left(\frac{\pi}{2}\right)\right| + \int_{(2m_x+1)\pi/2}^x \frac{ds}{s}$$
$$\leq \left|J_{m_x}\left(\frac{\pi}{2}\right)\right| + \frac{1}{2m_x+1},$$

where m_x is the largest integer satisfying $(2m+1)\pi \leq 2x$. Since the right-hand side converges to 0 as $x \to \infty$, the improper integral

$$\int_0^\infty \frac{\sin s}{s}\,ds,$$

exists and equals to $\pi/2$. Therefore

$$\left|\frac{\omega - \pi}{2} + \sum_{n=1}^m \frac{\sin n\omega}{n}\right| \leq |J_m(\eta)| + \left|I_m(\eta) - \frac{\pi}{2}\right|$$
$$= |J_m(\eta)| + \left|\int_{(2m+1)\eta}^\infty \frac{\sin s}{s}\,ds\right|.$$

The right-hand side clearly converges to 0 uniformly in $\eta \in [\delta/2, \pi - \delta/2]$ as $m \to \infty$. $\square$

SOLUTION 7.13

For any $x \in (0, 2\pi)$ let $s_n(x)$ be the nth partial sum of the Fourier series. Then

$$s_n(x) = \frac{a_0}{2} + \sum_{k=1}^{n} (a_k \cos kx + b_k \sin kx)$$

$$= \frac{1}{2\pi} \int_0^{2\pi} f(t) \left(1 + 2 \sum_{k=1}^{n} \cos k(t - x) \right) dt.$$

Now, as we have already seen in **SOLUTION 7.12**, the trigonometric sum in the big parentheses can be expressed as the ratio of sines; so we obtain

$$s_n(x) = \frac{1}{\pi} \int_{-x/2}^{\pi - x/2} f(x + 2y) \frac{\sin(2n + 1)y}{\sin y} \, dy.$$

When $f(x) = 1$, then obviously $a_0 = 2$ and $a_n = b_n = 0$ for all $n \geq 1$; hence

$$1 = \frac{1}{\pi} \int_{-x/2}^{\pi - x/2} \frac{\sin(2n + 1)y}{\sin y} \, dy.$$

Therefore

$$s_n(x) - f(x) = \frac{1}{\pi} \int_{-x/2}^{\pi - x/2} \phi_x(y) \sin(2n + 1)y \, dy$$

where

$$\phi_x(y) = \frac{f(x + 2y) - f(x)}{\sin y}.$$

Clearly $\phi_x \in \mathscr{C}(-x/2, \pi - x/2)$ and the improper integral

$$\int_{-x/2}^{\pi - x/2} |\phi_x(y)| \, dy$$

converges, because $f \in \mathscr{C}^1(0, 2\pi)$ and $\int_0^{2\pi} |f(x)| \, dx < \infty$. Therefore it follows from the remark after **SOLUTION 5.3** that

$$\lim_{n \to \infty} s_n(x) = f(x),$$

in view of

$$\int_{-x/2}^{\pi - x/2} \sin 2y \, dy = \int_{-x/2}^{\pi - x/2} \cos 2y \, dy = 0. \qquad \square$$

Chapter 8

Approximation by Polynomials

[Summary of Basic Points]

1. Any continuous function $f(x)$ on a bounded closed interval $[a, b]$ can be approximated uniformly by polynomials; that is, for any $\epsilon > 0$ there exists a polynomial $P(x)$ satisfying

$$|f(x) - P(x)| < \epsilon$$

for all x in $[a, b]$. This is known as Weierstrass' approximation theorem (1885) and may be easily shown in an elementary way by using the Bernstein polynomials (see **PROBLEM 8.1**).

2. Any continuous periodic function $f(x)$ with period 2π can be approximated uniformly by trigonometric polynomials; that is, for any $\epsilon > 0$ there exists a trigonometric polynomial

$$P(\theta) = \sum_{n=0}^{m} (a_n \cos n\theta + b_n \sin n\theta)$$

satisfying $|f(\theta) - P(\theta)| < \epsilon$ for all θ (see also **PROBLEM 8.3**). This theorem is referred to as the approximation theorem by trigonometric polynomials. This may be derived from Fejér's summability theorem, a very important result in the theory of Fourier series. Moreover it shows us a constructive way of such trigonometric polynomials, as follows:

$$P(\theta) = \frac{s_0(\theta) + s_1(\theta) + \cdots + s_{m-1}(\theta)}{m}$$

where $s_n(\theta)$ is the nth partial sum of the Fourier series for $f(x)$.

137

PROBLEM 8.1 ────────────────────────────────────

For any $f \in \mathscr{C}[0, 1]$ *the polynomial of degree n defined by*

$$B_n(f;x) = \sum_{k=0}^{n} f\left(\frac{k}{n}\right)\binom{n}{k}x^k(1-x)^{n-k}$$

is called the Bernstein polynomial. Show that $B_n(f;x)$ converges to $f(x)$ uniformly on $[0, 1]$.

Runge (1885ab) gave nearly simultaneous proof for Weierstrass' approximation theorem (1885), but his papers do not explicitly contain Weierstrass' theorem. This was pointed out by L. E. Phragmén in the paper of Mittag-Leffler (1900).

Many other proofs of the approximation theorem appeared shortly after Weierstrass. Picard (1891) used the Poisson integral and Volterra (1897) used Dirichlet's principle. Lebesgue (1898) used essentially the following uniformly convergent series (not Taylor series in x) on the interval $[-1, 1]$:

$$|x| = 1 - \frac{1}{2}(1 - x^2) - \frac{1}{2 \cdot 4}(1 - x^2)^2 - \frac{1 \cdot 3}{2 \cdot 4 \cdot 6}(1 - x^2)^3 - \cdots.$$

This is Lebesgue's first paper.

In a letter to É. Picard, Mittag-Leffler (1900) gave an elementary proof using the following discontinuity property:

$$\lim_{n \to \infty} \chi_n(x) = \begin{cases} 1 & (x > 0) \\ 0 & (x = 0) \\ -1 & (-2 < x < 0) \end{cases}$$

where $\chi_n(x) = 1 - 2^{1-(1+x)^n}$.

Fejér (1900) showed that the Fourier series for a bounded integrable function is uniformly Cesàro summable of the first order on an interval on which the function is continuous. Fejér was 20 years old when he wrote this paper and obtained the doctoral thesis at Univ. of Budapest under H. A. Schwarz 2 years later.

Lerch (1903) gave a proof using the Fourier series of special but simple piecewise linear functions. Landau (1908) used the formula:

$$\lim_{n \to \infty} \frac{\displaystyle\int_0^1 f(x)(1 - (x-y)^2)^n \, dx}{\displaystyle\int_0^1 (1 - x^2)^n \, dx} = f(y).$$

Bernstein (1912a) gave a probabilistic proof of Weierstrass' approximation theorem by introducing the Bernstein polynomials. Carleman (1927) showed that, for any $f \in \mathscr{C}(\mathbb{R})$, there exists a sequence of entire functions converging to $f(x)$ uniformly on $\mathbb{R}$.

PROBLEM 8.2

Show that there exists a sequence of polynomials with integral coefficients converging to $f \in \mathscr{C}[0,1]$ uniformly on $[0,1]$ if $f(0) = f(1) = 0$.

Pál (1914) noted that for any $\epsilon > 0$ there exists a sequence of polynomials with integral coefficients converging uniformly to $f \in \mathscr{C}[\epsilon - 1, 1 - \epsilon]$ if $f(0) = 0$. Kakeya (1914) found necessary and sufficient conditions on $f \in \mathscr{C}[-1,1]$ which can be approximated by polynomials with integer coefficients. Pál's work was extended to larger intervals by Kakeya (1914) and Okada (1923).

PROBLEM 8.3

From Weierstrass' approximation theorem by polynomials, deduce the approximation theorem by trigonometric polynomials.

PROBLEM 8.4

For any $f \in \mathscr{C}^1[0,1]$ show that $B_n'(f;x)$ converges to $f'(x)$ uniformly on $[0,1]$, where $B_n(f;x)$ is the Bernstein polynomial.

PROBLEM 8.5

Show that $f \in \mathscr{C}[0,1]$ satisfies

$$\int_0^1 x^n f(x)\, dx = 0$$

for all integers $n \geq 0$ if and only if $f(x)$ vanishes everywhere on $[0,1]$.

We can replace the condition $f \in \mathscr{C}[0,1]$ by a weaker one $f \in \mathscr{C}(0,1)$ supposing in addition that the improper integral

$$\int_0^1 f(x)\, dx$$

converges absolutely. To see this consider

$$F(x) = \int_0^x f(t)\, dt.$$

Plainly $F \in \mathscr{C}[0,1]$, $F(0) = 0$ and

$$\int_0^1 x^n F(x)\, dx = \frac{x^{n+1} - 1}{n+1} F(x)\Big|_0^1 - \frac{1}{n+1}\int_0^1 (x^{n+1} - 1) f(x)\, dx = 0$$

for all integer $n \geq 0$, which implies that f vanishes everywhere. It is easily seen that the interval $[0,1]$ can be replaced by any compact interval using a suitable affine transformation. However we cannot replace the interval $[0,1]$ by $[0,\infty)$ as the following shows.

PROBLEM 8.6

For all integers $n \geq 0$ show that

$$\int_0^\infty x^n (\sin x^{1/4}) \exp(-x^{1/4})\, dx = 0.$$

PROBLEM 8.7

Put $a_0 = 0$ and let $\{a_n\}_{n\geq 1}$ be a strictly increasing sequence of positive numbers such that

$$\frac{1}{a_1} + \frac{1}{a_2} + \cdots + \frac{1}{a_n} + \cdots$$

diverges. Then show that $f \in \mathscr{C}[0, 1]$ satisfies

$$\int_0^1 x^{a_n} f(x)\, dx = 0$$

for all $n \geq 0$ if and only if $f(x)$ vanishes everywhere on $[0, 1]$.

This is a question posed by S. N. Bernstein (1880–1968) and solved by Müntz(1914) affirmatively. Carleman(1922) gave an elegant another proof using the theory of functions of a complex variable.

◆ ❊ ◆

Solutions for Chapter 8

SOLUTION 8.1

For any $\epsilon > 0$ it follows from the uniform continuity of f that there is a $\delta > 0$ such that $|f(x) - f(y)| < \epsilon$ for any $x, y \in [0, 1]$ with $|x - y| < \delta$. Let M be the maximum of $|f(x)|$ over $[0, 1]$ and let $\{d_n\}$ be any monotone sequence of positive integers diverging to ∞ and satisfying $d_n = o(n)$ as $n \to \infty$. We divide the following difference into two parts:

$$B_n(f; x) - f(x) = \sum_{k=0}^{n} \left(f\left(\frac{k}{n}\right) - f(x) \right) \binom{n}{k} x^k (1 - x)^{n-k}$$

$$= S_0(x) + S_1(x),$$

where $S_0(x)$ is the sum over $k \in [0, n]$ satisfying $|x - k/n| \le d_n/n$ and $S_1(x)$ is the sum over the remaining values of k.

We consider any sufficiently large n satisfying $d_n/n < \delta$. For $S_0(x)$, using $|f(x) - f(k/n)| < \epsilon$ we get

$$|S_0(x)| \le \epsilon \sum_{k=0}^{n} \binom{n}{k} x^k (1 - x)^{n-k} = \epsilon.$$

Next, for $S_1(x)$ using $|k - nx| > d_n$ we obtain

$$|S_1(x)| < 2M \sum_{k=0}^{n} \left(\frac{k - nx}{d_n} \right)^2 \binom{n}{k} x^k (1 - x)^{n-k}$$

$$= \frac{2M}{d_n^2} (n^2 x^2 + nx(1 - x) - 2n^2 x^2 + n^2 x^2)$$

$$= \frac{2M}{d_n^2} nx(1 - x) \le \frac{nM}{2d_n^2},$$

where we used the fact that $B_n(1; x) = 1$, $B_n(x; x) = x$ and

$$B_n(x^2; x) = x^2 + \frac{x(1 - x)}{n}.$$

We now take $d_n = [n^{2/3}]$ so that $d_n/\sqrt{n}$ diverges as $n \to \infty$. Therefore we have $|S_0(x) + S_1(x)| < 2\epsilon$ for any sufficiently large n. $\qquad \square$

REMARK. Bohman (1952) and Korovkin (1953) gave an astonishing proof, using only the linearity, monotonicity and the uniform convergence for $1, x, x^2$ among basic properties of B_n as an operator.

Let us explain this. For any $\epsilon > 0$ take δ and M as in the above proof and put

$$c = \sup_{|x-y| \geq \delta} \frac{|f(x) - f(y)|}{(x-y)^2}.$$

We then have

$$-\epsilon - c(x-y)^2 \leq f(x) - f(y) \leq \epsilon + c(x-y)^2$$

for any (x, y) in the unit square $[0, 1]^2$. Applying the operator B_n to the function $\phi_y(x) = (x-y)^2$ with a parameter y, it follows from the linearity and monotonicity of B_n that

$$-\epsilon B_n(1; x) - c B_n(\phi_y; x) \leq B_n(f; x) - f(y) B_n(1; x)$$
$$\leq \epsilon B_n(1; x) + c B_n(\phi_y; x).$$

Let $\varphi_n(x)$ be the function obtained from $B_n(\phi_y; x)$ by carrying the substitution $y = x$. Since

$$B_n(\phi_y; x) = B_n(x^2; x) - 2y B_n(x; x) + y^2 B_n(1; x),$$

we have

$$\varphi_n(x) = (B_n(x^2; x) - x^2) - 2x(B_n(x; x) - x) + x^2(B_n(1; x) - 1),$$

which converges to 0 uniformly on the interval $[0, 1]$ as $n \to \infty$. Therefore

$$|B_n(f; x) - f(x)| \leq |B_n(f; x) - f(x) B_n(1; x)| + M |B_n(1; x) - 1|$$
$$\leq \epsilon + c |\varphi_n(x)| + (M + \epsilon) |B_n(1; x) - 1|,$$

which is less than 3ϵ uniformly in x for all sufficiently large n.

The reason why we do not use the exact expressions of $B_n(1; x)$, $B_n(x; x)$ and $B_n(x^2; x)$ in the above proof, is to clarify the role of B_n as an operator; hence the proof is indeed valid for any linear operators having the monotonicity and the uniform convergence for $1, x, x^2$.

SOLUTION 8.2

The Bernstein polynomial for f defined in **PROBLEM 8.1** is

$$B_n(f; x) = \sum_{k=1}^{n-1} f\left(\frac{k}{n}\right) \binom{n}{k} x^k (1-x)^{n-k}$$

because $f(0) = f(1) = 0$. If $n = p$ is a prime number, then the binomial coefficient $\binom{p}{k}$ is clearly a multiple of p for every integer $1 \leq k < p$; hence

$$\widetilde{B}_p(f;x) = \sum_{k=1}^{p-1} \frac{1}{p}\binom{p}{k}\left[f\left(\frac{k}{p}\right)p\right]x^k(1-x)^{p-k}$$

is a polynomial with integral coefficients. Since we have

$$\left|B_p(f;x) - \widetilde{B}_p(f;x)\right| \leq \frac{1}{p}\sum_{k=1}^{p-1}\binom{p}{k}x^k(1-x)^{p-k} < \frac{1}{p},$$

it is clear that $\widetilde{B}_p(f;x)$ converges also to $f(x)$ uniformly on $[0,1]$ as $p \to \infty$. □

SOLUTION 8.3

For any continuous periodic function with period 2π we can write

$$f(x) = f_+(x) + f_-(x) \quad \text{where} \quad f_\pm(x) = \frac{f(x) \pm f(-x)}{2}$$

respectively. The functions $f_\pm(x)$ are also continuous periodic functions with the same period satisfying $f_+(x) = f_+(-x)$ and $f_-(x) + f_-(-x) = 0$. Note that $f_-(k\pi) = 0$ for all integers k.

For any $\epsilon > 0$ we can take a continuous odd periodic function $\phi(x)$ with period 2π such that $|f_-(x) - \phi(x)| < \epsilon$ for any $x \in \mathbb{R}$ and that $\phi(x)$ vanishes on each small neighborhood of $k\pi$. Since $x = \arccos y$ maps the interval $[-1,1]$ onto $[0,\pi]$ homeomorphically, the functions

$$f_+(\arccos y) \quad \text{and} \quad \frac{\phi(\arccos y)}{\sin(\arccos y)}$$

are continuous on the interval $[-1,1]$. Thus applying Weierstrass' approximation theorem to these functions, we can find two polynomials $P(y)$ and $Q(y)$ satisfying

$$|f_+(x) - P(\cos x)| < \epsilon \quad \text{and} \quad |\phi(x) - Q(\cos x)\sin x| < \epsilon$$

for any $0 \leq x \leq \pi$. Moreover these inequalities hold for any $x \in \mathbb{R}$ by the evenness of $f_+(x)$ and $P(\cos x)$, by the oddness of $\phi(x)$ and $Q(\cos x)\sin x$ and, of course, by the periodicity of these functions. We therefore obtain

$$|f(x) - P(\cos x) - Q(\cos x)\sin x|$$
$$\leq |f_+(x) - P(\cos x)| + |f_-(x) - \phi(x)| + |\phi(x) - Q(\cos x)\sin x|$$
$$< 3\epsilon$$

for any $x \in \mathbb{R}$. Finally it is easily verified that each $\cos^k x$ can be written as a linear combination of $1, \cos x, \ldots, \cos kx$ and that each $\cos^k x \sin x$ can be expressed as a linear combination of $\sin x, \sin 2x, \ldots, \sin kx$. This completes the proof. $\qquad\square$

REMARK. The proof may be simpler than that of Achieser(1956) on p. 32.

SOLUTION 8.4

Since

$$B'_n(f;x) = \sum_{k=0}^{n} f\left(\frac{k}{n}\right)\binom{n}{k}(kx^{k-1}(1-x)^{n-k} - (n-k)x^k(1-x)^{n-k-1})$$

$$= n\sum_{k=0}^{n-1}\left(f\left(\frac{k+1}{n}\right) - f\left(\frac{k}{n}\right)\right)\binom{n-1}{k}x^k(1-x)^{n-k-1},$$

it follows from the mean value theorem that

$$B'_n(f;x) = \sum_{k=0}^{n-1} f'\left(\frac{k+\xi_k}{n}\right)\binom{n-1}{k}x^k(1-x)^{n-k-1}$$

for some $\xi_k \in (k/n, (k+1)/n)$. For any $\epsilon > 0$ there exists an integer N such that $|f'(x) - f'(y)| < \epsilon$ for all $x, y \in [0, 1]$ with $|x - y| \le 1/N$: so we have

$$\left|B'_n(f;x) - B_{n-1}(f';x)\right|$$

$$\le \sum_{k=0}^{n-1}\left|f'\left(\frac{k+\xi_k}{n}\right) - f'\left(\frac{k}{n}\right)\right|\binom{n-1}{k}x^k(1-x)^{n-k-1}$$

$$< \epsilon \sum_{k=0}^{n-1}\binom{n-1}{k}x^k(1-x)^{n-k-1} = \epsilon$$

for all $n > N$. $\qquad\square$

SOLUTION 8.5

For any $\epsilon > 0$ there exists a polynomial $P(x)$ satisfying $|f(x) - P(x)| < \epsilon$ on the interval $[0, 1]$ by Weierstrass' approximation theorem. Letting M be the maximum of $|f(x)|$ over $[0, 1]$ we get

$$\int_0^1 f^2(x)\,dx = \int_0^1 (f(x) - P(x))\,f(x)\,dx + \int_0^1 P(x)f(x)\,dx$$

$$\le \int_0^1 |f(x) - P(x)| \cdot |f(x)|\,dx < \epsilon M.$$

Since ϵ is arbitrary, we have $\int_0^1 f^2(x)\,dx = 0$; so $f(x)$ vanishes everywhere. $\quad\square$

<hr>

SOLUTION 8.6

For brevity put

$$I_n = \int_0^\infty x^n e^{-x} \sin x \, dx \quad \text{and} \quad J_n = \int_0^\infty x^n e^{-x} \cos x \, dx$$

for any integer $n \geq 0$. Substituting $t = x^{1/4}$ we see that the integral in the problem is equal to $4I_{4n+3}$. By partial integration we easily get

$$I_n = \frac{n}{2}(I_{n-1} + J_{n-1}) \quad \text{and} \quad J_n = \frac{n}{2}(J_{n-1} - I_{n-1})$$

for any $n \geq 1$. Solving these recursion formulae with the initial condition $I_0 = J_0 = 1/2$, we conclude that $I_n = 0$ for any integer n satisfying $n \equiv 3 \pmod{4}$. $\quad\square$

REMARK. We also get $J_n = 0$ for any integer n satisfying $n \equiv 1 \pmod{4}$. This means that

$$\int_0^\infty x^n \frac{(\cos x^{1/4}) \exp(-x^{1/4})}{\sqrt{x}} \, dx = 0$$

for all integers $n \geq 0$.

<hr>

SOLUTION 8.7

Since the case $\{a_n\} = \{n\}$ can be reduced to **PROBLEM 8.5**, we can assume that there exists an integer $m \geq 1$ satisfying $m \neq a_n$ for all $n \geq 1$. For such an m we first consider the definite integral

$$I_n = \int_0^1 (x^m - c_0 - c_1 x^{a_1} - c_2 x^{a_2} - \cdots - c_n x^{a_n})^2 \, dx$$

for any $n \geq 1$. Obviously I_n is a positive quadratic form in $c_0, c_1, \ldots, c_n$ and attains its minimum I_n^* at some point $(s_0, s_1, \ldots, s_n) \in \mathbb{R}^{n+1}$, which is a unique solution of the system of $n + 1$ linear equations:

$$\sum_{i=0}^n \frac{s_i}{a_i + a_j + 1} = \frac{1}{m + a_j + 1} \tag{8.1}$$

for $0 \leq j \leq n$ with $a_0 = 0$. Here the coefficient matrix

$$A = \left(\frac{1}{a_i + a_j + 1} \right)_{0 \leq i, j \leq n}$$

is symmetric and its determinant can be written explicitly as

$$\det A = \frac{\displaystyle\prod_{0 \le i < j \le n} (a_i - a_j)^2}{\displaystyle\prod_{0 \le i, j \le n} (a_i + a_j + 1)} \ne 0$$

by using the Cauchy determinant. It then follows from (8.1) that

$$I_n^* = \int_0^1 (x^m - s_0 - s_1 x^{a_1} - s_2 x^{a_2} - \cdots - s_n x^{a_n})^2 \, dx$$

$$= \frac{1}{2m+1} - 2 \sum_{j=0}^n \frac{s_j}{m + a_j + 1} + \sum_{i \ne j} \frac{s_i s_j}{a_i + a_j + 1}$$

$$= \frac{1}{2m+1} - \sum_{j=0}^n \frac{s_j}{m + a_j + 1}. \tag{8.2}$$

Combining (8.1) and (8.2) we thus have

$$\begin{pmatrix} & & 0 \\ A & & \vdots \\ & & 0 \\ a & & 1 \end{pmatrix} \begin{pmatrix} s_0 \\ \vdots \\ s_n \\ I_n^* \end{pmatrix} = \begin{pmatrix} {}^t a \\ \\ (2m+1)^{-1} \end{pmatrix}$$

where

$$a = \left(\frac{1}{m + a_0 + 1}, \frac{1}{m + a_1 + 1}, \ \cdots, \frac{1}{m + a_n + 1} \right).$$

Therefore we get

$$I_n^* = \frac{\det B}{\det A}$$

by Cramer's rule, where

$$B = \begin{pmatrix} A & {}^t a \\ \\ a & (2m+1)^{-1} \end{pmatrix}$$

is again symmetric and det B can be obtained from the Cauchy determinant by substituting a_{n+1} by m formally. Therefore we have

$$I_n^* = \frac{1}{2m+1} \prod_{k=0}^{n} \left(\frac{a_k - m}{a_k + m + 1} \right)^2$$

$$\leq \frac{1}{2m+1} \left(\frac{m}{m+1} \right)^{2\sigma_n} \exp\left(-2 \sum_{a_k > m} \frac{m}{a_k} \right),$$

because $\dfrac{m - a_k}{a_k + m + 1} < \dfrac{m}{m+1}$ for $a_k < m$ and $\dfrac{a_k - m}{a_k + m + 1} < \exp\left(-\dfrac{m}{a_k} \right)$ for $a_k > m$,

where σ_n is the number of k's satisfying $a_k < m$ for $0 \leq k \leq n$. If $\sigma_n \to \infty$ as $n \to \infty$, then clearly I_n^* converges to 0. On the other hand, if σ_n is bounded, then there exists an integer N such that $a_k > m$ for all $k > N$; so I_n^* also converges to 0, because $m \neq 0$ and $\sum 1/a_k = \infty$ by the assumption.

For any $\epsilon > 0$ we can take a sufficiently large n and $(c_0, c_1, ..., c_n) \in \mathbb{R}^{n+1}$ such that

$$\int_0^1 (x^m - c_0 - c_1 x^{a_1} - c_2 x^{a_2} - \cdots - c_n x^{a_n})^2 \, dx < \epsilon.$$

By the Cauchy-Schwarz inequality we have

$$\left(\int_0^1 x^m f(x) \, dx \right)^2 = \left(\int_0^1 (x^m - c_0 - c_1 x^{a_1} - c_2 x^{a_2} - \cdots - c_n x^{a_n}) \, f(x) \, dx \right)^2$$

$$\leq M \int_0^1 (x^m - c_0 - c_1 x^{a_1} - c_2 x^{a_2} - \cdots - c_n x^{a_n})^2 \, dx$$

$$< \epsilon M$$

where $M = \displaystyle\int_0^1 f^2(x) \, dx$. Since ϵ is arbitrary, $\displaystyle\int_0^1 x^m f(x) \, dx = 0$ for any integer m satisfying $m \neq a_n$ for all $n \geq 0$. This means that $\displaystyle\int_0^1 x^m f(x) \, dx = 0$ for all integers $m \geq 0$ and $f(x)$ vanishes everywhere on $[0, 1]$ by **Problem 8.5**. $\qquad \square$

Remark. For any integers $m, n \geq 1$ satisfying $a_k \neq m$ for all $k \geq 1$, define

$$d_{n,m} = \inf_{c_0, ..., c_n} \max_{0 \leq x \leq 1} |x^m - c_0 - c_1 x^{a_1} - \cdots - c_n x^{a_n}|$$

where the infimum ranges over all real numbers $c_0, ..., c_n$. The argument in the above proof can be used to show that $d_{n,m}$ converges to 0 as $n \to \infty$ for an arbitrarily fixed $m \geq 2$ if $a_1 \geq 1$. To see this, by using the Cauchy-Schwarz inequality,

$$|x^m - c_1 x^{a_1} - \cdots - c_n x^{a_n}|^2$$

$$= \left| \int_0^x (mt^{m-1} - a_1 c_1 t^{a_1-1} - \cdots - a_n c_n t^{a_n-1}) \, dt \right|^2$$

$$\leq m^2 \int_0^1 (t^{m-1} - c_1' t^{a_1-1} - \cdots - c_n' t^{a_n-1})^2 \, dt$$

where $c_k' = a_k c_k / m$. We can now take n and $c_1', \ldots, c_n'$ suitably so that the right-hand side becomes arbitrarily small.

◆ ❈ ◆

Chapter 9

Convex Functions

Some results in this chapter are due to J. L. W. V. Jensen (1859–1925) who introduced the notion of *convex* functions in Jensen (1906).

[Summary of Basic Points]

1. A function $f(x)$ defined on an interval I is said to be *convex in the sense of Jensen* provided that

$$f\left(\frac{x+y}{2}\right) \le \frac{f(x) + f(y)}{2}$$

for any x, y in I. Note that $f(x)$ is not necessarily continuous on I.

2. The set of all convex functions defined on I forms a positive cone; that is, if $f_1(x)$ and $f_2(x)$ are convex functions on I, then

$$c_1 f_1(x) + c_2 f_2(x)$$

are also convex for any constants $c_1, c_2 > 0$.

3. For example, the function $|x|$ is clearly convex on $\mathbb{R}$; hence a piecewise linear function

$$\sum_{k=1}^{n} c_k |x - x_k|$$

is also convex on $\mathbb{R}$ for any positive constants $c_1, c_2, \ldots, c_n$.

4. A function $g(x)$ on I is said to be *concave* if $-g(x)$ is convex on I.

5. A positive function $f(x)$ on I is said to be *logarithmically convex* if $\log f(x)$ is convex on I.

PROBLEM 9.1

Suppose that $f(x)$ is convex on an interval I. Show that

$$f\left(\frac{x_1 + x_2 + \cdots + x_n}{n}\right) \le \frac{f(x_1) + f(x_2) + \cdots + f(x_n)}{n}$$

for any n points $x_1, x_2, \ldots, x_n$ in I.

PROBLEM 9.2

Show that $f(x)$ is convex on an interval I if and only if $e^{\lambda f(x)}$ is convex on I for any $\lambda > 0$.

PROBLEM 9.3

Suppose that $f(x)$ is convex and bounded above on an open interval (a, b). Show then that $f(x)$ is continuous on (a, b).

PROBLEM 9.4

Suppose that $f \in \mathscr{C}(I)$ is convex. Show that

$$f\left(\frac{\lambda_1 x_1 + \cdots + \lambda_n x_n}{\lambda_1 + \cdots + \lambda_n}\right) \le \frac{\lambda_1 f(x_1) + \cdots + \lambda_n f(x_n)}{\lambda_1 + \cdots + \lambda_n}$$

for any n points $x_1, \ldots, x_n$ in I and any positive numbers $\lambda_1, \ldots, \lambda_n$.

PROBLEM 9.5

Suppose that $g, p \in \mathscr{C}[a, b]$, $p(x) \ge 0$ and

$$\sigma = \int_a^b p(x)\, dx > 0.$$

Let m and M be the minimum and the maximum of $g(x)$ on $[a, b]$ respectively. Suppose further that $f \in \mathscr{C}[m, M]$ is convex. Show then that

$$f\left(\frac{1}{\sigma} \int_a^b g(x) p(x)\, dx\right) \le \frac{1}{\sigma} \int_a^b f(g(x))\, p(x)\, dx.$$

PROBLEM 9.6

Show that any continuous convex function $f(x)$ on an open interval I possesses a finite derivative except for at most countable points.

PROBLEM 9.7

Show that $f \in \mathscr{C}[a, b]$ is convex if and only if

$$\mathscr{M}_{[s,t]}(f) \leq \frac{f(s) + f(t)}{2}$$

for any $s, t \in [a, b]$ with $s < t$.

PROBLEM 9.8

Suppose that $f(x)$ is twice differentiable in an open interval I. Show then that $f(x)$ is convex if and only if $f''(x) \geq 0$ on I.

Note that we do not suppose the continuity of $f''(x)$.

PROBLEM 9.9

Suppose that $f \in \mathscr{C}^2[0, \infty)$ is convex and bounded. Show that the improper integral

$$\int_0^\infty x f''(x) \, dx$$

converges.

PROBLEM 9.10

Let I be a closed interval of the form either $[0, a]$ or $[0, \infty)$. Suppose that $f \in \mathscr{C}(I)$ satisfies $f(0) = 0$. Show then that f is convex if and only if

$$\sum_{k=1}^n (-1)^{k-1} f(x_k) \geq f\left(\sum_{k=1}^n (-1)^{k-1} x_k \right)$$

for all integers $n \geq 2$ and any n points $x_1 \geq x_2 \geq \cdots \geq x_{n-1} \geq x_n$ in I.

Wright(1954) gave the simple proof and noted that this is a special case of Theorem 108 in Hardy, Littlewood and Pólya(1934).

PROBLEM 9.11

    Let $s > -1$ be a constant. Suppose that $f \in \mathscr{C}[0, \infty)$ is a convex function and satisfies $f(0) \geq 0$. Suppose further that $f(x)$ is differentiable on $(0, \infty)$ except for discrete points and $f'(x)$ is piecewise continuous. Suppose also that $f'(0+)$ exists when $f(0) = 0$. Then show that

$$\int_0^\infty x^s \exp\left(-\frac{f(x)}{x}\right) dx \leq \int_0^\infty x^s \exp\left(-f'\left(\frac{x}{e}\right)\right) dx.$$

Prove moreover that the constant e in the denominator of the right-hand side cannot be replaced by any smaller number in general.

    This is due to Carleson (1954). He used this inequality in the case $s = 0$ to show Carleman's inequality stated in **PROBLEM 2.11** in the following manner. Observe first that the given series can be arranged in decreasing order. He then defined the graph of $f(x)$ as the polygon whose vertices are the origin and the points

$$\left(n, \sum_{k=1}^n \log \frac{1}{a_k}\right)$$

for all integers $n \geq 1$.

                                   ✦ ✻ ✦

Solutions for Chapter 9

SOLUTION 9.1

Applying the convexity of f for m times, we easily get

$$f\left(\frac{x_1 + x_2 + \cdots + x_{2^m}}{2^m}\right) \leq \frac{f(x_1) + f(x_2) + \cdots + f(x_{2^m})}{2^m}$$

for any 2^m points $x_1, \dots, x_{2^m}$ in I. For any integer $n \geq 3$ we choose m satisfying $n < 2^m$. Now adding to $x_1, \dots, x_n$ the following new $2^m - n$ points

$$x_{n+1} = \cdots = x_{2^m} = \frac{x_1 + x_2 + \cdots + x_n}{n}$$

in I, we have

$$2^m f(y) \leq f(x_1) + \cdots + f(x_n) + (2^m - n)f\left(\frac{x_1 + x_2 + \cdots + x_n}{n}\right)$$

where

$$y = \frac{1}{2^m}\left(x_1 + x_2 + \cdots + x_n + (2^m - n)\frac{x_1 + x_2 + \cdots + x_n}{n}\right)$$

$$= \frac{x_1 + x_2 + \cdots + x_n}{n};$$

namely

$$f\left(\frac{x_1 + x_2 + \cdots + x_n}{n}\right) \leq \frac{f(x_1) + f(x_2) + \cdots + f(x_n)}{n}. \qquad \square$$

SOLUTION 9.2

Suppose first that $f(x)$ is convex on the interval I. Since $e^{\lambda x}$ is convex and monotone increasing on $\mathbb{R}$, we have

$$\exp\left(\lambda f\left(\frac{x+y}{2}\right)\right) \leq \exp\left(\lambda \frac{f(x) + f(y)}{2}\right) \leq \frac{e^{\lambda f(x)} + e^{\lambda f(y)}}{2}$$

for any x, y in I. Hence $e^{\lambda f(x)}$ is convex on I for any $\lambda > 0$.

Conversely suppose that $e^{\lambda f(x)}$ is convex on I for any $\lambda > 0$. The asymptotic expansions as $\lambda \to 0+$ of both sides of

$$\exp\left(\lambda f\left(\frac{x+y}{2}\right)\right) \leq \frac{e^{\lambda f(x)} + e^{\lambda f(y)}}{2}$$

give

$$1 + \lambda f\left(\frac{x+y}{2}\right) + O(\lambda^2) \leq 1 + \lambda \frac{f(x) + f(y)}{2} + O(\lambda^2)$$

for any x, y in I, which implies the convexity of $f(x)$ on I. $\qquad \square$

SOLUTION 9.3

Suppose that $f(x) < K$ for some constant K. For an arbitrarily fixed y in the interval (a, b) and any integer $n \geq 1$ the points $y \pm n\delta$ belong to (a, b) for all sufficiently small $\delta > 0$. Of course δ depends on n. Applying the inequality described in **PROBLEM 9.1** to the n points $x_1 = y \pm n\delta, x_2 = \cdots = x_n = y$, we get

$$f(y \pm \delta) \leq \frac{f(y \pm n\delta) + (n-1)f(y)}{n}$$

respectively. Hence

$$f(y+\delta) - f(y) \leq \frac{f(y+n\delta) - f(y)}{n} \leq \frac{K - f(y)}{n}$$

and

$$f(y) - f(y-\delta) \geq \frac{f(y) - f(y-n\delta)}{n} \geq \frac{f(y) - K}{n}.$$

Since $f(y) - f(y-\delta) \leq f(y+\delta) - f(y)$ and n is arbitrary, we conclude that $f(x)$ is continuous at the point y. $\qquad \square$

SOLUTION 9.4

By the inequality in **PROBLEM 9.1** it is easily seen that

$$f\left(\frac{k_1 x_1 + \cdots + k_n x_n}{k_1 + \cdots + k_n}\right) \leq \frac{k_1 f(x_1) + \cdots + k_n f(x_n)}{k_1 + \cdots + k_n}$$

for any points $x_1, \ldots, x_n$ in I and any positive integers $k_1, \ldots, k_n$. For any sufficiently large integer N we take

$$k_j = \left[\frac{\lambda_j N}{\lambda_1 + \cdots + \lambda_n}\right]$$

for each $1 \leq j \leq n$. Since

$$\frac{k_j}{k_1 + \cdots + k_n} \to \frac{\lambda_j}{\lambda_1 + \cdots + \lambda_n}$$

as $N \to \infty$, the required inequality follows from the continuity of f. $\qquad\square$

SOLUTION 9.5

We divide equally the interval $[a, b]$ into n parts and put

$$\lambda_k = \int_{t_{k-1}}^{t_k} p(x)\, dx \geq 0$$

for any subinterval $[t_{k-1}, t_k]$ so that $\sigma = \sum_{k=1}^{n} \lambda_k$. It follows from the first mean value theorem that

$$\int_{t_{k-1}}^{t_k} g(x) p(x)\, dx = \lambda_k g(\xi_k)$$

for some ξ_k in (t_{k-1}, t_k). Applying the inequality in **PROBLEM 9.4** to n points $x_k = g(\xi_k) \in [m, M]$, we obtain

$$f\left(\frac{1}{\sigma} \int_a^b g(x) p(x)\, dx \right) = f\left(\frac{1}{\sigma} \sum_{k=1}^{n} \lambda_k g(\xi_k) \right)$$

$$\leq \frac{1}{\sigma} \sum_{k=1}^{n} \lambda_k f(g(\xi_k))$$

$$= \frac{b-a}{\sigma n} \sum_{k=1}^{n} f(g(\xi_k))\, p(\eta_k)$$

for some η_k in (t_{k-1}, t_k). By the uniform continuity of $p(x)$, the difference between the expression on the right-hand side and one with $p(\eta_k)$ replaced by $p(\xi_k)$ is sufficiently small whenever n is sufficiently large. Therefore the right-hand side converges to

$$\frac{1}{\sigma} \int_a^b f(g(x))\, p(x)\, dx$$

as $n \to \infty$. $\qquad\square$

SOLUTION 9.6

Let $f'_+(x)$ and $f'_-(x)$ be the right and left derivatives of f at x respectively. First we will prove that $f'_\pm(x)$ exist at every point x in the open interval I. Let $\Delta(x, y)$ denote the difference quotient

$$\frac{f(x) - f(y)}{x - y}$$

for any $x \neq y$ in I. Let $x < y < z$ be arbitrary three points in I. Applying the inequality stated in **PROBLEM 9.4** with $\lambda_1 = z - y$, $x_1 = x$ and $\lambda_2 = y - x$, $x_2 = z$, we get $y = (\lambda_1 x_1 + \lambda_2 x_2)/(\lambda_1 + \lambda_2)$ and

$$f(y) \leq \frac{z-y}{z-x} f(x) + \frac{y-x}{z-x} f(z).$$

Thus we have $\Delta(x, y) \leq \Delta(y, z)$. Note that the above inequality can also be written as $\Delta(x, y) \leq \Delta(x, z)$ or as $\Delta(x, z) \leq \Delta(y, z)$. In particular, the quotients $\Delta(x - h, x)$ and $\Delta(x, x + h)$ are monotone increasing with respect to $h > 0$ and $\Delta(x - h, x) \leq \Delta(x, x + h)$ holds for any $h > 0$ satisfying $x \pm h \in I$. This means that both $f'_+(x)$ and $f'_-(x)$ certainly exist and satisfy $f'_-(x) \leq f'_+(x)$.

Next for any points $x < y$ in I we take a sufficiently small $h > 0$ satisfying $x + h < y - h$. Then

$$\Delta(x, x + h) \leq \Delta(x + h, y - h) \leq \Delta(y - h, y);$$

so, letting $h \to 0+$ we obtain $f'_+(x) \leq f'_-(y)$. Therefore if $f(x)$ does not possess a finite differential coefficient at x_0, then it follows that $f'_-(x_0) < f'_+(x_0)$. Moreover if we assign the point x_0 to the open interval $(f'_-(x_0), f'_+(x_0))$, then such intervals are disjoint mutually. Hence we can enumerate all such open intervals by labeling them, for example, in such a way that we count ones contained in $(-n, n)$ and having the length greater than $1/n$ for each positive integer n. $\qquad\square$

SOLUTION 9.7

First assume that a continuous function $f(x)$ is convex on the interval $[a, b]$. We divide a subinterval $[s, t]$ equally into n parts and put

$$x_k = \frac{n-k}{n} s + \frac{k}{n} t$$

for $0 \leq k \leq n$. It follows from the inequality in **PROBLEM 9.1** that

$$f(x_k) \leq \frac{n-k}{n} f(s) + \frac{k}{n} f(t).$$

Therefore

$$\mathcal{M}_{[s,t]}(f) = \lim_{n \to \infty} \frac{1}{n} \sum_{k=0}^{n} f(x_k)$$

$$\leq \limsup_{n \to \infty} \frac{n(n+1)}{2n^2} (f(s) + f(t)) = \frac{f(s) + f(t)}{2}.$$

Conversely assume that

$$\mathcal{M}_{[s,t]}(f) \leq \frac{f(s) + f(t)}{2}.$$

for any $s < t$ in $[a, b]$. Suppose, on the contrary, that there are two points $s < t$ in $[a, b]$ satisfying

$$f\left(\frac{s+t}{2}\right) > \frac{f(s) + f(t)}{2}.$$

Then we consider the set

$$E = \left\{ x \in [s, t] \ \middle| \ f(x) > f(s) + \frac{f(t) - f(s)}{t - s}(x - s) \right\}.$$

The set E is clearly open by the continuity of f and $E \neq \emptyset$ because it contains the point $(s + t)/2$. Note that E is the set of points on the interval $[s, t]$ at which the graph of $f(x)$ is situated in the upper side of the straight line through the two points $(s, f(s))$ and $(t, f(t))$. Let (u, v) be the connected component of E containing the point $(s + t)/2$. Since the end points $(u, f(u))$ and $(v, f(v))$ must lie on that line, we have

$$f(u) = f(s) + \frac{f(t) - f(s)}{t - s}(u - s),$$

$$f(v) = f(s) + \frac{f(t) - f(s)}{t - s}(v - s),$$

which imply that

$$\frac{f(u) + f(v)}{2} = f(s) + \frac{f(t) - f(s)}{t - s}\left(\frac{u + v}{2} - s\right).$$

Therefore

$$\mathscr{M}_{[u,v]}(f) > \frac{1}{v - u} \int_u^v \left(f(s) + \frac{f(t) - f(s)}{t - s}(x - s) \right) dx$$

$$= f(s) + \frac{f(t) - f(s)}{t - s}\left(\frac{u + v}{2} - s\right)$$

$$= \frac{f(u) + f(v)}{2},$$

contrary to the assumption. $\qquad \square$

SOLUTION 9.8

Suppose that $f(x)$ is convex on I. We have already seen in the proof of **PROBLEM 9.6** that $f'_+(x) \le f'_-(y)$ for any points $x < y$ in I. Since $f(x)$ is twice differentiable, the derivative $f'(x)$ is monotone increasing on I, whence $f''(x) \ge 0$.

Conversely suppose that $f''(x) \ge 0$. Let x and y be any points in I. By Taylor's formula centered at $(x + y)/2$ we obtain

$$f(x) = f\left(\frac{x+y}{2}\right) + f'\left(\frac{x+y}{2}\right)\frac{x-y}{2} + f''(c)\frac{(x-y)^2}{8},$$

$$f(y) = f\left(\frac{x+y}{2}\right) + f'\left(\frac{x+y}{2}\right)\frac{y-x}{2} + f''(c')\frac{(x-y)^2}{8}$$

for some c and c'. Adding these equalities we get

$$f(x) + f(y) = 2f\left(\frac{x+y}{2}\right) + \left(f''(c) + f''(c')\right)\frac{(x-y)^2}{8}$$

$$\geq 2f\left(\frac{x+y}{2}\right).$$

Hence $f(x)$ is convex on I. $\qquad\square$

SOLUTION 9.9

Suppose $\delta = f'(x_0) > 0$ for some $x_0 > 0$. Since $f'(x)$ is monotone increasing, we have $f'(x) \geq \delta$ for all $x \geq x_0$ and hence

$$f(x) = f(x_0) + \int_{x_0}^{x} f'(t)\,dt$$

$$\geq f(x_0) + \delta(x - x_0),$$

contrary to the assumption that f is bounded. Thus $f'(x) \leq 0$ for all $x > 0$ and it follows that $f(x)$ is monotone decreasing and converges to some λ as $x \to \infty$. Therefore

$$\lambda - f(x) = \int_{x}^{\infty} f'(t)\,dt$$

and in particular, $f'(x)$ converges to 0 as $x \to \infty$. By the Cauchy criterion we have, for any $\epsilon > 0$,

$$0 < -\int_{\alpha}^{\beta} f'(t)\,dt < \epsilon$$

for all sufficiently large α and $\beta > \alpha$. Since $f'(x)$ is non-positive and monotone increasing, we obtain $(\beta - \alpha)|f'(\beta)| < \epsilon$; therefore

$$\beta|f'(\beta)| < \epsilon + \alpha|f'(\beta)|.$$

The right-hand side can be smaller than 2ϵ if we take β sufficiently large. Hence $xf'(x)$ converges to 0 as $x \to \infty$. Thus

$$\int_{0}^{x} tf''(t)\,dt = xf'(x) - f(x) + f(0)$$

converges to $f(0) - \lambda$ as $x \to \infty$. $\square$

Suppose first that f is convex on the closed interval I. Let $x_1 > x_2 > x_3$ be arbitrary three points in I and define a positive number λ with $x_2 = \lambda x_1 + (1-\lambda)x_3$. Since f is convex, it follows from **PROBLEM 9.4** that

$$f(x_2) = f(\lambda x_1 + (1-\lambda)x_3) \le \lambda f(x_1) + (1-\lambda)f(x_3)$$

and

$$f(x_1 - x_2 + x_3) = f((1-\lambda)x_1 + \lambda x_3) \le (1-\lambda)f(x_1) + \lambda f(x_3);$$

so we have

$$f(x_2) + f(x_1 - x_2 + x_3) \le f(x_1) + f(x_3),$$

which is also valid for any $x_1 \ge x_2 \ge x_3$ in I by the continuity of f. If we take $x_3 = 0$, then clearly $f(x_1 - x_2) \le f(x_1) - f(x_2)$ by virtue of $f(0) = 0$. We thus have the inequality in the problem for $n = 2$ and 3. Suppose now the inequality holds for $n = m \ge 2$. Then for any $m + 2$ points $x_1 \ge x_2 \ge \cdots \ge x_{m+2}$ in I,

$$\sum_{k=1}^{m+2} (-1)^{k-1} f(x_k) \ge f(x_1) - f(x_2) + f\left(\sum_{k=3}^{m+2} (-1)^{k-1} x_k\right)$$
$$\ge f\left(\sum_{k=1}^{m+2} (-1)^{k-1} x_k\right),$$

which means that the inequality in the problem holds for $n = m + 2$; therefore for every $n \ge 2$.

Conversely suppose that the inequality in the problem for the case $n = 3$ holds:

$$f(x_2) + f(x_1 - x_2 + x_3) \le f(x_1) + f(x_3)$$

for any points $x_1 \ge x_2 \ge x_3$ in I. By taking $x_2 = (x_1 + x_3)/2$ we get

$$f\left(\frac{x_1 + x_3}{2}\right) \le \frac{f(x_1) + f(x_3)}{2},$$

which means that f is convex on I. $\square$

As is shown in **SOLUTION 9.6**, the difference quotient

$$\Delta(x + h, x) = \frac{f(x + h) - f(x)}{h}$$

is monotone increasing for $h > 0$; therefore

$$\Delta(\alpha x, x) \geq \Delta(x + h, x)$$

for any $\alpha > 1$ and $x > 0$ if $(\alpha - 1)x \geq h$ is fulfilled. Thus, letting $h \to 0+$, we get $\Delta(\alpha x, x) \geq f'(x)$ if the derivative exists; in other words,

$$f(\alpha x) \geq f(x) + (\alpha - 1)x f'(x).$$

For brevity put

$$F(x) = x^s \exp\left(-\frac{f(x)}{x}\right) \quad \text{and} \quad G(x) = x^s \exp(-f'(x)).$$

Note that the improper integral

$$\int_0^\infty F(x)\,dx$$

converges at $x = 0$ if $f(0) > 0$. When $f(0) = 0$ and $f'(0+)$ exists, it also converges by virtue of $f'(0+) \leq f(x)/x$. Substituting $x = \alpha t$ we have

$$\int_0^L F(x)\,dx = \alpha^{s+1} \int_0^{L/\alpha} t^s \exp\left(-\frac{f(\alpha t)}{\alpha t}\right) dt$$

$$\leq \alpha^{s+1} \int_0^{L/\alpha} F^{1/\alpha}(t) G^{(\alpha-1)/\alpha}(t)\,dt$$

for any $L > 0$. By Hölder's inequality the right-hand side is less than or equal to

$$\alpha^{s+1} \left(\int_0^{L/\alpha} F(x)\,dx\right)^{1/\alpha} \left(\int_0^{L/\alpha} G(x)\,dx\right)^{(\alpha-1)/\alpha}.$$

Replacing L/α by L we thus have

$$\int_0^L F(x)\,dx < \phi^{s+1}(\alpha) \int_0^L G(x)\,dx$$

where $\phi(\alpha) = \alpha^{\alpha/(\alpha-1)}$ is strictly monotone increasing on $(1, \infty)$. Letting $\alpha \to 1+$ we get

$$\int_0^L F(x)\,dx \leq e^{s+1} \int_0^L G(x)\,dx$$

$$= \int_0^{eL} t^s \exp\left(-f'\left(\frac{t}{e}\right)\right) dt.$$

We obtain the desired inequality by letting $L \to \infty$.

To see that the constant e is best possible, we take $f(x) = x^\beta$ for $\beta > 1$. Then it is not hard to see that

$$\int_0^\infty F(x)\, dx = \frac{1}{\beta - 1} \Gamma\left(\frac{s+1}{\beta - 1}\right)$$

and

$$\int_0^\infty G(x)\, dx = \frac{1}{(\beta - 1)\beta^{(s+1)/(\beta-1)}} \Gamma\left(\frac{s+1}{\beta - 1}\right)$$

where $\Gamma(s)$ is the Gamma function. Then the ratio of the two integrals converges to e^{s+1} as $\beta \to 1+$. $\qquad\square$

REMARK. Carleson (1954) obtained the inequality under the condition $f(0) = 0$. He also claimed that 'the sign of equality is excluded, but we shall not insist on this detail'.

* ❋ ◆

Chapter 10

Various Proofs of $\zeta(2) = \pi^2/6$

[Summary of Basic Points]

1. The standard way to evaluate the sum

$$\zeta(2) = \sum_{n=1}^{\infty} \frac{1}{n^2}$$

may be the usage of the partial fraction expansion for $\cot \pi z$:

$$\pi \cot \pi z = \frac{1}{z} + \sum_{n=1}^{\infty} \frac{2z}{z^2 - n^2}.$$

This method has an advantage of evaluating all $\zeta(2n)$ at the same time, but one needs certain justification for representations of meromorphic functions by partial fractions or factorizations, as described in Ahlfors (1979) in detail. See also Kanemitsu and Tsukada (2007), where it is shown that the partial fraction expansion of the cotangent function is a form of the functional equation for the Riemann zeta function.

2. There is also a way of finding the sum $\zeta(2)$ using the Fourier series for suitable periodic continuous functions. For example, the following trigonometric series appeared in **Problem 7.13**:

$$\sum_{n=1}^{\infty} \frac{\sin n\theta}{n},$$

which converges boundedly to $(\pi - \theta)/2$ in $(0, 2\pi)$. Hence, integrating by parts we get

$$\sum_{n=1}^{\infty} \frac{1 - (-1)^n}{n^2} = \int_0^{\pi} \frac{\pi - \theta}{2} \, d\theta = \frac{\pi^2}{4},$$

which implies $\zeta(2) = \pi^2/6$. But one needs some justification for good convergence of the Fourier series to apply partial integration. In general, using the trigonometric series of the Bernoulli polynomials one can obtain the closed form of $\zeta(2n)$, as described in Dieudonné (1971) in detail.

3. Using hypergeometric series Choi and Rathie (1997) gave an evaluation for $\zeta(2)$. See also Choi, Rathie and Srivastava (1999).

In this chapter we are to evaluate the series

$$\sum_{n=1}^{\infty} \frac{1}{n^2}.$$

Various easier proofs for $\zeta(2) = \pi^2/6$ are collected here, so you can enjoy them.

PROBLEM 10.1

It follows from the Gregory-Leibniz series

$$\sum_{n=0}^{\infty} \frac{(-1)^n}{2n+1} = \frac{\pi}{4}$$

that the sequence

$$a_n = \sum_{k=-n}^{n} \frac{(-1)^k}{2k+1}$$

converges to $\pi/2$ as $n \to \infty$. Then square the a_n.

This is due to J. M. Borwein and P. B. Borwein (1987). A similar method using the Gregory-Leibniz series was already found by Denquin (1912) and by Estermann (1947).

PROBLEM 10.2

The reciprocal of the function $\sin\theta$ is denoted by $\operatorname{cosec}\theta$. Use

$$\frac{1}{\theta^2} < \operatorname{cosec}^2\theta < 1 + \frac{1}{\theta^2}$$

for $0 < \theta < \pi/2$ and the formula

$$\operatorname{cosec}^2\theta = \frac{1}{4}\left(\operatorname{cosec}^2\frac{\theta}{2} + \operatorname{cosec}^2\frac{\theta+\pi}{2}\right).$$

This is due to Hofbauer (2002).

PROBLEM 10.3

The reciprocal of the function of $\tan\theta$ *is denoted by* $\cot\theta$. *Use*

$$\cot^2\theta < \frac{1}{\theta^2} < 1 + \cot^2\theta$$

for $0 < \theta < \pi/2$ *and the formula*

$$\cot^2\frac{\pi}{2n+1} + \cdots + \cot^2\frac{n\pi}{2n+1} = \frac{n(2n-1)}{3}.$$

This is due to A. M. Yaglom and I. M. Yaglom (1953). The same argument can be found in Holme (1970) and in Papadimitriou (1973). A similar but slightly complicated proof was given by Kortram (1996). The same method was also discussed in Arratia (1999).

PROBLEM 10.4

Multiply the following formula by θ *and integrate from 0 to* $\pi/2$:

$$\frac{1}{2} + \cos 2\theta + \cos 4\theta + \cdots + \cos 2n\theta = \frac{\sin(2n+1)\theta}{2\sin\theta}.$$

This is due to Giesy (1972). A similar proof was given by Stark (1969), who multiplied the so-called Fejér kernel:

$$\frac{\sin^2(n+1)\theta}{2(n+1)\sin^2\theta} = \frac{1}{2} + \sum_{k=1}^{n}\left(1 - \frac{k}{n+1}\right)\cos 2k\theta$$

by θ and integrated from 0 to $\pi/2$.

PROBLEM 10.5

Carry out the substitution $x = \sin\theta$ *in the Taylor series*

$$\arcsin x = x + \sum_{n=1}^{\infty}\frac{1\cdot 3\cdot 5\cdots(2n-1)}{2\cdot 4\cdot 6\cdots(2n)}\cdot\frac{x^{2n+1}}{2n+1}$$

valid for $|x| \le 1$ *and use the formula*

$$\int_0^{\pi/2}\sin^{2n+1}\theta\,d\theta = \frac{2\cdot 4\cdots(2n)}{3\cdot 5\cdots(2n+1)}.$$

This is due to Choe (1987), who used substantially the value of $(\arcsin 1)^2$. But this is very close to Euler's proof described in Ayoub (1974). Kimble (1987) reproduced it without words.

PROBLEM 10.6

Show first that

$$\zeta(2) = 3 \sum_{n=1}^{\infty} \frac{(n-1)!^2}{(2n)!}.$$

To evaluate the series use the Taylor series

$$\arcsin^2 x = \sum_{n=1}^{\infty} \frac{2^{2n-1}(n-1)!^2}{(2n)!} x^{2n}.$$

This is due to Knopp and Schur (1918).

PROBLEM 10.7

Put

$$I_n = \int_0^{\pi/2} \cos^{2n} \theta \, d\theta,$$

$$J_n = \int_0^{\pi/2} \theta^2 \cos^{2n} \theta \, d\theta$$

for all integer $n \geq 0$. Show the recursion formula

$$I_n = n(2n-1)J_{n-1} - 2n^2 J_n$$

and use

$$\int_0^{\pi/2} \cos^{2n} \theta \, d\theta = \frac{1 \cdot 3 \cdots (2n-1)}{2 \cdot 4 \cdots (2n)} \cdot \frac{\pi}{2}.$$

This is due to Matsuoka (1961). See the comment after **PROBLEM 18.6**.

PROBLEM 10.8

Carry out the substitution

$$2 \cos \theta = e^{\theta i} + e^{-\theta i}$$

in the improper integral

$$\int_0^{\pi/2} \log(2 \cos \theta) \, d\theta.$$

This is due to Russell (1991).

PROBLEM 10.9

For an arbitrarily fixed $x \in (0, 1]$ carry out the substitution

$$\cos\theta = y + \frac{x}{2}(y^2 - 1)$$

in the improper repeated integral

$$\int_0^1 \int_{-1}^1 \frac{1}{1 + xy}\, dy dx.$$

This is due to Goldscheider(1913) who answered the question posed by Stäckel(1913) as Aufgabe 208.

PROBLEM 10.10

Interchange the order of integration of the improper repeated integral

$$\int_0^\infty \int_0^1 \frac{x}{(1 + x^2)(1 + x^2 y^2)}\, dy dx.$$

This is due to Harper(2003).

PROBLEM 10.11

Carry out the affine transformation

$$u = \frac{x + y}{\sqrt{2}}, \quad v = \frac{y - x}{\sqrt{2}}$$

in the improper double integral

$$\zeta(2) = \iint_S \frac{dx dy}{1 - xy},$$

where S is the unit square $[0, 1) \times [0, 1)$, in order to show that

$$\zeta(2) = 4\int_0^{1/\sqrt{2}} \arctan\frac{t}{\sqrt{2 - t^2}}\frac{dt}{\sqrt{2 - t^2}}$$
$$+ 4\int_{1/\sqrt{2}}^{\sqrt{2}} \arctan\frac{\sqrt{2} - t}{\sqrt{2 - t^2}}\frac{dt}{\sqrt{2 - t^2}}.$$

Then substitute $t = \sqrt{2}\sin\theta$ in the first integral and $t = \sqrt{2}\cos 2\theta$ in the second integral.

This is due to Apostol(1983).

PROBLEM 10.12

Carry out the transformation

$$x = \frac{\sin\theta}{\cos\varphi}, \quad y = \frac{\sin\varphi}{\cos\theta}$$

on the triangular region $\{(\theta,\varphi) \mid \theta,\varphi > 0, \ \theta + \varphi < \pi/2\}$ *in the improper double integral*

$$\frac{3}{4}\zeta(2) = \iint\limits_{S} \frac{dxdy}{1 - x^2 y^2},$$

where S is the unit square $[0,1) \times [0,1)$.

According to Kalman(1993) this proof was given in a lecture by Don B. Zagier in 1989, who mentioned that it was shown to him by a colleague who had learned of it through the grapevine. Elkies(2003) has reported that this proof, as well as the higher dimensional generalization, is due to Calabi and that the only paper containing the proof is Beukers, Calabi and Kolk(1993). See **PROBLEM 18.8**.

PROBLEM 10.13

The power series

$$D(z) = \sum_{n=1}^{\infty} \frac{z^n}{n^2}$$

converges absolutely on $|z| \le 1$ with the radius of convergence 1 and satisfies the relations $D(1) = \zeta(2)$ and $D(-1) = -\zeta(2)/2$. We have

$$D(z) = -\int_0^z \frac{\log(1-z)}{z}\, dz,$$

from which $D(z)$ has an analytic continuation to the whole complex plane except for the half line $[1,\infty)$ by taking the principal branch of the logarithm $\log(1-z)$. Show then the functional equation

$$D\left(-\frac{1}{z}\right) + D(-z) = 2D(-1) - \frac{1}{2}\log^2 z$$

and consider the limit as $z \to -1$ in the upper half plane.

The function $D(z)$ is known as the dilogarithm. This is taken from Levin's book(1981).

Solutions for Chapter 10

SOLUTION 10.1

For brevity put

$$b_n = \sum_{k=-n}^{n} \frac{1}{(2k+1)^2}.$$

Then we have

$$
\begin{aligned}
a_n^2 - b_n &= \sum_{-n \leq k \neq m \leq n} \frac{(-1)^{k+m}}{(2k+1)(2m+1)} \\
&= \sum_{-n \leq k \neq m \leq n} \frac{(-1)^{k+m}}{2(m-k)} \left(\frac{1}{2k+1} - \frac{1}{2m+1} \right) \\
&= \sum_{-n \leq k \neq m \leq n} \frac{(-1)^{k+m}}{(m-k)(2k+1)}.
\end{aligned}
$$

The last expression can be written as

$$\sum_{k=-n}^{n} \frac{(-1)^k}{2k+1} c_{k,n}$$

where

$$c_{k,n} = \sum \frac{(-1)^m}{m-k}$$

and in the summation m runs through $[-n, n]$ except for k. The sum $c_{k,n}$ contains several terms to be canceled. Indeed,

$$c_{k,n} = (-1)^{k+1} \sum_{\ell=n-k+1}^{k+n} \frac{(-1)^\ell}{\ell}$$

for positive k. It is also clear that $c_{0,n} = 0$ and $c_{-k,n} = -c_{k,n}$; so

$$|c_{k,n}| \leq \frac{1}{n+|k|+1}.$$

Hence

$$\left|a_n^2 - b_n\right| \le \sum_{k=-n}^{n} \frac{1}{|2k+1|\,(n-|k|+1)}$$
$$< \frac{1}{n} + \frac{6}{2n+1}\left(1 + \frac{1}{2} + \cdots + \frac{1}{2n+1}\right).$$

The right-hand side converges to 0 as $n \to \infty$ and thus b_n converges to $\pi^2/4$, which is equal to $3\zeta(2)/2$. □

SOLUTION 10.2

Applying repeatedly the formula described in the problem, starting with $1 = \operatorname{cosec}^2(\pi/2)$, we get

$$1 = \frac{1}{4}\left(\operatorname{cosec}^2 \frac{\pi}{4} + \operatorname{cosec}^2 \frac{3\pi}{4}\right)$$
$$= \frac{1}{16}\left(\operatorname{cosec}^2 \frac{\pi}{8} + \operatorname{cosec}^2 \frac{3\pi}{8} + \operatorname{cosec}^2 \frac{5\pi}{8} + \operatorname{cosec}^2 \frac{7\pi}{8}\right)$$
$$\vdots$$
$$= \frac{1}{4^n} \sum_{k=0}^{2^n-1} \operatorname{cosec}^2 \frac{2k+1}{2^{n+1}}\pi.$$

On the other hand, we have

$$\theta > \sin\theta > \theta - \frac{\theta^3}{6} > \frac{\theta}{\sqrt{1+\theta^2}}$$

for $0 < \theta < \pi/2$, which implies the inequality on 'cosec' stated in the problem. Using that inequality we have

$$4^{n+1} s_n < 2^{2n-1} < 2^n + 4^{n+1} s_n$$

where

$$s_n = \frac{1}{\pi^2} \sum_{k=0}^{2^{n-1}-1} \frac{1}{(2k+1)^2}.$$

Dividing both sides by 4^{n+1} and letting $n \to \infty$, we find that the sequence s_n converges to $1/8$, which is equal to $\frac{3}{4\pi^2}\zeta(2)$. □

SOLUTION 10.3

Putting

$$\phi(\theta) = \sin\theta - \theta\cos\theta$$

for $0 < \theta < \pi/2$, we have $\phi(0+) = 0$, $\phi'(\theta) = \theta\sin\theta > 0$; so $\theta < \tan\theta$ and hence

$$\cot^2\theta < \frac{1}{\theta^2} < 1 + \cot^2\theta.$$

Put

$$\xi_k = \cot^2\frac{k\pi}{2n+1}$$

for integer $k \geq 1$. Then

$$\frac{c_n}{(2n+1)^2} < \sum_{k=1}^{n}\frac{1}{(k\pi)^2} < \frac{n}{(2n+1)^2} + \frac{c_n}{(2n+1)^2}$$

where $c_n = \xi_1 + \xi_2 + \cdots + \xi_n$. Thus it suffices to show that c_n/n^2 converges to $2/3$ as $n \to \infty$.

To see this we introduce a polynomial of degree n vanishing at ξ_k for $1 \leq k \leq n$. We now have

$$\sin(2n+1)\theta = \operatorname{Im} e^{(2n+1)\theta i} = \operatorname{Im} \sum_{k=0}^{2n+1} i^k \binom{2n+1}{k} \cos^{2n+1-k}\theta \sin^k\theta$$

$$= \sum_{\ell=0}^{n} (-1)^\ell \binom{2n+1}{2\ell+1} \cos^{2n-2\ell}\theta \sin^{2\ell+1}\theta.$$

Dividing both sides by $\sin^{2n+1}\theta$ we see that

$$\frac{\sin(2n+1)\theta}{\sin^{2n+1}\theta} = \sum_{\ell=0}^{n} (-1)^\ell \binom{2n+1}{2\ell+1} \cot^{2n-2\ell}\theta.$$

The right-hand side is a polynomial in $\cot^2\theta$ of degree n, which we denote by $Q(\cot^2\theta)$. Solving the equation $\sin(2n+1)\theta = 0$ with $\sin\theta \neq 0$, we see that ξ_1, ..., ξ_n are n real simple zeros of $Q(x)$. Since

$$Q(x) = (2n+1)x^n - \frac{n}{3}(4n^2-1)x^{n-1} + \cdots,$$

we have $c_n = n(2n-1)/3$, whence $c_n/n^2 \to 2/3$ as $n \to \infty$, as required. $\qquad\square$

SOLUTION 10.4

The formula in the problem will be easily shown in the same way as in the proof of **PROBLEM 1.7**. Using

$$\int_0^{\pi/2} \theta \cos 2k\theta \, d\theta = -\frac{1}{2k} \int_0^{\pi/2} \sin 2k\theta \, d\theta = \frac{(-1)^k - 1}{4k^2}$$

it can be seen that the resulting value of the left-hand side in the problem for $n = 2m + 1$ is equal to

$$\frac{\pi^2}{16} - \frac{1}{2}\left(\frac{1}{1^2} + \frac{1}{3^2} + \cdots + \frac{1}{(2m+1)^2}\right),$$

which converges to $\pi^2/16 - 3\zeta(2)/8$ as $m \to \infty$. On the other hand, it follows from **PROBLEM 5.3** that

$$\frac{1}{2} \int_0^{\pi/2} \frac{\theta}{\sin\theta} \sin(4m + 3)\theta \, d\theta$$

converges to 0 as $m \to \infty$. Note that the function $\theta/\sin\theta$ can be extended to $\theta = 0$ continuously, while the function $1/\sin\theta$ is not integrable on $(0, \pi/2]$. □

SOLUTION 10.5

The Taylor series of the function $(1 - x^2)^{-1/2}$ about $x = 0$ is

$$1 + \frac{1}{2}x^2 + \frac{1\cdot3}{2\cdot4}x^4 + \frac{1\cdot3\cdot5}{2\cdot4\cdot6}x^6 + \cdots,$$

whose radius of convergence is equal to 1. The termwise integration yields the Taylor series for $\arcsin x$ as follows:

$$\arcsin x = x + \sum_{n=1}^{\infty} \frac{1\cdot3\cdot5\cdots(2n-1)}{2\cdot4\cdot6\cdots(2n)} \cdot \frac{x^{2n+1}}{2n+1},$$

which is valid for $|x| < 1$. Moreover this converges uniformly on the interval $[-1, 1]$, because Stirling's approximation implies that

$$\frac{1\cdot3\cdot5\cdots(2n-1)}{2\cdot4\cdot6\cdots(2n)} = \binom{2n}{n} 4^{-n} = O\left(\frac{1}{\sqrt{n}}\right)$$

as $n \to \infty$. Substituting $x = \sin\theta$ in the Taylor series of $\arcsin x$ and integrating from 0 to $\pi/2$ we get

$$\int_0^{\pi/2} \theta \, d\theta = 1 + \sum_{n=1}^{\infty} \frac{1}{(2n+1)^2},$$

which implies that $\pi^2/8 = 3\zeta(2)/4$. □

SOLUTION 10.6

First we put

$$\sigma_m = \sum_{n \geq m} \frac{m!}{n^2(n+1)\cdots(n+m)}$$

for any integer $m \geq 1$. The first difference of this sequence is

$$\sigma_m - \sigma_{m+1} = \frac{(m-1)!^2}{(2m)!}$$
$$+ \sum_{n>m} \left(\frac{m!}{n^2(n+1)\cdots(n+m)} - \frac{(m+1)!}{n^2(n+1)\cdots(n+m+1)} \right).$$

It is easily seen that the sum on the right-hand side can be written as

$$\sum_{n>m} \frac{m!}{n(n+1)\cdots(n+m+1)},$$

which is transformed into

$$\frac{1}{m+1} \sum_{n>m} \sum_{k=0}^{m+1} (-1)^k \binom{m+1}{k} \frac{1}{n+k} = \frac{1}{m+1} \int_0^1 t^m (1-t)^m \, dt$$
$$= \frac{m!^2}{(m+1)(2m+1)!}.$$

On the other hand, since

$$\sigma_1 = \sum_{n=1}^{\infty} \frac{1}{n^2(n+1)} = \sum_{n=1}^{\infty} \frac{1}{n}\left(\frac{1}{n} - \frac{1}{n+1}\right) = \zeta(2) - 1,$$

we obtain

$$\zeta(2) - \sigma_k = 1 + \sigma_1 - \sigma_k = 1 + \sum_{m=1}^{k-1} (\sigma_m - \sigma_{m+1})$$
$$= 1 + \sum_{m=1}^{k-1} \left(\frac{(m-1)!^2}{(2m)!} + \frac{m!^2}{(m+1)(2m+1)!} \right),$$

which is equal to

$$\sum_{m=1}^{k-1} \frac{(m-1)!^2}{(2m)!} + 2 \sum_{m=1}^{k} \frac{(m-1)!^2}{(2m)!}$$

for any integer $k \geq 1$. Therefore, noting that

$$0 < \sigma_k < \int_0^1 t^{k-1}(1-t)^{k-1} \, dt$$

and that the right-hand side converges to 0 as $k \to \infty$, we have

$$\zeta(2) = 3 \sum_{n=1}^{\infty} \frac{(n-1)!^2}{(2n)!}.$$

To evaluate the series we note that the power series stated in the problem satisfies the differential equation

$$(1 - x^2)\, y'' - xy' - 2 = 0$$

with the initial conditions $y(0) = y'(0) = 0$ as well as the function $\arcsin^2 x$; hence they coincide with each other. We thus have

$$\zeta(2) = 6 \arcsin^2 \frac{1}{2} = \frac{\pi^2}{6}. \qquad \square$$

| SOLUTION 10.7 |

For any integer $n \geq 1$ it follows from integration by parts that

$$
\begin{aligned}
I_n &= \theta \cos^{2n} \theta \Big|_0^{\pi/2} + 2n \int_0^{\pi/2} \theta \sin \theta \cos^{2n-1} \theta \, d\theta \\
&= n\theta^2 \sin \theta \cos^{2n-1} \theta \Big|_0^{\pi/2} \\
&\quad - n \int_0^{\pi/2} \theta^2 \left(\cos^{2n} \theta - (2n-1)\sin^2 \theta \cos^{2n-2} \theta \right) d\theta;
\end{aligned}
$$

therefore

$$I_n = n(2n-1) J_{n-1} - 2n^2 J_n.$$

Multiplying this by $2^{2n-1}(n-1)!^2/(2n)!$ we get

$$\frac{\pi}{4n^2} = \frac{4^{n-1}(n-1)!^2}{(2n-2)!} J_{n-1} - \frac{4^n n!^2}{(2n)!} J_n.$$

Adding these formulae from $n = 1$ to $n = m$ we obtain

$$\frac{\pi}{4} \sum_{n=1}^{m} \frac{1}{n^2} = J_0 - \frac{4^m m!^2}{(2m)!} J_m.$$

Since $J_0 = \pi^3/24$, it suffices to show that the second term on the right-hand side of the above expression converges to 0 as $m \to \infty$. To see this, using Jordan's inequality $\pi \sin \theta \geq 2\theta$ valid on $[0, \pi/2]$ we get

$$J_m < \frac{\pi^2}{4} \int_0^{\pi/2} \sin^2 \theta \cos^{2m} \theta \, d\theta = \frac{\pi^2}{4}(I_m - I_{m-1})$$

$$= \frac{\pi^2}{4(m+1)} I_m.$$

Hence

$$0 < \frac{4^m m!^2}{(2m)!} J_m < \frac{4^m m!^2}{(2m)!} \cdot \frac{\pi^2}{4(m+1)} \cdot \frac{(2m)!\pi}{m!^2 2^{2m+1}}$$

$$= \frac{\pi^3}{8(m+1)}$$

and the right-hand side converges to 0 as $m \to \infty$, as required. $\square$

SOLUTION 10.8

Let I be the improper integral stated in the problem. Since both complex numbers $e^{i\theta}$ and $1 + e^{i\theta}$ do not touch the negative real axis as θ varies in the interval $[0, \pi/2)$, we can write

$$\log(2\cos\theta) = i\theta + \log(1 + e^{-2i\theta})$$

by taking the principal value of the logarithm. Therefore we have

$$I = \frac{\pi^2}{8}i + \int_0^{\pi/2} \log(1 + e^{-2i\theta}) \, d\theta.$$

The Taylor series of $\log(1 + z)$ converges on compact sets in $\{|z| = 1 \mid z \neq -1\}$, which is the unit circle centered at the origin removing one point $z = -1$. Thus for any sufficiently small $\epsilon > 0$ we have

$$\int_0^{\pi/2 - \epsilon} \log(1 + e^{-2i\theta}) \, d\theta = \sum_{n=1}^{\infty} \frac{(-1)^{n-1}}{n} \cdot \frac{1 - (-1)^n e^{2n\epsilon i}}{2ni}.$$

Since the right-hand side now converges absolutely to $-3\zeta(2)i/4$ as $\epsilon \to 0+$, we get

$$I = \frac{i}{8}(\pi^2 - 6\zeta(2)).$$

However I is obviously real and so we have $I = 0$; that is, $\zeta(2) = \pi^2/6$. $\square$

SOLUTION 10.9

For any x in the interval $(-1, 1)$ we put

$$\phi(x) = \int_{-1}^{1} \frac{dy}{1 + xy} = \frac{1}{x} \log \frac{1 + x}{1 - x}.$$

Expanding the function $1/(1 + xy)$ into the Taylor series with respect to y, we get

$$\phi(x) = \sum_{n=0}^{\infty} (-1)^n x^n \int_{-1}^{1} y^n \, dy = 2 \sum_{n=0}^{\infty} \frac{x^{2n}}{2n + 1}.$$

Therefore we have

$$I_\epsilon = \int_{0}^{1-\epsilon} \phi(x) \, dx = 2 \sum_{n=0}^{\infty} \frac{(1 - \epsilon)^{2n+1}}{(2n + 1)^2}$$

for any ϵ in $(0, 1)$. The series on the right-hand side clearly converges to $3\zeta(2)/2$ as $\epsilon \to 0+$.

On the other hand, the relation

$$\cos \theta = y + \frac{x}{2}(y^2 - 1)$$

gives a smooth one-to-one correspondence between the interval $[-1, 1]$ in y and the interval $[0, \pi]$ in θ. Hence we obtain

$$\phi(x) = \int_{\pi}^{0} \frac{1}{1 + xy} \cdot \frac{dy}{d\theta} \, d\theta = \int_{0}^{\pi} \frac{\sin \theta}{(1 + xy)^2} \, d\theta$$

$$= \int_{0}^{\pi} \frac{\sin \theta}{1 + 2x \cos \theta + x^2} \, d\theta.$$

Since

$$\frac{\sin \theta}{1 + 2x \cos \theta + x^2} = \frac{d}{dx} \left(\arctan \frac{x + \cos \theta}{\sin \theta} \right),$$

it follows from interchanging the order of integration that

$$I_\epsilon = \int_{0}^{\pi} \left(\arctan \frac{1 + \cos \theta - \epsilon}{\sin \theta} - \arctan \cot \theta \right) d\theta$$

$$= \int_{0}^{\pi} \arctan \frac{(1 - \epsilon) \sin \theta}{1 + (1 - \epsilon) \cos \theta} \, d\theta,$$

which converges to

$$\int_{0}^{\pi} \arctan \frac{\sin \theta}{1 + \cos \theta} \, d\theta = \int_{0}^{\pi} \frac{\theta}{2} \, d\theta = \frac{\pi^2}{4}$$

as $\epsilon \to 0+$, because the integrand is uniformly bounded. $\square$

| SOLUTION 10.10 |

Let I be the value of the improper repeated integral in the problem. By integrating first in y we have

$$I = \int_0^\infty \frac{1}{1 + x^2} \arctan xy \Big|_{y=0}^{y=1} dx = \int_0^\infty \frac{\arctan x}{1 + x^2} dx,$$

which is equal to

$$\frac{1}{2} \arctan^2 x \Big|_0^\infty = \frac{\pi^2}{8}.$$

Next by integrating first in x we have

$$I = \frac{1}{2} \int_0^1 \frac{dy}{1 - y^2} \int_0^\infty \left(\frac{2x}{1 + x^2} - \frac{2xy^2}{1 + x^2 y^2} \right) dx$$

$$= \frac{1}{2} \int_0^1 \frac{1}{1 - y^2} \log \frac{1 + x^2}{1 + x^2 y^2} \Big|_{x=0}^{x=\infty} dy,$$

which is equal to

$$-\int_0^1 \frac{\log y}{1 - y^2} dy.$$

Then by the termwise integration we obtain

$$I = -\sum_{n=0}^\infty \int_0^1 y^{2n} \log y \, dy = \sum_{n=0}^\infty \frac{1}{(2n + 1)^2} = \frac{3}{4} \zeta(2). \qquad \square$$

| SOLUTION 10.11 |

The affine transformation described in the problem clearly rotates the unit square $(0, 1) \times (0, 1)$ with -45 degrees so that the four vertices $(0, 0)$, $(1, 0)$, $(1, 1)$ and $(0, 1)$ are transformed to $(0, 0)$, $(1/\sqrt{2}, -1/\sqrt{2})$, $(\sqrt{2}, 0)$ and $(1/\sqrt{2}, 1/\sqrt{2})$ respectively. Since

$$1 - xy = 1 - \frac{u^2 - v^2}{2}$$

is an even function with respect to v, we have

$$\frac{\zeta(2)}{4} = \int_0^{1/\sqrt{2}} \int_0^u \frac{dv \, du}{2 - u^2 + v^2} + \int_{1/\sqrt{2}}^{\sqrt{2}} \int_0^{\sqrt{2}-u} \frac{dv \, du}{2 - u^2 + v^2}. \qquad (10.1)$$

Hence we obtain the required formula given in the problem by using

$$\int_0^x \frac{dv}{2 - u^2 + v^2} = \frac{1}{\sqrt{2 - u^2}} \arctan \frac{x}{\sqrt{2 - u^2}}.$$

Substituting u by $\sqrt{2} \sin \theta$ and $\sqrt{2} \cos 2\theta$ in the first and the second integrals in (10.1) respectively, we get

$$\frac{\zeta(2)}{4} = \int_0^{\pi/6} \theta \, d\theta + 2 \int_0^{\pi/6} \theta \, d\theta = \frac{\pi^2}{24}. \qquad \square$$

SOLUTION 10.12

Let Φ and Δ be the transformation and the triangular region stated in the problem respectively. Obviously Φ maps Δ into the unit open square $(0, 1) \times (0, 1)$ in the xy-plane. Conversely for any x and y in the interval $(0, 1)$ we have

$$\sin^2 \theta = \frac{x^2(1 - y^2)}{1 - x^2 y^2} \quad \text{and} \quad \sin^2 \varphi = \frac{y^2(1 - x^2)}{1 - x^2 y^2},$$

from which we can determine θ and φ in $(0, \pi/2)$ respectively. Moreover

$$\cos(\theta + \varphi) = \frac{\sqrt{(1 - x^2)(1 - y^2)}}{1 + xy} > 0$$

implies that $\theta + \varphi < \pi/2$; hence $(\theta, \varphi) \in \Delta$ and so Φ is onto. Since the Jacobian

$$\begin{vmatrix} \dfrac{\partial x}{\partial \theta} & \dfrac{\partial x}{\partial \varphi} \\[2mm] \dfrac{\partial y}{\partial \theta} & \dfrac{\partial y}{\partial \varphi} \end{vmatrix} = \begin{vmatrix} \dfrac{\cos \theta}{\cos \varphi} & \dfrac{\sin \theta}{\cos \varphi} \tan \varphi \\[2mm] \dfrac{\sin \varphi}{\cos \theta} \tan \theta & \dfrac{\cos \varphi}{\cos \theta} \end{vmatrix}$$

is equal to $1 - x^2 y^2 > 0$, we conclude that

$$\frac{3}{4} \zeta(2) = \iint_\Delta d\theta \, d\varphi = \frac{\pi^2}{8}. \qquad \square$$

SOLUTION 10.13

Since

$$\left(D\left(-\frac{1}{z}\right) + D(-z) \right)' = \frac{\log(1 + 1/z) - \log(1 + z)}{z} = -\frac{\log z}{z}$$

for any complex $z \in \mathbb{C} \setminus (-\infty, 0]$, we have the functional equation

$$D\left(-\frac{1}{z}\right) + D(-z) = c - \frac{1}{2} \log^2 x$$

for some constant c. To determine the value of c we simply put $z = 1$ in the above formula so that

$$c = 2D(-1) = -\zeta(2).$$

On the other hand, putting $z = -1 + \epsilon i$ and taking the limit of $D(-1/z)$ as $\epsilon \to 0+$, we see that $D(-1/z)$, together with $D(-z)$, converges to $D(1) = \zeta(2)$. Hence it follows from the functional equation that

$$2\zeta(2) = -\zeta(2) - \frac{1}{2}(-\pi i)^2;$$

namely $\zeta(2) = \pi^2/6$, as required. □

✦ ❋ ✦

Chapter 11

Functions of Several Variables

[Summary of Basic Points]

1. A function $f(x, y)$ defined on an open set $U \subset \mathbb{R}^2$ is (totally) differentiable at $(a, b) \in U$ if and only if there exist two functions $\alpha(x, y), \beta(x, y)$ on U, which are continuous at (a, b), satisfying

$$f(x, y) = f(a, b) + \alpha(x, y)(x - a) + \beta(x, y)(y - b).$$

It is then obvious that

$$\frac{\partial f}{\partial x}(a, b) = \alpha(a, b) \quad \text{and} \quad \frac{\partial f}{\partial y}(a, b) = \beta(a, b).$$

This definition of 'differentiability' is in Carathéodory's way. See the first basic point in Chapter 4. Unlike the one variable case the functions $\alpha(x, y), \beta(x, y)$ are *not* determined uniquely, because one can replace $\alpha(x, y)$ and $\beta(x, y)$ by $\alpha(x, y) + (y - b)\phi(x, y)$ and $\beta(x, y) - (x - a)\phi(x, y)$ respectively for any ϕ continuous at (a, b).

2. Let $f(x, y)$ and $\frac{\partial f}{\partial y}(x, y)$ be continuous on $[a, b] \times (c, d)$. Then it follows that

$$\frac{d}{dy} \int_a^b f(x, y)\, dx = \int_a^b \frac{\partial f}{\partial y}(x, y)\, dx$$

for any $y \in (c, d)$. This means that we can interchange the order of differentiation and integration; in other words, we can differentiate under the integral sign.

3. If $f(x, y)$ and $\frac{\partial f}{\partial y}(x, y)$ are continuous on $[a, \infty) \times (c, d)$,

$$\int_a^\infty f(x, y)\, dx$$

exists, and if

$$\int_a^\infty \frac{\partial f}{\partial y}(x, y)\, dx$$

converges uniformly on compact sets in y, then it holds that

$$\frac{d}{dy} \int_a^\infty f(x, y)\, dx = \int_a^\infty \frac{\partial f}{\partial y}(x, y)\, dx$$

for any y in (c, d).

4. For n-tuple of non-negative integers $\boldsymbol{m} = (m_1, m_2, \ldots, m_n)$ the partial differential operator

$$\frac{\partial^{|\boldsymbol{m}|}}{\partial x_1^{m_1} \partial x_2^{m_2} \cdots \partial x_n^{m_n}}$$

is denoted by $D^{\boldsymbol{m}}$ where $|\boldsymbol{m}| = m_1 + m_2 + \cdots + m_n$ is called the order of $D^{\boldsymbol{m}}$.

5. Let f have continuous partial derivatives of order s at each point of an open set U in $\mathbb{R}^n$. If the line segment joining two points $\boldsymbol{x} = (x_1, \ldots, x_n)$ and $\boldsymbol{a} = (a_1, \ldots, a_n)$ is contained in U, then there exists a θ in the interval $(0, 1)$ such that

$$f(\boldsymbol{x}) = f(\boldsymbol{a}) + \sum_{k=1}^{s-1} \frac{1}{k!} \left(\sum_{j=1}^n (x_j - a_j) \frac{\partial}{\partial x_j} \right)^k f(\boldsymbol{a})$$

$$+ \frac{1}{s!} \left(\sum_{j=1}^n (x_j - a_j) \frac{\partial}{\partial x_j} \right)^s f((1 - \theta)\boldsymbol{a} + \theta \boldsymbol{x}),$$

known as Taylor's formula for functions of several variables.

6. For any polynomial $P(x_1, \ldots, x_n)$ it may be convenient to express

$$P(\boldsymbol{x}) = \sum_{k_1=0}^{r_1} \cdots \sum_{k_n=0}^{r_n} \frac{1}{\boldsymbol{k}!} D^{\boldsymbol{k}} P(\boldsymbol{a})(\boldsymbol{x} - \boldsymbol{a})^{\boldsymbol{k}}$$

where $\boldsymbol{k} = (k_1, \ldots, k_n)$, $\boldsymbol{k}! = k_1! k_2! \cdots k_n!$,

$$(\boldsymbol{x} - \boldsymbol{a})^{\boldsymbol{k}} = (x_1 - a_1)^{k_1} \cdots (x_n - a_n)^{k_n}$$

and r_j is the degree of P with respect to x_j.

7. Let $u_k(x_1, \ldots, x_n)$, $1 \le k \le n$ be a smooth transformation from an open region

U onto V in $\mathbb{R}^n$. Suppose that the Jacobian

$$J = \begin{vmatrix} \dfrac{\partial u_1}{\partial x_1} & \cdots & \dfrac{\partial u_1}{\partial x_n} \\ \vdots & \ddots & \vdots \\ \dfrac{\partial u_n}{\partial x_1} & \cdots & \dfrac{\partial u_n}{\partial x_n} \end{vmatrix}$$

does not vanish on U. Then

$$\int_V \!\!\cdots\!\! \int f(u_1, \ldots, u_n)\, du_1 \cdots du_n = \int_U \!\!\cdots\!\! \int f(u_1(\boldsymbol{x}), \ldots, u_n(\boldsymbol{x})) \, |J| \, dx_1 \cdots dx_n$$

for any $f \in \mathscr{C}(V)$, if the integral on the left-hand side exists.

8. After Apéry's discovery (1978) of the irrationality of $\zeta(3)$, Beukers (1978) gave another elegant proof using the following improper triple integral:

$$\zeta(3) = \iiint_B \frac{dx\,dy\,dz}{1 - (1 - xy)z}$$

where B is the unit cube $(0, 1)^3$.

✦ ⌒ ✳ ⌒ ✦

PROBLEM 11.1

Show that

$$\iint_{0 < x < y < \pi} \log|\sin(x - y)| \, dx\,dy = -\frac{\pi^2}{2} \log 2.$$

PROBLEM 11.2

Let $f(x, y)$ be a function defined on an open region $U \subset \mathbb{R}^2$. Assume that both $\dfrac{\partial f}{\partial x}(x, y)$ and $\dfrac{\partial f}{\partial y}(x, y)$ exist on U and they are totally differentiable at a point $(a, b) \in U$. Show then that

$$\frac{\partial^2 f}{\partial x \partial y}(a, b) = \frac{\partial^2 f}{\partial y \partial x}(a, b).$$

This is due to Young (1909) and is called the fundamental theorem of differentials. Note that no continuity of partial derivatives are required.

PROBLEM 11.3

Let $\phi(x, y)$ be a bounded $\mathscr{C}^\infty$-function defined on a punctured region $U = \{(x, y) \mid 0 < x^2 + y^2 < 1\}$. Suppose that $\phi(h, 0)$ and $\phi(0, h)$ converge to α and β as $h \to 0$ respectively. Suppose also that

$$\lim_{(x,y)\to(0,0)} xy \frac{\partial \phi}{\partial x}(x, y) = \lim_{(x,y)\to(0,0)} xy \frac{\partial \phi}{\partial y}(x, y) = 0.$$

Show then that

$$f(x, y) = \begin{cases} xy\phi(x, y) & ((x, y) \in U) \\ 0 & (x = y = 0) \end{cases}$$

is differentiable at $(0, 0)$ and satisfies

$$\frac{\partial^2 f}{\partial x \partial y}(0, 0) = \alpha \quad and \quad \frac{\partial^2 f}{\partial y \partial x}(0, 0) = \beta.$$

Peano gave an example

$$\phi(x, y) = \frac{x^2 - y^2}{x^2 + y^2}$$

in Genocchi (1884), p. 174 with $\alpha = 1, \beta = -1$. This book that was based on Genocchi's lecture at the university of Turin was edited by Peano with many important additions by Peano himself. One may observe that the function

$$\phi(x, y) = \arctan \frac{x^2}{y^2}$$

also satisfies the assumptions in the problem with $\alpha = \dfrac{\pi}{2}, \beta = 0$.

PROBLEM 11.4

Show that

$$\int \cdots \int_{[0,1]^n} \frac{n}{\dfrac{1}{x_1} + \dfrac{1}{x_2} + \cdots + \dfrac{1}{x_n}} \, dx_1 dx_2 \cdots dx_n$$

converges as $n \to \infty$.

The integrand is known as the harmonic mean of $x_1, x_2, \ldots, x_n$ and satisfies

$$\frac{n}{\dfrac{1}{x_1} + \cdots + \dfrac{1}{x_n}} \le \sqrt[n]{x_1 \cdots x_n} \le \frac{x_1 + \cdots + x_n}{n}$$

for any $x_i > 0$. These inequalities are merely the convexity of the functions e^{-x} and $-\log x$ respectively.

PROBLEM 11.5

Let Δ be the set of all points $(x_1, x_2, ..., x_n)$ satisfying $0 \leq x_1 \leq x_2 \leq \cdots \leq x_n \leq 1$. For any $f \in \mathscr{C}[0,1]$ show that

$$\int \cdots \int_{\Delta} f(x_1) \cdots f(x_n) dx_1 \cdots dx_n = \frac{1}{n!} \left(\int_0^1 f(x) \, dx \right)^n.$$

PROBLEM 11.6

Suppose that $\phi_n(x_1, ..., x_n) \in \mathscr{C}[0,1]^n$ satisfies $0 \leq \phi_n(x_1, ..., x_n) \leq 1$ on $[0,1]^n$ for all integers $n \geq 1$ and that

$$\int \cdots \int_{[0,1]^n} \phi_n(x_1, ..., x_n) \, dx_1 \cdots dx_n \quad and \quad \int \cdots \int_{[0,1]^n} \phi_n^2(x_1, ..., x_n) \, dx_1 \cdots dx_n$$

converge to α and α^2 as $n \to \infty$ respectively. Show then that

$$\lim_{n \to \infty} \int \cdots \int_{[0,1]^n} f(\phi_n(x_1, ..., x_n)) \, dx_1 \cdots dx_n = f(\alpha)$$

for any $f \in \mathscr{C}[0,1]$.

Arithmetic mean case $\phi_n = (x_1 + x_2 + \cdots + x_n)/n$ with $\alpha = 1/2$ is given in Kac's book (1959). The geometric mean $\phi_n = \sqrt[n]{x_1 x_2 \cdots x_n}$ also satisfies the above assumptions with $\alpha = 1/e$.

PROBLEM 11.7

Let $n \geq 1$ be an integer. We assign a complex number λ_m to each n-tuple of non-negative integers $\boldsymbol{m} = (m_1, ..., m_n)$ arbitrarily. Show that there exists an $f(x_1, ..., x_n) \in \mathscr{C}^\infty(\mathbb{R}^n)$ satisfying

$$D^m f(\boldsymbol{0}) = \lambda_m$$

for all $\boldsymbol{m}$, where $\boldsymbol{0} = (0, ..., 0)$.

One variable case $n = 1$ was shown by Borel(1895). Later Rosenthal(1953) gave a simpler proof by considering

$$g(x) = \sum_{n=0}^{\infty} a_n e^{-|a_n| n! x^2} x^n$$

where a_n is determined according to the given $\{g^{(k)}(0)\}$. Mirkil(1956) gave a proof for the n-dimensional case.

PROBLEM 11.8

For all integers $n \geq 2$ show that

$$\int \cdots \int_{[0,1]^n} \prod_{1 \leq j < k \leq n} \sin^2 \pi(x_k - x_j) \, dx_1 \cdots dx_n = \frac{n!}{2^{n(n-1)}}.$$

Since the integrand

$$f(x_1, \ldots, x_n) = \prod_{1 \leq j < k \leq n} \sin^2 \pi(x_k - x_j)$$

satisfies $f(x_1, \ldots, x_n) = f(x_{\sigma(1)}, \ldots, x_{\sigma(n)})$ for any permutation σ, we can apply the same way as stated in **SOLUTION 11.5** to the above integral so that

$$\int \cdots \int_\Delta f(x_1, \ldots, x_n) \, dx_1 \cdots dx_n = \frac{1}{2^{n(n-1)}}$$

where $\Delta = \{(x_1, x_2, \ldots, x_n) \mid 0 \leq x_1 \leq x_2 \leq \cdots \leq x_n \leq 1\}$.

❖　✳　❖

Solutions for Chapter 11

SOLUTION 11.1

The affine transformation

$$\begin{pmatrix} x \\ y \end{pmatrix} = \begin{pmatrix} -1/2 & 1/2 \\ 1/2 & 1/2 \end{pmatrix} \begin{pmatrix} u \\ v \end{pmatrix}$$

maps $D = \{0 < u < v < 2\pi - u\}$ onto the triangular region $\{0 < x < y < \pi\}$ and its Jacobian is $-1/2$. Hence the double integral in the problem is

$$\frac{1}{2} \iint_D \log|\sin u| \, dudv = \int_0^\pi (\pi - u) \log|\sin u| \, du, \quad \text{say } J.$$

Since $(\pi/2 - u) \log|\sin u|$ is an odd function with respect to $u = \pi/2$, the corresponding integral over the interval $(0, \pi)$ vanishes; therefore

$$J = \pi \int_0^{\pi/2} \log|\sin u| \, du$$

and we get

$$\frac{2}{\pi} J = \int_0^{\pi/2} \log|\sin u| \, du + \int_0^{\pi/2} \log|\cos u| \, du$$

$$= \int_0^{\pi/2} \log|\sin 2u| \, du - \frac{\pi}{2} \log 2 = \frac{J}{\pi} - \frac{\pi}{2} \log 2,$$

which implies that $J = -\dfrac{\pi^2}{2} \log 2.$ $\qquad\square$

REMARK. This evaluation of J is due to Euler. On the other hand, Pólya and Szegö (1972) used the so-called Vandermonde determinant

$$\begin{vmatrix} 1 & 1 & 1 & \cdots & 1 \\ 1 & z & z^2 & \cdots & z^{n-1} \\ 1 & z^2 & z^4 & \cdots & z^{2(n-1)} \\ \vdots & \vdots & \vdots & \ddots & \vdots \\ 1 & z^{n-1} & z^{2(n-1)} & \cdots & z^{(n-1)^2} \end{vmatrix} = \prod_{0 \le j < k < n} (z^k - z^j)$$

to evaluate the value of J. Computing the product of the determinant with $z = \exp(2\pi i/n)$ and its conjugate, they obtained

$$\prod_{0 \leq j < k < n} \left(2 \sin \frac{j-k}{n} \pi \right)^2 = n^n.$$

The right-hand side comes from computing the product of two matrices directly. Therefore

$$\frac{\pi^2}{n^2} \sum_{0 \leq j < k < n} \log \left| \sin \left(\frac{j\pi}{n} - \frac{k\pi}{n} \right) \right| = \frac{\pi^2}{2n} \log n - \left(1 - \frac{1}{n} \right) \frac{\pi^2}{2} \log 2.$$

The left-hand side may be the 2-dimensional Riemann sum divided equally for the function $\log |\sin(x - y)|$, although we must verify the convergence of the Riemann sum to the improper double integral.

SOLUTION 11.2

Since $\dfrac{\partial f}{\partial x}$ is totally differentiable at (a, b), we have

$$\frac{\partial f}{\partial x}(a + \epsilon, b + \eta) = \frac{\partial f}{\partial x}(a, b) + \epsilon\alpha(a + \epsilon, b + \eta) + \eta\beta(a + \epsilon, b + \eta) \qquad (11.1)$$

for some functions $\alpha(x, y), \beta(x, y)$ continuous at (a, b) and satisfying

$$\alpha(a, b) = \frac{\partial^2 f}{\partial x^2}(a, b) \quad \text{and} \quad \beta(a, b) = \frac{\partial^2 f}{\partial y \partial x}(a, b).$$

Similarly we get

$$\frac{\partial f}{\partial y}(a + \epsilon, b + \eta) = \frac{\partial f}{\partial y}(a, b) + \epsilon\gamma(a + \epsilon, b + \eta) + \eta\delta(a + \epsilon, b + \eta)$$

for some functions $\gamma(x, y), \delta(x, y)$ continuous at (a, b) and satisfying

$$\gamma(a, b) = \frac{\partial^2 f}{\partial x \partial y}(a, b) \quad \text{and} \quad \delta(a, b) = \frac{\partial^2 f}{\partial y^2}(a, b).$$

Now we consider the double increment

$$\Delta = f(a + \epsilon, b + \eta) - f(a + \epsilon, b) - f(a, b + \eta) + f(a, b).$$

Since $\Delta = \phi(a + \epsilon) - \phi(a)$ with $\phi(x) = f(x, b + \eta) - f(x, b)$, it follows from the mean value theorem and (11.1) that

$$\Delta = \epsilon\phi'(a + \theta\epsilon) = \epsilon \left(\frac{\partial f}{\partial x}(a + \theta\epsilon, b + \eta) - \frac{\partial f}{\partial x}(a + \theta\epsilon, b) \right)$$

$$= \epsilon\eta\beta(a + \theta\epsilon, b + \eta) + \theta\epsilon^2(\alpha(a + \theta\epsilon, b + \eta) - \alpha(a + \theta\epsilon, b))$$

for some $\theta \in (0, 1)$. Similarly, since $\Delta = \psi(b + \eta) - \psi(b)$ with $\psi(y) = f(a + \epsilon, y) - f(a, y)$, we get

$$\Delta = \eta \psi'(b + \tilde{\theta}\eta) = \eta \left(\frac{\partial f}{\partial y}(a + \epsilon, b + \tilde{\theta}\eta) - \frac{\partial f}{\partial y}(a, b + \tilde{\theta}\eta) \right)$$

$$= \epsilon \eta \gamma(a + \epsilon, b + \tilde{\theta}\eta) + \tilde{\theta}\eta^2 (\delta(a + \epsilon, b + \tilde{\theta}\eta) - \delta(a, b + \tilde{\theta}\eta))$$

for some $\tilde{\theta} \in (0, 1)$. Note that $\theta, \tilde{\theta}$ depend on ϵ and on η. Taking $\epsilon = \eta$ we thus obtain

$$|\beta(a + \theta\epsilon, b + \epsilon) - \gamma(a + \epsilon, b + \tilde{\theta}\epsilon)| \leq |\alpha(a + \theta\epsilon, b + \epsilon) - \alpha(a + \theta\epsilon, b)|$$

$$+ |\delta(a + \epsilon, b + \tilde{\theta}\epsilon) - \delta(a, b + \tilde{\theta}\epsilon)|.$$

Therefore

$$|\beta(a, b) - \gamma(a, b)| \leq |\beta(a + \theta\epsilon, b + \epsilon) - \beta(a, b)|$$

$$+ |\gamma(a + \epsilon, b + \tilde{\theta}\epsilon) - \gamma(a, b)|$$

$$+ |\alpha(a + \theta\epsilon, b + \epsilon) - \alpha(a + \theta\epsilon, b)|$$

$$+ |\delta(a + \epsilon, b + \tilde{\theta}\epsilon) - \delta(a, b + \tilde{\theta}\epsilon)|$$

and the right-hand side converges to 0 as $\epsilon \to 0$ by the continuity of α, β, γ and δ at (a, b). This completes the proof. $\square$

SOLUTION 11.3

One can write $f(x, y) = x\gamma(x, y)$ where

$$\gamma(x, y) = \begin{cases} y\phi(x, y) & ((x, y) \in U) \\ 0 & (x = y = 0) \end{cases}.$$

By the boundedness of ϕ we see that $\gamma(x, y)$ tends to 0 as $(x, y) \to (0, 0)$; hence, $\gamma(x, y)$ is continuous at $(0, 0)$. By the definition in Carathéodory's way $f(x, y)$ is differentiable at $(0, 0)$ and

$$\frac{\partial f}{\partial x}(0, 0) = \frac{\partial f}{\partial y}(0, 0) = 0.$$

Since $\frac{\partial f}{\partial x}(x, y) = \gamma(x, y) + xy\frac{\partial \phi}{\partial x}(x, y)$ in U, we see that $\frac{\partial f}{\partial x}$ is continuous at $(0, 0)$, as well as $\frac{\partial f}{\partial y}$. Therefore

$$\frac{\partial^2 f}{\partial y \partial x}(0, 0) = \lim_{h \to 0} \frac{1}{h} \left(\frac{\partial f}{\partial x}(0, h) - \frac{\partial f}{\partial x}(0, 0) \right) = \lim_{h \to 0} \phi(0, h) = \beta$$

and

$$\frac{\partial^2 f}{\partial x \partial y}(0,0) = \lim_{h \to 0} \frac{1}{h}\left(\frac{\partial f}{\partial y}(h,0) - \frac{\partial f}{\partial y}(0,0)\right) = \lim_{h \to 0} \phi(h,0) = \alpha. \qquad \square$$

SOLUTION 11.4

It is easily seen that

$$\frac{(m+n)^2}{s+t} \le \frac{m^2}{s} + \frac{n^2}{t}$$

holds for any $s, t > 0$ and any integers m, n. This means that $a_{m+n} \le a_m + a_n$ where

$$a_n = \int\cdots\int_{[0,1]^n} \frac{n^2}{\dfrac{1}{x_1} + \cdots + \dfrac{1}{x_n}}\, dx_1 \cdots dx_n.$$

Hence the positive sequence $\{a_n/n\}$ converges as $n \to \infty$ by **PROBLEM 1.5**. $\qquad \square$

SOLUTION 11.5

The hypercube $[0,1]^n$ can be divided into $n!$ polytopes:

$$\Delta_\sigma = \{(x_1, \ldots, x_n) \mid 0 \le x_{\sigma(1)} \le \cdots \le x_{\sigma(n)} \le 1\}$$

for every permutation σ in the symmetric group $\mathfrak{S}_n$ of degree n. Since

$$\int\cdots\int_{\Delta_\sigma} f(x_1)\cdots f(x_n)dx_1\cdots dx_n = \int\cdots\int_{\Delta_\sigma} f(x_{\sigma(1)})\cdots f(x_{\sigma(n)})dx_{\sigma(1)}\cdots dx_{\sigma(n)}$$

$$= \int\cdots\int_{\Delta} f(x_1)\cdots f(x_n)dx_1\cdots dx_n,$$

the integral over Δ_σ is independent of σ; so we get

$$n! \int\cdots\int_{\Delta} f(x_1)\cdots f(x_n)dx_1\cdots dx_n = \sum_\sigma \int\cdots\int_{\Delta_\sigma} f(x_1)\cdots f(x_n)dx_1\cdots dx_n$$

$$= \int\cdots\int_{[0,1]^n} f(x_1)\cdots f(x_n)dx_1\cdots dx_n = \left(\int_0^1 f(x)\,dx\right)^n. \qquad \square$$

SOLUTION 11.6

We put $B_n = [0,1]^n$ for brevity. Note that

$$c_n = \int\cdots\int_{B_n} (\phi_n(x_1, \ldots, x_n) - \alpha)^2\, dx_1 \cdots dx_n$$

converges to 0 as $n \to \infty$. If $c_n = 0$, then $\phi_n = \alpha$ is a constant function; so, we have

$$\int_{B_n} \cdots \int f(\phi_n(x_1, ..., x_n)) \, dx_1 \cdots dx_n = f(\alpha).$$

Hence we can assume that $c_n > 0$ hereafter. Let A_n be the set of points $(x_1, ..., x_n)$ in B_n satisfying

$$|\phi_n(x_1, ..., x_n) - \alpha| \geq c_n^{1/3}.$$

Then

$$c_n \geq \int_{A_n} \cdots \int (\phi_n(x_1, ..., x_n) - \alpha)^2 \, dx_1 \cdots dx_n$$

$$\geq c_n^{2/3} \int_{A_n} \cdots \int dx_1 \cdots dx_n,$$

which implies that

$$\int_{A_n} \cdots \int dx_1 \cdots dx_n \leq c_n^{1/3}.$$

For any $\epsilon > 0$ we can find a sufficiently large integer N such that

$$\left| f(\phi_n(x_1, ..., x_n)) - f(\alpha) \right| < \epsilon$$

on $B_n \backslash A_n$ for all $n > N$, because

$$|\phi_n(x_1, ..., x_n) - \alpha| < c_n^{1/3}$$

on the set $B_n \backslash A_n$. Thus,

$$\left| \int_{B_n} \cdots \int f(\phi_n(x_1, ..., x_n)) \, dx_1 \cdots dx_n - f(\alpha) \right|$$

$$\leq \int_{A_n} \cdots \int + \int_{B_n \backslash A_n} \cdots \int \left| f(\phi_n(x_1, ..., x_n)) - f(\alpha) \right| dx_1 \cdots dx_n$$

$$\leq 2Mc_n^{1/3} + \epsilon,$$

where M is the maximum of $|f(x)|$ on the interval $[0, 1]$. We can take a larger N if necessary so that $2Mc_n^{1/3} < \epsilon$ holds for all $n > N$. $\qquad\square$

SOLUTION 11.7

Let $\|x\|$ be the Euclidean norm of a vector $x = (x_1, \ldots, x_n) \in \mathbb{R}^n$. Take a $\mathscr{C}^\infty$-function $\phi(x)$ defined on $\mathbb{R}^n$ satisfying

$$\phi(x) = \begin{cases} 1 & (\|x\| \le 1) \\ 0 & (\|x\| \ge 2) \end{cases}.$$

For any integer $k \ge 0$ we put

$$f_k(x) = \sum_{|m|=k} \frac{\lambda_m}{m!} x^m \phi(x),$$

where $m! = m_1! m_2! \cdots m_n!$, $x^m = x_1^{m_1} x_2^{m_2} \cdots x_n^{m_n}$ and the summation runs through $m = (m_1, \ldots, m_n)$ satisfying $|m| = m_1 + m_2 + \cdots + m_n = k$. Then it can be seen that

$$D^\ell f_k(x) = \sum_{|m|=k} \frac{\lambda_m}{m!} \sum_{i+j=\ell} \binom{\ell}{i} D^i(x^m) \, D^j\phi(x)$$

for any n-tuple of non-negative integers $\ell = (\ell_1, \ldots, \ell_n)$, where

$$\binom{\ell}{i} = \frac{\ell!}{i!(\ell - i)!}.$$

If $m_s \ge i_s$ for all $1 \le s \le n$, then

$$D^i(x^m) = \frac{m!}{(m-i)!} x^{m-i}.$$

On the other hand, if $m_s < i_s$ for some s, then clearly $D^i(x^m)$ vanishes. Hence $D^i(x^m)(0)$ does not vanish if and only if $i = m$, whence we obtain

$$D^m(x^m)(0) = m!.$$

Since $\phi(x)$ is constant on $\|x\| \le 1$, it is obvious that $D^j\phi(0)$ does not vanish if and only if $j = 0$. Thus $D^0\phi(0) = \phi(0) = 1$ and we have $D^\ell f_k(0) = 0$ for all $|\ell| \ne k$ and $D^\ell f_k(0) = \lambda_\ell$ for all $|\ell| = k$.

Put next

$$M_k = 2^k \max_{|\ell| < k} \max_{\|x\| \le 2} \left| D^\ell f_k(x) \right|$$

and

$$g_k(x) = \frac{1}{M_k^k} f_k(M_k x_1, M_k x_2, \ldots, M_k x_n).$$

Since

$$D^\ell g_k(x) = M_k^{|\ell|-k} D^\ell f_k(M_k x),$$

we have $D^\ell g_k(0) = 0$ for all $|\ell| \neq k$ and $D^\ell g_k(0) = \lambda_\ell$ for any $|\ell| = k$. So $f_k(x)$ and $g_k(x)$ have the same partial derivatives at the origin. Moreover, since

$$\max_x \left| D^\ell g_k(x) \right| = M_k^{|\ell|-k} \max_x \left| D^\ell f_k(x) \right| \leq \frac{1}{2^k}$$

for all $|\ell| < k$, this implies that every series obtained by termwise partial differentiation from

$$g(x) = \sum_{k=0}^{\infty} g_k(x)$$

converges uniformly in $\mathbb{R}^n$. Therefore $g(x) \in \mathscr{C}^\infty(\mathbb{R}^n)$ and

$$D^\ell g(0) = \sum_{k=0}^{\infty} D^\ell g_k(0) = D^\ell g_{|\ell|}(0) = \lambda_\ell$$

for all n-tuple of non-negative integers ℓ. $\square$

SOLUTION 11.8

Let I_n be the integral in the problem. Using $4\sin^2(\alpha - \beta) = \left| e^{2i\alpha} - e^{2i\beta} \right|^2$ and substituting $\theta_j = 2\pi x_j$, we have

$$I_n = \frac{1}{2^{n(n-1)}} \cdot \frac{1}{(2\pi)^n} \int_0^{2\pi} \cdots \int_0^{2\pi} \prod_{1 \leq j < k \leq n} \left| e^{i\theta_k} - e^{i\theta_j} \right|^2 d\theta_1 \cdots d\theta_n$$

$$= \frac{1}{2^{n(n-1)}} \cdot \frac{1}{(2\pi i)^n} \int_C \cdots \int_C \prod_{1 \leq j < k \leq n} |z_k - z_j|^2 \frac{dz_1}{z_1} \cdots \frac{dz_n}{z_n}$$

where C is the unit circle $|z| = 1$. Since $\bar{z} = 1/z$ on C, we have

$$I_n = \frac{1}{2^{n(n-1)}} \cdot \frac{1}{(2\pi i)^n} \int_C \cdots \int_C \prod_{1 \leq j < k \leq n} (z_k - z_j)\left(\frac{1}{z_k} - \frac{1}{z_j} \right) \frac{dz_1}{z_1} \cdots \frac{dz_n}{z_n}$$

$$= \frac{(-1)^{n(n-1)/2}}{2^{n(n-1)}} \cdot \frac{1}{(2\pi i)^n} \int_C \cdots \int_C \prod_{1 \leq j < k \leq n} (z_k - z_j)^2 \frac{dz_1}{z_1^n} \cdots \frac{dz_n}{z_n^n}$$

$$= \frac{(-1)^{n(n-1)/2}}{2^{n(n-1)}} c_n.$$

Here c_n is the coefficient of $(z_1 \cdots z_n)^{n-1}$ in the expansion of

$$\prod_{1 \le j < k \le n} (z_k - z_j)^2 = D^2(z_1, \ldots, z_n)$$

where

$$D(z_1, \ldots, z_n) = \begin{vmatrix} 1 & z_1 & \cdots & z_1^{n-1} \\ 1 & z_2 & \cdots & z_2^{n-1} \\ \vdots & \vdots & \ddots & \vdots \\ 1 & z_n & \cdots & z_n^{n-1} \end{vmatrix}$$

is the Vandermonde determinant. Hence we have

$$c_n = \frac{1}{(n-1)!^n} \frac{\partial^{n(n-1)}}{\partial z_1^{n-1} \cdots \partial z_n^{n-1}} D^2(z_1, \ldots, z_n) \Big|_{z_1 = \cdots = z_n = 0}.$$

Note that the partial derivative in the right-hand side is constant itself, because the total degree of each term in the expansion of D^2 is $n(n-1)$. However it follows from Leibniz' rule that

$$\frac{\partial^{n(n-1)}}{\partial z_1^{n-1} \cdots \partial z_n^{n-1}} D^2 = \sum_{k_1=0}^{n-1} \cdots \sum_{k_n=0}^{n-1} \binom{n-1}{k_1} \cdots \binom{n-1}{k_n}$$
$$\times \, \Delta(k_1, \ldots, k_n) \Delta(n-1-k_1, \ldots, n-1-k_n)$$

where

$$\Delta(k_1, \ldots, k_n) = \begin{vmatrix} v_{k_1}(z_1) \\ \vdots \\ v_{k_n}(z_n) \end{vmatrix}$$

and

$$v_k(t) = \frac{d^k}{dt^k}(1, t, \ldots, t^{n-1})$$
$$= k! \left(0, \ldots, 0, 1, \binom{k+1}{1} t, \ldots, \binom{n-1}{n-1-k} t^{n-1-k} \right).$$

So if we put $v_k(0) = k! \, e(k)$, then $e(k)$ is the unit row vector with 1 in the $(k+1)$th position and 0 elsewhere. Thus, putting $z_1 = \cdots = z_n = 0$ we obtain

$$c_n = \frac{1}{(n-1)!^n} \sum_{k_1=0}^{n-1} \cdots \sum_{k_n=0}^{n-1} \binom{n-1}{k_1} \cdots \binom{n-1}{k_n}$$
$$\times \, \Delta_0(k_1, \ldots, k_n) \Delta_0(n-1-k_1, \ldots, n-1-k_n),$$

where

$$\Delta_0(k_1, \ldots, k_n) = \Delta(k_1, \ldots, k_n)\Big|_{z_1 = \cdots = z_n = 0} = k_1! \cdots k_n! \begin{vmatrix} e(k_1) \\ \vdots \\ e(k_n) \end{vmatrix}.$$

Therefore we get

$$c_n = \sum_{k_1=0}^{n-1} \cdots \sum_{k_n=0}^{n-1} \begin{vmatrix} e(k_1) \\ \vdots \\ e(k_n) \end{vmatrix} \cdot \begin{vmatrix} e(n-1-k_1) \\ \vdots \\ e(n-1-k_n) \end{vmatrix}.$$

Now it is easily seen that the determinants in the right-hand side do not vanish if and only if $\{k_1, \ldots, k_n\} = \{0, 1, \ldots, n-1\}$; that is, if $(k_1, \ldots, k_n)$ is a permutation of n symbols $0, 1, \ldots, n-1$. In such a case each determinant is either 1 or -1. Obviously such permutations exist in $n!$ ways. Moreover the matrix

$$\begin{pmatrix} e(n-1-k_1) \\ \vdots \\ e(n-1-k_n) \end{pmatrix}$$

is transformed to

$$\begin{pmatrix} e(k_1) \\ \vdots \\ e(k_n) \end{pmatrix}$$

by interchanging the first column and the nth column, the second column and the $(n-1)$th column, etc. Thus we have

$$\begin{vmatrix} e(n-1-k_1) \\ \vdots \\ e(n-1-k_n) \end{vmatrix} = \operatorname{sgn} \sigma \begin{vmatrix} e(k_1) \\ \vdots \\ e(k_n) \end{vmatrix}$$

where σ is the permutation $\begin{pmatrix} 0 & 1 & \cdots & n-1 \\ n-1 & n-2 & \cdots & 0 \end{pmatrix}$. Since it can be easily seen that $\operatorname{sgn} \sigma = (-1)^{n(n-1)/2}$, we finally get

$$c_n = (-1)^{n(n-1)/2} n!.$$

This completes the proof. $\qquad\qquad\qquad\qquad\qquad\qquad\qquad\qquad\qquad\qquad$ □

Chapter 12

Uniform Distribution

H. Weyl(1885–1955) introduced the notion of 'uniform distributions' in a general form in Weyl(1916).

[Summary of Basic Points]

1. Given a sequence $\{a_n\}_{n \geq 1}$ in the unit interval $I = [0, 1]$, let $N_n(J)$ be the number of a_k in J for $1 \leq k \leq n$, and $|J|$ be the length of J for any subinterval J. The sequence $\{a_n\}_{n \geq 1}$ is said to be uniformly distributed, or equidistributed, on I provided that

$$\mu_n(J) = \frac{N_n(J)}{n}$$

converges to $|J|$ as $n \to \infty$ for any subinterval J. In other words, the probability of finding a_n in J is proportional to $|J|$.

2. Obviously we have $\mu_n(I) = 1$ and

$$\mu_n(J \cup J') = \mu_n(J) + \mu_n(J')$$

for any disjoint subintervals J and J' in I.

PROBLEM 12.1 ─────────────────────────────────

*For any integer $m \geq 1$ and any interval $J \subset [0,1]$ satisfying $|J| < 1/m$
suppose that*

$$\limsup_{n \to \infty} \mu_n(J) \leq \frac{1}{m}.$$

Show that the sequence $\{a_n\}$ is uniformly distributed in $[0,1]$.

PROBLEM 12.2 ─────────────────────────────────

*For any irrational number α show that the fractional parts of αn, which is
denoted by $\{\alpha n\}$, are uniformly distributed in $[0,1]$.*

The above two facts are due to Callahan(1964). His argument is very simple
because he makes use of neither continued fractions nor exponential sums. This
will be proved as a simple application of **PROBLEM 12.1**.

PROBLEM 12.3 ─────────────────────────────────

*Suppose that $f \in \mathscr{C}^1(0, \infty)$ satisfies $f(x) > 0$, $f'(x) > 0$ for $x > 0$ and
$f(x)/x$ diverges to ∞ as $x \to \infty$. Suppose further that*

$$\lim_{k \to \infty} \frac{M_k}{m_k} = 1$$

*where M_k and m_k are the maximum and the minimum of $f'(x)$ on the interval
$[k, k+1]$ respectively. Let f^{-1} be the inverse function of f. Show then that
the fractional parts of $f^{-1}(n)$ are uniformly distributed in $[0,1]$.*

For example, the fractional parts of n^α and $\log^\beta n$ are uniformly distributed on
$[0,1]$ when $0 < \alpha < 1$ and $\beta > 1$ respectively.

If f^{-1} grows as fast as or faster than polynomials, the problem becomes much
more difficult. For example, it is not known whether the fractional parts of e^n
and $(3/2)^n$ are uniformly distributed or not. It is worthwhile to note that an
upper bound for the fractional part of $(3/2)^n$ is closely related to that of the
form of $g(k)$ in Waring problem.

However we know that if θ is the Pisot number, then the distance of θ^n from
$\mathbb{Z}$ converges to 0 as $n \to \infty$. The Pisot numbers were studied by Pisot (1938,
1946) and by Vijayaraghavan (1940–42, 1948) independently and are sometimes called the 'PV numbers'.

PROBLEM 12.4

A sequence $\{a_n\}$ contained in $[0, 1]$ is uniformly distributed if and only if

$$\lim_{n\to\infty} \frac{1}{n} \sum_{k=1}^{n} f(a_k) = \int_0^1 f(x)\,dx$$

for any $f \in \mathscr{C}[0, 1]$.

PROBLEM 12.5

Show that the fractional parts of a given sequence $\{a_n\}$ is uniformly distributed if and only if

$$\lim_{n\to\infty} \frac{1}{n} \sum_{k=1}^{n} \exp(2\pi i m a_k) = 0$$

for every integer $m \geq 1$.

This is due to Weyl (1916) and known as the *Weyl criterion*.

PROBLEM 12.6

Let α be an arbitrary positive irrational number. Show that the unit circle $|z| = 1$ is the natural boundary of the power series

$$f(z) = \sum_{n=1}^{\infty} [\alpha n]\, z^n.$$

This is due to Hecke (1921). The function $f(z)$ is related to the 'Hecke-Mahler series'

$$M_\alpha(w, z) = \sum_{n=1}^{\infty} \sum_{m=1}^{[\alpha n]} w^m z^n$$

by the relation $M_\alpha(1, z) = f(z)$. Various arithmetical properties of the values of $f(z)$ can be studied by virtue of certain functional equations satisfied by $M_\alpha(w, z)$, $M_{1/\alpha}(z, w)$ and $M_{k+\alpha}(w, z)$. The author (1982) encountered this function in the study of a mathematical neuron model as a special case of Caianiello's equation (1961). Also the author (2015) studied the rational approximations to $M_\alpha(w, z)$ by using Farey series and some one-dimensional discrete dynamical system.

Solutions for Chapter 12

SOLUTION 12.1

Let $\epsilon > 0$ be an arbitrary number. For any subinterval $J \subset I$ with $|J| < 1$ take a rational number $p/q < 1$ satisfying $|J| < p/q < |J| + \epsilon$. We divide J into p equal parts and name them $J_1, J_2, ..., J_p$ from left to right. Since $|J_k| < 1/q$ we have

$$\limsup_{n\to\infty} \mu_n(J) \le \sum_{k=1}^{p} \limsup_{n\to\infty} \mu_n(J_k)$$

$$\le \frac{p}{q} < |J| + \epsilon;$$

therefore $\limsup_{n\to\infty} \mu_n(J) \le |J|$, because ϵ is arbitrary.

On the other hand, the set $I \setminus J$ is either an interval or a union of two disjoint intervals; so let K_0 and K_1 be such intervals (the latter may be empty). Then

$$\liminf_{n\to\infty} \mu_n(J) = 1 - \limsup_{n\to\infty} \mu_n(K_0 \cup K_1)$$

$$\ge 1 - \limsup_{n\to\infty} \mu_n(K_0) - \limsup_{n\to\infty} \mu_n(K_1)$$

$$\ge 1 - |K_0| - |K_1| = |J|,$$

which implies that $\lim_{n\to\infty} \mu_n(J) = |J|$, as required.   □

SOLUTION 12.2

Let J be any subinterval in $[0, 1]$ with $|J| < 1/m$ for some integer $m \ge 2$. Since $\{\alpha n\}$ forms a dense set in $[0, 1]$ (this follows easily from the pigeon hole principle), there exists a sufficiently large integer k satisfying

$$|J| < (\alpha k) < \frac{1 - |J|}{m - 1}.$$

Put now $J_0 = J$ and let J_1 be the shifted interval of J_0 by $\{\alpha k\}$. We continue this process up to J_{m-1}; that is,

$$J_j \equiv J + j\{\alpha k\} \pmod 1$$

for $0 \le j < m$. Since the distance between the left end point of J_0 and the right end point of J_{m-1} measured in the positive direction is

$$m |J| + (m-1)(\{\alpha k\} - |J|) < |J| + (m-1)\frac{1-|J|}{m-1} = 1,$$

the intervals $J_0, \ldots, J_{m-1}$ are mutually disjoint; therefore

$$\mu_n(J_0) + \mu_n(J_1) + \cdots + \mu_n(J_{m-1}) \le 1.$$

Moreover, since $j\{\alpha k\} \equiv \{\alpha j k\}$ (mod 1), it can be easily seen that $\{\alpha \ell\} \in J$ if and only if $\{\alpha(\ell + jk)\} \in J_j$ for all $0 \le j < m$. We thus have

$$|N_n(J) - N_n(J_j)| \le 2jk,$$

which implies that

$$1 \ge \sum_{j=0}^{m-1} \mu_n(J_j) \ge \sum_{j=0}^{m-1}\left(\mu_n(J) - \frac{2jk}{n}\right)$$

$$= m\mu_n(J) - \frac{km(m-1)}{n}.$$

Since m and k are independent of n, the superior limit of $\mu_n(J)$ as $n \to \infty$ is less than or equal to $1/m$. $\qquad\square$

SOLUTION 12.3

We first show that

$$\lim_{x\to\infty} \frac{f(x+1)}{f(x)} = 1.$$

To see this, by virtue of l'Hôpital's rule, it suffices to show

$$\lim_{x\to\infty} \frac{f'(x+1)}{f'(x)} = 1.$$

For any $\epsilon > 0$ take a large integer k_0 satisfying $M_k/m_k < 1 + \epsilon$ for all $k \ge k_0$. Then for any $x \ge k_0$,

$$\frac{f'(x+1)}{f'(x)} \le \frac{M_{[x]+1}}{m_{[x]}} < (1+\epsilon)\frac{m_{[x]+1}}{m_{[x]}} \le (1+\epsilon)\frac{M_{[x]}}{m_{[x]}} < (1+\epsilon)^2.$$

Similarly we can see that the left-hand side is larger than $(1 + \epsilon)^{-2}$. This implies the limit exists and is equal to 1, because ϵ is arbitrary.

Now, for any subinterval $J = [a, b)$ contained in $[0, 1]$, we give an upper estimate for $N_n(J)$. For any $\epsilon > 0$ we take a large integer k_1 satisfying $k < \epsilon f(k)$,

$$\frac{M_k}{m_k} < 1 + \epsilon \quad \text{and} \quad \frac{f(k+1)}{f(k)} < 1 + \epsilon$$

for all $k \geq k_1$. For any integer $n > f(k_1 + a)$ we can take a unique integer $K_n \geq k_1$ satisfying

$$f(K_n + a) \leq n < f(K_n + a + 1).$$

Let $\nu(k)$ be the number of ℓ's satisfying

$$f(k + a) \leq \ell < f(k + b),$$

which is equivalent to $f^{-1}(\ell) - k \in J$; therefore

$$N_n(J) \leq f(k_1 + a) + \sum_{k=k_1}^{K_n} \nu(k).$$

On the other hand, it follows from the mean value theorem that

$$\nu(k) \leq [f(b + k)] - [f(a + k)] + 1$$
$$\leq f(b + k) - f(a + k) + 2$$
$$= (b - a)f'(\xi_k) + 2$$

for some ξ_k in $(a + k, b + k)$. Hence

$$\nu(k) \leq (b - a)M_k + 2$$
$$< (1 + \epsilon)(b - a)m_k + 2$$
$$\leq (1 + \epsilon)(b - a)(f(k + 1) - f(k)) + 2;$$

thus

$$N_n(J) \leq 2K_n + c(k_1, f) + (1 + \epsilon)(b - a)f(K_n + 1)$$

where $c(k_1, f)$ is a constant depending only on k_1 and f. Using $K_n < \epsilon f(K_n) \leq \epsilon n$ and $f(K_n + 1) < (1 + \epsilon)f(K_n) \leq (1 + \epsilon)n$, which follow from the choice of K_n, we get

$$\mu_n(J) < 2\epsilon + (1 + \epsilon)^2(b - a) + O\left(\frac{1}{n}\right),$$

which implies that the superior limit of $\mu_n(J)$ is less than or equal to $|J|$. $\square$

SOLUTION 12.4

For any subinterval J in $\mathbb{R}$ the discontinuous function of the first kind

$$\chi_J(x) = \begin{cases} 1 & (x \in J) \\ 0 & (x \notin J) \end{cases},$$

is called the *characteristic function* of J. It follows from the definition that $\{a_n\}$ is uniformly distributed in I if and only if

$$\lim_{n \to \infty} \frac{1}{n} \sum_{k=1}^{n} \chi_J(a_k) = \int_0^1 \chi_J(x)\,dx \qquad (12.1)$$

for any subinterval J in $[0, 1]$.

Since every step function is a finite combination of characteristic functions, (12.1) holds also for any step functions. It is now clear from the definition of uniform continuity that there exists a step function $\phi(x)$ satisfying

$$\sup_{x \in I} |f(x) - \phi(x)| < \epsilon$$

for any $f \in \mathscr{C}(I)$ and any $\epsilon > 0$. Thus we obtain

$$\left| \frac{1}{n} \sum_{k=1}^{n} f(a_k) - \int_0^1 f(x)\,dx \right| \le \left| \frac{1}{n} \sum_{k=1}^{n} \phi(a_k) - \int_0^1 \phi(x)\,dx \right| + 2\epsilon.$$

Since the superior limit as $n \to \infty$ of the left-hand side is less than or equal to 2ϵ,

$$\lim_{n \to \infty} \frac{1}{n} \sum_{k=1}^{n} f(a_k) = \int_0^1 f(x)\,dx \qquad (12.2)$$

holds for any $f \in \mathscr{C}(I)$.

Conversely suppose that (12.2) holds for any $f \in \mathscr{C}(I)$. For any subinterval J in I and any $\epsilon > 0$ consider two piecewise linear trapezoidal functions $f_-(x)$ and $f_+(x)$ such that

$$f_-(x) \le \chi_J(x) \le f_+(x)$$

for all $x \in I$ and that $f_\pm(x)$ differ from $\chi_J(x)$ only in some ϵ-neighborhoods of two end points of J. Then obviously

$$\frac{1}{n} \sum_{k=1}^{n} f_-(a_k) \le \frac{1}{n} \sum_{k=1}^{n} \chi_J(a_k) \le \frac{1}{n} \sum_{k=1}^{n} f_+(a_k)$$

and the both sides converge to $\int_0^1 f_\pm(x)\,dx$ as $n \to \infty$ respectively. Since

$$\left| \int_0^1 f_\pm(x)\,dx - \int_0^1 \chi_J(x)\,dx \right| \le 2\epsilon,$$

we can conclude that (12.1) holds for any subinterval J in I. $\square$

SOLUTION 12.5

By virtue of the previous problem it suffices to show the sufficiency. Suppose that

$$\lim_{n\to\infty} \frac{1}{n} \sum_{k=1}^n \exp(2\pi i m a_k) = 0$$

for any integer $m \ge 1$. Then clearly we have

$$\lim_{n\to\infty} \frac{1}{n} \sum_{k=1}^n T(a_k) = d_0 = \int_0^1 T(x)\,dx$$

for any trigonometric polynomial

$$T(x) = \sum_{j=1}^p c_j \sin(2\pi j x) + \sum_{j=0}^q d_j \cos(2\pi j x).$$

In the previous proof one can assume in addition that $f_\pm(0) = f_\pm(1)$ respectively. Therefore such $f_\pm$ can be approximated uniformly by some trigonometric polynomials and in consequence (12.1) holds for any subinterval J. This completes the proof. $\square$

SOLUTION 12.6

We will show that

$$z_m = e^{2\pi i \alpha m} = e^{2\pi i \{\alpha m\}}$$

on the unit circle $|z| = 1$ is a singular point of $f(z)$ for any integer $m \ge 1$. As is seen in **PROBLEM 12.2** such points form a dense subset on the unit circle. Putting $a_n = \{\alpha n\}\, e^{2\pi i \alpha m n}$ and $\sigma_n = a_1 + a_2 + \cdots + a_n$ we have

$$f(r z_m) = \sum_{n=1}^\infty \{\alpha n\}\, r^n e^{2\pi i \alpha m n} = \sum_{n=1}^\infty a_n r^n$$

for $0 < r < 1$ and

$$\sigma_n = \sum_{k=1}^{n} \{\alpha k\} r^n e^{2\pi i m \{\alpha k\}} = \sum_{k=1}^{n} \phi(\{\alpha k\}),$$

where $\phi(x) = x e^{2\pi i m x}$. By virtue of **PROBLEM 12.2** and of **PROBLEM 12.4** the sequence σ_n/n converges to

$$\int_0^1 \phi(x)\,dx = \frac{1}{2\pi i m}$$

as $n \to \infty$. We put

$$\sigma_n = \frac{n}{2\pi i m} + \tau_n.$$

For any $\epsilon > 0$ we take a sufficiently large integer n_0 satisfying $|\tau_n| < \epsilon n$ for any $n \geq n_0$. Hence

$$\frac{f(rz_m)}{1 - r} = \sum_{n=1}^{\infty} a_n r^n \sum_{k=0}^{\infty} r^k = \sum_{n=1}^{\infty} \sigma_n r^n$$

$$= \sum_{n=1}^{\infty} \left(\frac{n}{2\pi i m} + \tau_n \right) r^n$$

$$= \frac{1}{2\pi i m} \cdot \frac{r}{(1 - r)^2} + \sum_{n=1}^{\infty} \tau_n r^n.$$

Multiplying the both sides by $(1 - r)^2$ we get

$$\left| (1 - r) f(rz_m) - \frac{1}{2\pi i m} \right| \leq \frac{1 - r}{2\pi m} + (1 - r)^2 \sum_{n=1}^{\infty} |\tau_n| r^n$$

$$< \frac{1 - r}{2\pi m} + (1 - r)^2 \sum_{n < n_0} |\tau_n|$$

$$+ \epsilon (1 - r)^2 \sum_{n \geq n_0} n r^n.$$

Since the third term on the right-hand side is clearly less than ϵ, the radial limit of $(z - z_m) f(z)$, as $z \to z_m$ within the unit disk, is equal to $z_m/(2\pi i m) \neq 0$. This implies that z_m is a singular point of $f(z)$, as required. $\qquad\square$

Chapter 13

Rademacher Functions

[Summary of Basic Points]

1. Any real number x in the interval $[0, 1)$ can be expanded into a series

$$x = \frac{s_1(x)}{2} + \frac{s_2(x)}{2^2} + \cdots + \frac{s_n(x)}{2^n} + \cdots,$$

known as a dyadic or binary expansion of x. Each numerator $s_n(x)$ is a discontinuous function of the first kind taking the value 0 or 1 only. To ensure uniqueness we adopt that the expansion in which all digits are 0 after some place for any irreducible fraction whose denominator is a power of 2. Note that each function $s_n(x)$ is left-continuous.

2. The functions defined by

$$r_n(x) = 1 - 2s_n(x)$$

are called 'Rademacher function', which form an orthogonal system on the interval $[0, 1]$. This system is not complete but very useful as a sample space of the probability events of coin tossing. Rademacher functions were studied in Rademacher(1922). According to Grosswald(1980), Rademacher wrote a sequel containing the completion of the system, but he decided not to publish taking Schur's opinion. Soon after Walsh(1923) published some closely relevant results which he obtained independently. Walsh's complete system of orthogonal functions $\{w_n(x)\}$ is defined by

$$w_n(x) = r_{v_1+1}(x) r_{v_2+1}(x) \cdots r_{v_k+1}(x)$$

where $n = 2^{v_1} + 2^{v_2} + \cdots + 2^{v_k}$ with integers $0 \le v_1 < v_2 < \cdots < v_k$.

3. We have

$$1 - 2x = \sum_{n=1}^{\infty} \frac{r_n(x)}{2^n}. \tag{13.1}$$

207

Since the series in (13.1) has a majorant converging absolutely, it follows from termwise integration that

$$2x(1-x) = \sum_{n=0}^{\infty} \frac{1}{4^n} \psi(2^n x) \qquad (13.2)$$

for $0 \le x \le 1$, where $\psi(x)$ is the continuous piecewise linear function extended to $\mathbb{R}$ defined in **PROBLEM 4.8** by virtue of

$$\int_0^x r_n(s)\,ds = \frac{1}{2^{n-1}} \psi(2^{n-1}x).$$

Note that the series (13.2) becomes the Takagi function $T(x)$, which is continuous but nowhere differentiable, if we replace the coefficient 4^{-n} to 2^{-n}. The expansion (13.2) is nothing but the *method of exhaustion* used by Archimedes to show that the area surrounded by the parabola $y = 2x(1-x)$ and the x-axis is equal to $4/3$ of that of the triangle $y = \psi(x)$.

4. If we define

$$r(x) = \begin{cases} 1 & (0 \le x < 1) \\ -1 & (1 \le x < 2) \end{cases}$$

and extend it to $\mathbb{R}$ periodically with period 2, then the Rademacher functions can be written as $r_n(x) = r(2^n x)$, or since $r(x) = (-1)^{[x]}$, one can write

$$r_n(x) = (-1)^{[2^n x]}.$$

The Rademacher functions are also defined by

$$r_n(x) = \text{sgn}(\sin 2^n \pi x)$$

in some books, which differs from ours only at a set of countable points; so there is no influence on integrals.

5. We use the following notation

$$I_n[f(x)] = \int_0^1 f\left(\sum_{k=1}^n r_k(x)\right) dx$$

for any integer $n \ge 1$ and any function $f(x)$ defined on $\mathbb{Z}$.

The first seven problems are given in Kac's fruitful monograph (1959).

PROBLEM 13.1

For any integers $1 \le k_1 < k_2 < \cdots < k_n$ show that

$$\int_0^1 r_{k_1}(x) r_{k_2}(x) \cdots r_{k_n}(x)\, dx = 0.$$

If $k_1, k_2, \ldots, k_n$ are arbitrary positive integers, then the corresponding integral is equal to 1 if and only if the number of j satisfying $k_\ell = k_j$ is even for all $1 \le \ell \le n$; otherwise the integral vanishes. For example, we have for any $j \ne k$,

$$\int_0^1 (r_j(x) + r_k(x))^{2m}\, dx = \binom{2m}{0} + \binom{2m}{2} + \cdots + \binom{2m}{2m}$$

$$= \left. \frac{(1+x)^{2m} + (1-x)^{2m}}{2} \right|_{x=1}$$

$$= 2^{2m-1}.$$

PROBLEM 13.2

For any real numbers $c_1, c_2, \ldots, c_n$ show that

$$\int_0^1 \cos\left(\sum_{k=1}^n c_k r_k(x) \right) dx = \prod_{k=1}^n \cos c_k.$$

PROBLEM 13.3

Compute $I_n[x^2]$ and $I_n[x^4]$.

PROBLEM 13.4

For any integer $m \ge 0$ show that $I_n[x^{2m}]$ is a polynomial in n of degree m with integer coefficients and the leading coefficient is

$$\frac{(2m)!}{2^m m!} = 1 \cdot 3 \cdot 5 \cdots (2m - 1).$$

PROBLEM 13.5

Let s be an arbitrary real number. Prove that

$$I_n[e^{s|x|}] < 2\left(\frac{e^s + e^{-s}}{2} \right)^n.$$

PROBLEM 13.6

Show that

$$\lim_{n \to \infty} \frac{I_n[|x|]}{\sqrt{n}} = \sqrt{\frac{2}{\pi}}.$$

Hint: Use

$$|s| = \frac{2}{\pi} \int_0^\infty \frac{1 - \cos(sx)}{x^2} \, dx$$

to obtain

$$I_n[|x|] = \frac{2}{\pi} \int_0^\infty \frac{1 - \cos^n x}{x^2} \, dx.$$

PROBLEM 13.7

For any $\epsilon > 0$ show that the series

$$\sum_{n=1}^\infty \frac{1}{n^{2+\epsilon}} \exp\left(\sqrt{\frac{2 \log n}{n}} \left| \sum_{k=1}^n r_k(x) \right| \right)$$

converges almost everywhere.

PROBLEM 13.8

For any integer $n \geq 0$ and $f \in \mathscr{C}[0, 1]$ we put

$$a_n = \int_0^1 f(x) r_n(x) \, dx$$

where $r_0(x) = 1$. Show that

$$\sum_{n=0}^\infty a_n^2 \leq \int_0^1 f^2(x) \, dx \tag{13.3}$$

and the equality holds if and only if f is a linear function.

The inequality (13.3), known as Bessel's inequality, holds for any system of orthonormal functions. The equality in (13.3) means that f belongs to the closure of the space spanned by Rademacher functions in L^2-sense. Thus (13.1) gives an essentially unique Rademacher series representing a non-constant continuous function.

PROBLEM 13.9

For any $f \in \mathscr{C}^1[0, 1]$ show that

$$\lim_{n \to \infty} 2^n \int_0^1 f(x) r_n(x)\, dx = \frac{f(0) - f(1)}{2}.$$

PROBLEM 13.10

Let $m, n \geq 1$ be integers and $\nu \geq 0$ be the largest integer satisfying $2^\nu \mid m$. Show that

$$\int_0^1 e^{2\pi i m x} r_n(x)\, dx = \begin{cases} \dfrac{2^n i}{\pi m} & (\nu = n - 1) \\[2mm] 0 & (\text{otherwise}) \end{cases}.$$

Since the Rademacher function $r_n(x)$ is of bounded variation, it can be expanded to the Fourier series as

$$r_n(x) = \frac{\pi}{4} \sum_{k=1}^{\infty} \frac{1}{2k - 1} \sin(2^n(2k - 1)\pi x)$$

except for $x = \ell/2^n$, $\ell \in \mathbb{Z}$ by the Dirichlet-Jordan test. At these points one can observe the Gibbs phenomena as is illustrated in Fig. 13.1.

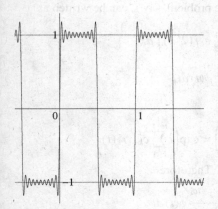

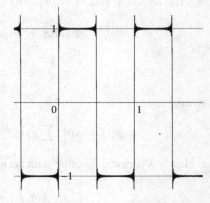

(a) The graph of partial sum up to the first 10 terms of the Fourier series of $r_1(x)$.

(b) The graph of partial sum up to the first 50 terms of the Fourier series of $r_1(x)$.

Fig. 13.1

Solutions for Chapter 13

SOLUTION 13.1

The left-hand side of the equality in the problem, say J, can be written as

$$J = \int_0^1 r(2^{k_1}x)\, r(2^{k_2}x) \cdots r(2^{k_n}x)\, dx$$

$$= \frac{1}{2^{k_1}} \int_0^{2^{k_1}} r(t)\phi(t)\, dt,$$

where

$$\phi(t) = r(2^{k_2-k_1}t) \cdots r(2^{k_n-k_1}t)$$

is a periodic function with period 1; hence

$$2^{k_1} J = \sum_{j=0}^{2^{k_1-1}-1} \int_{2j}^{2j+2} r(t)\phi(t)\, dt = 0. \qquad \Box$$

SOLUTION 13.2

The left-hand side of the equality in the problem, say I, can be written as

$$I = \mathrm{Re} \int_0^1 \exp\!\left(i \sum_{k=1}^n c_k r(2^k x)\right) dx$$

$$= \frac{1}{2}\, \mathrm{Re} \int_0^2 e^{ic_1 r(t)} \phi(t)\, dt,$$

where

$$\phi(t) = \exp\!\left(i \sum_{k=2}^n c_k r(2^{k-1}t)\right) = \exp\!\left(i \sum_{k=1}^{n-1} c_{k+1} r_k(t)\right)$$

is clearly a periodic function with period 1. Therefore

$$I = \mathrm{Re} \int_0^1 \frac{e^{ic_1} + e^{-ic_1}}{2} \phi(t)\, dt$$

$$= (\cos c_1) \times \mathrm{Re} \int_0^1 \phi(t)\, dt.$$

By repeating this process we may get the desired equality. $\qquad \Box$

SOLUTION 13.3

The integral of $r_i(x)r_j(x)$ over the interval $[0, 1]$ vanishes if $i \neq j$ and is equal to 1 if $i = j$. Expanding

$$(r_1(x) + r_2(x) + \cdots + r_n(x))^2$$

we get immediately $I_n[x^2] = n$. Similarly the integral of

$$r_i(x)r_j(x)r_k(x)r_\ell(x)$$

over $[0, 1]$ is equal to 1 if and only if either $i = j = k = \ell$ (n kinds of 'four cards') or $\binom{n}{2}$ kinds of 'two pairs'; otherwise the integral vanishes. Therefore

$$I_n[x^4] = n + \binom{4}{2}\binom{n}{2}. \qquad \square$$

REMARK. From the expression of $I_n[x^4]$ we can conclude that

$$\phi_n(x) = \frac{1}{n} \sum_{k=1}^{n} r_k(x)$$

converges to 0 almost everywhere as $n \to \infty$, because

$$\sum_{n=1}^{\infty} \int_0^1 \phi_n^4(x)\, dx < \infty$$

and so $\sum \phi_n^4(x)$ has a finite sum almost everywhere by the monotone convergence theorem of Beppo Levi.

SOLUTION 13.4

It is clear that

$$r_1(x) + r_2(x) + \cdots + r_n(x) = n - 2k$$

where k is the number of r_j satisfying $r_j(x) = -1$. Then the number of subintervals of length 2^{-n} on which $r_j(x) = -1$ is equal to the number of ways of picking k unordered -1's from n possibilities. Hence we obtain in general

$$I_n[f(x)] = \frac{1}{2^n} \sum_{k=0}^{n} \binom{n}{k} f(n - 2k);$$

so we put

$$A_m(n) = I_n[x^{2m}] = \frac{1}{2^n} \sum_{k=0}^{n} \binom{n}{k} (n - 2k)^{2m}.$$

We now introduce the rational function

$$R_n(x) = \left(x + \frac{1}{x}\right)^n = \sum_{k=0}^{n} \binom{n}{k} x^{n-2k}$$

and the differential operator $\delta = x\dfrac{d}{dx}$ so that

$$A_m(n) = \frac{1}{2^n} \delta^{2m}(R_n)\Big|_{x=1}.$$

Since $\delta^2(R_n) = n^2 R_n - 4n(n-1)R_{n-2}$, we have

$$A_{m+1}(n) = \frac{1}{2^n} \delta^{2m}\Big(n^2 R_n - 4n(n-1)R_{n-2}\Big)\Big|_{x=1}$$

$$= n^2 A_m(n) - n(n-1)A_m(n-2)$$

for any integers $m \geq 0$ and $n \geq 3$. This recursion formula enables us to determine all $A_m(n)$ from the initial conditions $A_0(n) = 1$, $A_m(1) = 1$ and $A_m(2) = 2^{2m-1}$. This also gives a solution to **PROBLEM 13.3** but the proof given there is much simpler.

By the recursion formula it is easily seen that $A_m(n)$ is a polynomial in n of degree m with integer coefficients whose leading coefficient is

$$1 \cdot 3 \cdot 5 \cdots (2m - 1).$$

Moreover we see that every polynomial $A_m(n)$ has a factor n if $m \geq 1$. $\qquad\square$

SOLUTION 13.5

Obviously we have

$$I_n[e^{s|x|}] < I_n[e^{sx}] + I_n[e^{-sx}] = J_n.$$

Since

$$g_n(x) = \exp\left(s \sum_{k=1}^{n} r_k(x)\right)$$

is a periodic function with period 1, we get

$$J_n = \frac{1}{2}\left(\int_0^2 e^{sr(t)} g_{n-1}(t)\, dt + \int_0^2 \frac{e^{-sr(t)}}{g_{n-1}(t)}\, dt\right)$$

$$= \frac{e^s + e^{-s}}{2}(I_{n-1}[e^{sx}] + I_{n-1}[e^{-sx}]);$$

therefore

$$\frac{J_n}{J_{n-1}} = \frac{e^s + e^{-s}}{2}.$$

Solving this with $J_0 = 2$ we may get the desired estimate. $\qquad \square$

SOLUTION 13.6

To show the integral formula representing $|s|$ we can assume $s > 0$. Substituting $t = sx$ we get

$$\int_0^\infty \frac{1 - \cos(sx)}{x^2} \, dx = s \int_0^\infty \frac{1 - \cos t}{t^2} \, dt$$

$$= s \int_0^\infty \frac{\sin t}{t} \, dt = \frac{\pi s}{2},$$

where the evaluation of the last improper integral was given in SOLUTION 7.13. Interchanging the order of integrations and using PROBLEM 13.2 we have

$$I_n[|x|] = \frac{2}{\pi} \int_0^1 dx \int_0^\infty \frac{1 - \cos(s(r_1(x) + r_2(x) + \cdots + r_n(x)))}{s^2} \, ds$$

$$= \frac{2}{\pi} \int_0^\infty \frac{1 - I_n[\cos(sx)]}{s^2} \, ds$$

$$= \frac{2}{\pi} \int_0^\infty \frac{1 - \cos^n s}{s^2} \, ds,$$

as required. For a real parameter $0 \le \epsilon < 1$ we consider the function

$$\varphi_\epsilon(x) = \frac{x^2}{2(1 - \epsilon)} + \log \cos x$$

for $0 \le x < \pi/2$. Clearly $\varphi_\epsilon(0) = \varphi'_\epsilon(0) = 0$ and

$$\varphi''_\epsilon(x) = \frac{1}{1 - \epsilon} - \frac{1}{\cos^2 x}.$$

Hence it follows that $\varphi_0(x) < 0 < \varphi_\epsilon(x)$ on the interval $(0, \alpha(\epsilon))$ if $\epsilon > 0$, where

$$\alpha(\epsilon) = \arccos \sqrt{1 - \epsilon};$$

therefore

$$\exp\left(-\frac{x^2}{2(1 - \epsilon)}\right) < \cos x < \exp\left(-\frac{x^2}{2}\right)$$

for $0 < x < \alpha(\epsilon)$. We then have

$$K(2 - 2\epsilon) + \beta(\epsilon) < \frac{\pi}{2} I_n[|x|] < K(2) + \beta(\epsilon) \tag{13.4}$$

where

$$K(\sigma) = \int_0^{\alpha(\epsilon)} \frac{1 - \exp(-ns^2/\sigma)}{s^2}\, ds \quad \text{and} \quad \beta(\epsilon) = \int_{\alpha(\epsilon)}^{\infty} \frac{1 - \cos^n s}{s^2}\, ds.$$

Substituting $t = \sqrt{n/\sigma}\, s$ we get

$$\begin{aligned} K(\sigma) &= \int_0^{\alpha(\epsilon)} \frac{1 - \exp(-ns^2/\sigma)}{s^2}\, ds \\ &= \sqrt{\frac{n}{\sigma}} \int_0^{\tau(\epsilon)} \frac{1 - e^{-t^2}}{t^2}\, dt, \end{aligned}$$

which is asymptotic to

$$\sqrt{\frac{n}{\sigma}} \int_0^{\infty} \frac{1 - e^{-t^2}}{t^2}\, dt = \sqrt{\frac{\pi n}{\sigma}}$$

as $n \to \infty$, where $\tau(\epsilon) = \sqrt{n/\sigma}\, \alpha(\epsilon)$. Since $\beta(\epsilon) < 1/\alpha(\epsilon)$, it follows from (13.4) that

$$\sqrt{\frac{2}{\pi(1 - \epsilon)}} \le \liminf_{n \to \infty} \frac{I_n[|x|]}{\sqrt{n}} \le \limsup_{n \to \infty} \frac{I_n[|x|]}{\sqrt{n}} \le \sqrt{\frac{2}{\pi}}.$$

This completes the proof, because ϵ is arbitrary. $\square$

SOLUTION 13.7

Put

$$f_n(x) = \frac{1}{n^{2+\epsilon}} \exp\left(\sqrt{\frac{2 \log n}{n}} \left| \sum_{k=1}^{n} r_k(x) \right| \right).$$

It follows from **PROBLEM 13.5** that

$$\begin{aligned} \int_0^1 f_n(x)\, dx &< \frac{2}{n^{2+\epsilon}} \left(\frac{e^{\sqrt{(2 \log n)/n}} + e^{-\sqrt{(2 \log n)/n}}}{2} \right)^n \\ &= \frac{2}{n^{1+\epsilon}} + O(n^{-3/2}) \end{aligned}$$

as $n \to \infty$ for any $0 < \epsilon < 1/2$; therefore

$$\sum_{n=1}^{\infty} \int_0^1 f_n(x)\, dx < \infty.$$

By Beppo Levi's theorem $\displaystyle\sum_{n=1}^{\infty} f_n(x)$ has a finite value almost everywhere. $\square$

REMARK. In particular the above result implies that

$$\limsup_{n\to\infty} \frac{|r_1(x) + r_2(x) + \cdots + r_n(x)|}{\sqrt{n\log n}} \le \sqrt{2} \tag{13.5}$$

almost everywhere. To see this, note that almost every x in $(0, 1)$ satisfies

$$\frac{|r_1(x) + r_2(x) + \cdots + r_n(x)|}{\sqrt{n\log n}} < \frac{2+\epsilon}{\sqrt{2}}$$

for all sufficiently large n, because this inequality is equivalent to $f_n(x) < 1$. Now for any integer $k \ge 1$ let E_k be the set of points $x \in (0, 1)$ such that the superior limit of

$$\frac{|r_1(x) + r_2(x) + \cdots + r_n(x)|}{\sqrt{n\log n}}$$

as $n \to \infty$ exceeds $\sqrt{2} + 1/k$. The above result implies that the set E_k is a null set for every k and so is the infinite union

$$\bigcup_{k=1}^{\infty} E_k.$$

The inequality (13.5) was first shown by Hardy and Littlewood (1914) in a different way. Later Khintchine (1924) strengthened (13.5) to

$$\limsup_{n\to\infty} \frac{|r_1(x) + r_2(x) + \cdots + r_n(x)|}{\sqrt{n\log\log n}} = \sqrt{2}$$

almost everywhere, which is called the *law of the iterated logarithm*.

SOLUTION 13.8

For any $f \in \mathscr{C}[0, 1]$ we put $\phi_m(x) = \sum_{n=0}^{m} a_n r_n(x)$ for brevity. Then we have

$$0 \le \int_0^1 (f(x) - \phi_m(x))^2 \, dx = \int_0^1 f^2(x) \, dx - \sum_{n=0}^{m} a_n^2,$$

which gives (13.3). Suppose next that $f \in \mathscr{C}[0, 1]$ satisfies the equality in (13.3). We simply say that $f(x)$ is 'balanced' on an interval $I = [a, b]$ if

$$f(a + x) + f(a + b - x)$$

is a constant function on I. A subinterval in $[0, 1]$ of the form

$$\left[\frac{k}{2^n}, \frac{k+1}{2^n}\right], \quad 0 \le k < 2^n, \, n \ge 0$$

is called a *dyadic interval*. Observe that any Rademacher function $r_n(x)$, $n \geq 1$ is balanced on every dyadic interval except for a finite number of points.

We first show that $f(x)$ is balanced on $[0, 1]$. For any $\epsilon > 0$ we take a sufficiently large integer m satisfying

$$\sum_{n > m} a_n^2 < \epsilon.$$

Then we have

$$\int_0^1 (f(x) - \phi_m(x))^2 \, dx = \int_0^1 f^2(x) \, dx - \sum_{n=0}^m a_n^2$$

$$= \sum_{n > m} a_n^2 < \epsilon,$$

which yields that

$$\left(\int_0^1 (f(x) + f(1 - x) - 2a_0)^2 \, dx \right)^{1/2} \leq \left(\int_0^1 (f(x) - \phi_m(x))^2 \, dx \right)^{1/2}$$

$$+ \left(\int_0^1 (f(1 - x) - \phi_m(1 - x))^2 \, dx \right)^{1/2}$$

$$< 2\sqrt{\epsilon},$$

because $r_n(x) + r_n(1 - x) = 0$ except for a finite number of points for each $n \geq 1$. Here we used the Minkowski inequality. Since $f(x) + f(1 - x)$ is continuous, we see that $f(x)$ is balanced on $[0, 1]$, as required. Note that $a_0 = f(1/2)$.

We next consider the function $g(x) = f\left(\dfrac{x}{2}\right) \in \mathscr{C}[0, 1]$ and put

$$a_n' = \int_0^1 g(x) r_n(x) \, dx$$

for $n \geq 0$. Then it follows that

$$a_n' = \int_0^1 f\left(\frac{x}{2}\right) r_n(x) \, dx = 2 \int_0^{1/2} f(s) r_{n+1}(s) \, ds$$

$$= \int_0^{1/2} f(s) r_{n+1}(s) \, ds + \int_0^{1/2} (f(1 - s) - 2a_0) r_{n+1}(1 - s) \, ds$$

$$= a_{n+1} + \begin{cases} a_0 & (n = 0) \\ 0 & (n \geq 1) \end{cases}$$

and therefore

$$\int_0^1 g^2(x)\,dx = \int_0^1 f^2\left(\frac{x}{2}\right) dx = \int_0^{1/2} f^2(s)\,ds + \int_{1/2}^1 (f(s) - 2a_0)^2\,ds$$

$$= \int_0^1 f^2(s)\,ds - 4a_0 \int_{1/2}^1 f(s)\,ds + 2a_0^2$$

$$= \int_0^1 f^2(s)\,ds - 2a_0 \int_0^1 f(s)(r_0(s) - r_1(s))\,ds + 2a_0^2$$

$$= \int_0^1 f^2(s)\,ds + 2a_0 a_1$$

$$= (a_0 + a_1)^2 + \sum_{n=2}^\infty a_n^2 = \sum_{n=0}^\infty a_n'^2.$$

Similarly we can see that $g(x) = f\left(\dfrac{1+x}{2}\right)$ has the same property.

Repeating the above argument we conclude that $f(x)$ is balanced on all dyadic intervals. Put

$$\varphi(x) = f(x) - f(0)(1-x) - f(1)x \in \mathscr{C}[0, 1].$$

Since any linear function is balanced on any interval, the function $\psi(x)$ is also balanced on every dyadic interval. Thus we have $\varphi(x) + \varphi(1 - x) = c$ for some constant c; therefore $c = 0$ and $\varphi(1/2) = 0$, because $\varphi(0) = \varphi(1) = 0$. Repeating this argument we obtain that

$$\varphi\left(\frac{k}{2^n}\right) = 0$$

for all $0 \le k < 2^n$ and $n \ge 0$. Since $\varphi(x)$ is continuous, we see that $\varphi(x) = 0$ everywhere, and so $f(x)$ is a linear function.

Conversely every linear function satisfies the equality in (13.3), because

$$\int_0^1 x r_n(x)\,dx = -\frac{1}{2^{n+1}}$$

for all $n \ge 1$ by virtue of (13.1). $\qquad\qquad\square$

SOLUTION 13.9

We have

$$a_n = \int_0^1 f(x) r_n(x)\,dx = 2 \sum_{j=0}^{2^{n-1}-1} \int_{2j/2^n}^{(2j+1)/2^n} f(x)\,dx - \int_0^1 f(x)\,dx.$$

For any $\epsilon > 0$ take a sufficiently large integer n such that

$$|f'(x) - f'(y)| < \epsilon$$

whenever $|x - y| \le 2^{-n}$. Then, for any $0 \le a < b \le 1$ with $b - a = 2^{-n}$ it follows from Taylor's formula that

$$f(x) = f(a) + (f'(a) + \xi(x))(x - a), \quad |\xi(x)| < \epsilon$$
$$f(x) = f(b) + (f'(a) + \eta(x))(x - b), \quad |\eta(x)| < \epsilon$$

for every $x \in [a, b]$. Hence

$$\mathcal{M}_{[a,b]}(f) = f(a) + \frac{f'(a) + \xi_0}{2^{n+1}}, \quad |\xi_0| < \epsilon$$

$$\mathcal{M}_{[a,b]}(f) = f(b) - \frac{f'(a) + \eta_0}{2^{n+1}}, \quad |\eta_0| < \epsilon$$

for some constants ξ_0, η_0; so, we get

$$\left| 2\mathcal{M}_{[a,b]}(f) - f(a) - f(b) \right| < \frac{\epsilon}{2^n}.$$

Therefore, applying this estimate to $a = 0, 2/2^n, ..., 2(2^{n-1} - 1)$ respectively, we have

$$\left| 2^n a_n - \sum_{j=0}^{2^n - 1} f\left(\frac{j}{2^n} \right) + 2^n \int_0^1 f(x)\,dx \right| < \epsilon.$$

We obtain the limit of $2^n a_n$ as $n \to \infty$ by applying **PROBLEM 5.7** for $\theta = 0$. $\qquad \square$

ANOTHER SOLUTION. Let $\psi(x)$ is the continuous piecewise linear periodic function defined in **PROBLEM 4.8**. Since

$$2^n \int_0^x r_n(s)\,ds = 2\psi(2^{n-1}x),$$

it follows from integration by parts and **PROBLEM 5.3** that

$$2^n \int_0^1 f(x) r_n(x)\,dx = -2 \int_0^1 f'(x)\psi(2^{n-1}x)\,dx$$

$$\to -2\mathcal{M}_{[0,1]}(\psi) \int_0^1 f'(x)\,dx$$

$$= \frac{f(0) - f(1)}{2}$$

as $n \to \infty$.

SOLUTION 13.10

Put $e(x) = e^{2\pi i x}$ for brevity. We have

$$I = \int_0^1 e(mx) r_n(x) \, dx$$

$$= \frac{1}{2\pi i m} \sum_{k=0}^{2^{n-1}-1} \left(e\left(\frac{m(2k+1)}{2^n}\right) - e\left(\frac{mk}{2^{n-1}}\right) \right.$$

$$\left. - e\left(\frac{m(2k+2)}{2^n}\right) + e\left(\frac{m(2k+1)}{2^n}\right) \right)$$

$$= -\frac{(1 - e(m/2^n))^2}{2\pi i m} \sum_{k=0}^{2^{n-1}-1} X^k,$$

where $X = e(m/2^{n-1})$. Note that $X = 1$ if and only if $\nu \geq n - 1$. If $\nu < n - 1$, then $X \neq 1$ and so

$$I = -\frac{(1 - e(m/2^n))^2}{2\pi i m} \cdot \frac{1 - X^{2^{n-1}}}{1 - X} = 0.$$

If $\nu \geq n - 1$, then we can write $m = \ell \cdot 2^{n-1}$ for some integer $\ell \geq 1$ and

$$I = -\frac{(1 - e(m/2^n))^2}{2\pi i m} 2^{n-1} = -\frac{(1 - (-1)^\ell)^2}{2\pi i \ell};$$

thus $I = 0$ if ℓ is even. If ℓ is odd, then $\nu = n - 1$ and we have

$$I = \frac{2i}{\pi \ell} = \frac{2^n i}{\pi m}. \qquad \square$$

❖ ✳ ❖

Chapter 14

Legendre Polynomials

[Summary of Basic Points]

1. A. M. Legendre (1752–1833) introduced an orthogonal system of polynomials with weight function 1 on the interval $[-1, 1]$ in 1784 in his papers on celestial mechanics. We have already encountered this polynomial in **PROBLEM 4.7** with

$$P_n(x) = \frac{1}{\pi} \int_0^\pi \left(x + i\sqrt{1 - x^2} \cos\theta \right)^n d\theta, \tag{14.1}$$

known as the Laplace-Mehler integral.

2. O. Rodrigues (1794–1851) gave the formula:

$$P_n(x) = \frac{1}{2^n n!} \frac{d^n}{dx^n} (x^2 - 1)^n.$$

3. The nth Legendre polynomial $P_n(x)$ is a polynomial of degree n with rational coefficients. The first seven polynomials are as follows:

$$P_0(x) = 1$$
$$P_1(x) = x$$
$$P_2(x) = \frac{1}{2}(3x^2 - 1)$$
$$P_3(x) = \frac{1}{2}(5x^3 - 3x)$$
$$P_4(x) = \frac{1}{8}(35x^4 - 30x^2 + 3)$$
$$P_5(x) = \frac{1}{8}(63x^5 - 70x^3 + 15x)$$
$$P_6(x) = \frac{1}{16}(231x^6 - 315x^4 + 105x^2 - 5).$$

4. It follows from the Rodrigues formula that

$$P_n(x) = \frac{1}{2^n n!} \sum_{k=0}^{[n/2]} (-1)^k \binom{n}{k} \frac{(2n-2k)!}{(n-2k)!} x^{n-2k}.$$

Obviously $P_n(x)$ is even or odd according to the parity of n.

5. Let $\delta_{m,n}$ be the Kronecker delta; namely, $\delta_{m,n} = 0$ for any $m \neq n$ and $\delta_{n,n} = 1$. We have

$$\int_{-1}^{1} P_m(x)P_n(x)\,dx = \frac{2}{2n+1}\,\delta_{m,n}.$$

6. [*Recursion Formula*]

$$(n+1)P_{n+1}(x) = (2n+1)xP_n(x) - nP_{n-1}(x).$$

7. [*Differential Equation*]

$$(1-x^2)\,P_n''(x) - 2xP_n'(x) + n(n+1)P_n(x) = 0.$$

PROBLEM 14.1

Show that

$$\int_{-1}^{1} (P_n(x))^2\,dx = \frac{2}{2n+1} \quad and \quad \int_{-1}^{1} (P_n'(x))^2\,dx = n(n+1).$$

PROBLEM 14.2

Show that

$$\frac{1}{\sqrt{1-2xy+x^2}} = \sum_{n=0}^{\infty} P_n(y)x^n$$

for any $|y| \leq 1$ and $|x| < 1$.

Compare with **PROBLEM 15.1**.

PROBLEM 14.3

Show that

$$(n-1)n(n+1)(n+2) \int_1^x \int_1^t P_n(s)\,ds\,dt = (1-x^2)^2 P_n''(x).$$

PROBLEM 14.4

Show that the minimum of the integral

$$\int_0^1 (x^n + a_1 x^{n-1} + \cdots + a_n)^2\,dx$$

as $a_1, a_2, \ldots, a_n$ range over all real numbers, is equal to $\dfrac{1}{2n+1}\dbinom{2n}{n}^{-2}$ and that the minimum is attained by the monic polynomial

$$\binom{2n}{n}^{-1} P_n(2x-1).$$

Compare with **PROBLEM 15.7**.

PROBLEM 14.5

Show that

$$|P_n(\cos\varphi)| < \frac{c}{\sqrt{n}\,\sin\varphi}$$

for any $0 < \varphi < \pi$ with $c = \sqrt{\pi/2}$.

This estimate is not sharp at all. It was Stieltjes (1890) who showed that

$$|P_n(\cos\varphi)| < \frac{c'}{\sqrt{n}\,\sin\varphi}$$

for any $0 < \varphi < \pi$ with some constant c'. Gronwall (1913) gave a proof with $c' = 2\sqrt{2/\pi}$. Fejér (1925) gave a worse inequality with $c' = 4\sqrt{2/\pi}$ but in an elementary way. Finally Bernstein (1931) obtained the inequality with $c' = \sqrt{2/\pi}$, which cannot be replaced by any smaller constant, because

$$P_{2n}(0) = \frac{(-1)^n}{2^{2n}}\binom{2n}{n} \sim \frac{(-1)^n}{\sqrt{\pi n}}$$

as $n \to \infty$. Bernstein's proof will be shown in the remark after **SOLUTION 14.5**.

PROBLEM 14.6

For any $0 < \varphi < \pi$ show that

$$P_n(\cos \varphi) = \frac{\sqrt{2}}{\pi} \int_0^\varphi \frac{\cos(n + 1/2)\theta}{\sqrt{\cos \theta - \cos \varphi}} \, d\theta.$$

This is due to Mehler (1872), known as the Dirichlet-Mehler integral.

PROBLEM 14.7

Show that

$$P_n^2(x) \geq P_{n+1}(x) P_{n-1}(x)$$

for any $|x| \leq 1$ with equality only at $x = \pm 1$.

This is due to P. Turán (1910–1976). Szegö (1948) gave four different proofs for Turán's inequality. Later Turán published his original proof in Turán (1950).

PROBLEM 14.8

Show that

$$(P_n'(x))^2 > P_{n+1}'(x) P_{n-1}'(x)$$

for any $|x| \leq 1$.

PROBLEM 14.9

Suppose that $\{p_n(x)\}_{n \geq 0}$ is a system of orthogonal polynomials on the interval $[a, b]$ with positive integrable weight $\rho(x)$; namely, $\deg p_n = n$ and

$$\int_a^b p_m(x) p_n(x) \rho(x) \, dx = 0$$

for all $m \neq n$. Show that each $p_n(x)$ possesses n simple roots in (a, b) and that there exists an exactly one root of $p_{n-1}(x)$ between any consecutive roots of $p_n(x)$. Note that $p_0(x)$ is a non-zero constant.

This is one of various properties satisfied by a general system of orthogonal polynomials. For details see Szegö (1934).

PROBLEM 14.10

Find a polynomial $q_n(x)$ of degree $n \geq 1$ with integer coefficients having the property that the coefficient of x^k in the Taylor series of

$$q_n(x) \log(1 + x)$$

at the origin vanishes for every $k \in (n, 2n]$.

For example,

$$(x + 2) \log(1 + x) = 2x + \frac{1}{6}x^3 + \cdots,$$

$$(x^2 + 6x + 6) \log(1 + x) = 6x + 3x^2 + \frac{1}{30}x^5 + \cdots,$$

$$(x^3 + 12x^2 + 30x + 20) \log(1 + x) = 20x + 20x^2 + \frac{11}{3}x^3 + \frac{1}{140}x^7 + \cdots.$$

The approximation formula

$$q_n(x) \log(1 + x) - p_m(x) = O(x^{n+m+1}) \tag{14.2}$$

with non-zero polynomials $p_m(x), q_n(x)$ satisfying $\deg p_m \leq m$ and $\deg q_n \leq n$ is called a Padé approximation to $\log(1 + x)$. The existence of such Padé approximants follows easily from the general theory of systems of homogeneous linear equations by comparing the number of equations with that of unknowns. The expression (14.2) means that the value of $\log(1 + x)$ may be closely approximated by the rational function $p_m(x)/q_n(x)$ when x is small. To verify this we need a detailed information about the error term.

✦ ❋ ✦

Solutions for Chapter 14

SOLUTION 14.1

Since the leading coefficient of $P_n(x)$ is $\dfrac{1}{2^n}\dbinom{2n}{n}$ and since $P_n(x)$ is orthogonal to any polynomial of degree less than n, we have

$$\int_{-1}^{1} (P_n(x))^2 \, dx = \frac{1}{2^n}\binom{2n}{n} \int_{-1}^{1} x^n P_n(x) \, dx.$$

It then follows from the integration by parts that the right-hand side is equal to

$$\frac{1}{2^n}\binom{2n}{n}\left(-\frac{1}{2}\right)^n \int_{-1}^{1} (x^2 - 1)^n \, dx.$$

Substituting $x = 2s - 1$ we see that this is

$$2\binom{2n}{n}\int_{0}^{1} s^n (1 - s)^n \, ds$$

and by the definition of the Beta function we finally obtain

$$2\binom{2n}{n} B(n+1, n+1) = \frac{2}{2n+1}.$$

Similarly integrating by parts we have

$$\int_{-1}^{1} (P_n'(x))^2 \, dx = P_n(x) P_n'(x) \Big|_{-1}^{1} - \int_{-1}^{1} P_n(x) P_n''(x) \, dx$$

$$= P_n'(1) - (-1)^n P_n'(-1),$$

because $P_n(1) = 1$ and $P_n(-1) = (-1)^n$. To determine the values of the derivative we use Leibniz' rule so that

$$P_n'(x) = \frac{1}{2^n n!} \frac{d^{n+1}}{dx^{n+1}} ((x+1)^n (x-1)^n)$$

$$= \frac{1}{2^n} \sum_{k=1}^{n} \binom{n+1}{k}\binom{n}{k} k (x+1)^{k-1} (x-1)^{n-k};$$

therefore

$$P_n'(1) = \frac{n(n+1)}{2} \quad \text{and} \quad P_n'(-1) = (-1)^{n-1}\frac{n(n+1)}{2},$$

which implies

$$\int_{-1}^{1} (P_n'(x))^2 \, dx = n(n+1). \qquad \Box$$

SOLUTION 14.2

We first notice that $|P_n(y)| \le 1$ for all $|y| \le 1$ by the Laplace-Mehler integral. Hence the radius of convergence of the series on the right-hand side is greater than or equal to 1 as a power series of complex variable x. Thus it suffices to show the given formula for any sufficiently small $x > 0$.

Let $G(x)$ be the generating function of the Legendre polynomials; that is,

$$G(x) = \sum_{n=0}^{\infty} P_n(y)x^n.$$

Applying Cauchy's integral formula to the Rodrigues formula we obtain

$$G(x) = \sum_{n=0}^{\infty} \frac{1}{2\pi i} \int_{C_y} \left(\frac{x(z^2-1)}{2(z-y)} \right)^n \frac{dz}{z-y}$$

$$= -\frac{1}{\pi i} \int_{C_y} \frac{dz}{xz^2 - 2z + 2y - x},$$

where C_y is the oriented unit circle centered at y, that is, $C_y = \{z \in \mathbf{C} \mid |z-y| = 1\}$. Let $\phi(z-y)$ be the denominator of the integrand of the last integral; namely,

$$\phi(w) = xw^2 - 2(1-xy)w - x(1-y^2),$$

which is a quadratic polynomial with real coefficients. It is easily verified that $\phi(w)$ has two real roots $\alpha_\pm$ in the intervals $(1, \infty)$ and $(-1, 0)$ respectively for all sufficiently small $x > 0$. Thus, if we put

$$z_\pm = y + \alpha_\pm = \frac{1 \pm \sqrt{1 - 2xy + x^2}}{x}$$

respectively, the point z_+ lies outside of C_y and z_- lies inside of C_y. Since

$$\phi(z-y) = x(z-z_+)(z-z_-),$$

it follows from the residue theorem that

$$G(x) = -\frac{2}{x(z_- - z_+)} = \frac{1}{\sqrt{1 - 2xy + x^2}}. \qquad \square$$

SOLUTION 14.3

Using the differential equation

$$(1 - x^2)\, P_n'' = 2x P_n' - n(n + 1) P_n$$

we have

$$\begin{aligned}
\left((1 - x^2)^2\, P_n''\right)' &= -2x\left(2x P_n' - n(n + 1) P_n\right) \\
&\quad + (1 - x^2)(2P_n' + 2x P_n'' - n(n + 1) P_n') \\
&= -(n - 1)(n + 2)(1 - x^2) P_n'.
\end{aligned}$$

Hence the derivative of the left-hand side is equal to $(n - 1)n(n + 1)(n + 2) P_n$. Therefore

$$(n - 1)n(n + 1)(n + 2) \int_1^x \!\!\int_1^t P_n(s)\, ds\, dt = (1 - x^2)^2 P_n''(x) + \alpha x + \beta$$

for some constants α and β. We then have $\alpha = \beta = 0$ because the left-hand side has the factor $(x - 1)^2$. $\qquad \square$

SOLUTION 14.4

It is shown in **SOLUTION 8.7** that the minimum of

$$\int_0^1 \left(x^n - c_0 - c_1 x^1 - c_2 x^2 - \cdots - c_{n-1} x^{n-1}\right)^2 dx$$

is

$$\frac{1}{2n + 1} \prod_{k=0}^{n-1} \left(\frac{n - k}{n + k + 1}\right)^2 = \frac{1}{2n + 1}\binom{2n}{n}^{-2}.$$

On the other hand, it follows from **PROBLEM 14.1** that

$$\int_0^1 P_n^2(2x - 1)\, dx = \frac{1}{2n + 1}$$

and $\binom{2n}{n}^{-1} P_n(2x - 1)$ is a monic polynomial of degree n. $\qquad \square$

It follows from the Laplace-Mehler integral that

$$|P_n(\cos\varphi)| \le \frac{2}{\pi} \int_0^{\pi/2} (1 - \sin^2\varphi \sin^2\theta)^{n/2} \, d\theta.$$

Using Jordan's inequality $\sin\theta \ge 2\theta/\pi$, the right-hand side is estimated from above by

$$\frac{2}{\pi} \int_0^{\pi/2} \left(1 - \frac{4\theta^2}{\pi^2} \sin^2\varphi\right)^{n/2} \, d\theta.$$

Substituting $t = 2\theta/\pi$ we see that it is equal to

$$\int_0^1 (1 - t^2 \sin^2\varphi)^{n/2} \, dt.$$

Moreover using the inequality $1 - s \le e^{-s}$, it is less than or equal to

$$\int_0^1 \exp\left(-\frac{n}{2} t^2 \sin^2\varphi\right) \, dt.$$

Finally, substituting $\tau = \sqrt{n/2}\,(\sin\varphi)t$ we obtain

$$|P_n(\cos\varphi)| < \int_0^\infty \exp\left(-\frac{n}{2} t^2 \sin^2\varphi\right) \, dt$$

$$= \frac{\sqrt{2/n}}{\sin\varphi} \int_0^\infty e^{-\tau^2} \, d\tau.$$

Since the integral on the right-hand side is equal to $\sqrt{\pi}/2$, we obtain

$$|P_n(\cos\varphi)| < \frac{\sqrt{\pi/2}}{\sqrt{n}\,\sin\varphi}. \qquad \square$$

REMARK. Bernstein's tricky proof is as follows. Put

$$f(\theta) = \sqrt{\sin\theta}\, P_n(\cos\theta)$$

for $0 < \theta < \pi$, which satisfies the differential equation $f''(\theta) + A(\theta)f(\theta) = 0$ where

$$A(\theta) = \frac{1}{4\sin^2\theta} + \left(n + \frac{1}{2}\right)^2.$$

We then put

$$F(\theta) = f^2(\theta) + \frac{(f'(\theta))^2}{A(\theta)};$$

hence

$$F'(\theta) = -\left(\frac{f'(\theta)}{A(\theta)}\right)^2 A'(\theta).$$

Since

$$2A'(\theta) = -\frac{\cos\theta}{2\sin^3\theta},$$

the function $F(\theta)$ is monotone increasing on $(0, \pi/2]$ and monotone decreasing on $[\pi/2, \pi)$. Moreover, since $f(0+) = 0$ and $|F(\theta)| = |F(\pi - \theta)|$, it follows that

$$f^2(\theta) \le F(\theta) \le F\left(\frac{\pi}{2}\right) = P_n^2(0) + \frac{(P_n'(0))^2}{n^2 + n + 1/2}.$$

If $n = 2m$, then $P_n'(0) = 0$ and $|f(\theta)| \le |P_n(0)|$. If $n = 2m + 1$, then $P_n(0) = 0$ and

$$|f(\theta)| < \frac{|P_n'(0)|}{\sqrt{n^2 + n + 1/2}} = \frac{n|P_{n-1}(0)|}{\sqrt{n^2 + n + 1/2}}.$$

Now it follows from the Rodrigues formula or the Laplace-Mehler integral that

$$|P_{2m}(0)| = \frac{1}{2^{2m}}\binom{2m}{m}, \quad \text{say } c_m.$$

The desired inequality follows from the following properties on c_m:

$$\sqrt{2m}\,c_m \quad \text{and} \quad \frac{(2m+1)^{3/2}}{\sqrt{4m^2 + 6m + 5/2}}c_m$$

are monotone increasing sequences which converge to $\sqrt{2/\pi}$ as $m \to \infty$. This limit may be obtained by Stirling's approximation mentioned after SOLUTION 16.7.

SOLUTION 14.6

For an arbitrarily fixed $x \in (-1, 1)$ the point

$$z = x + i\sqrt{1 - x^2}\cos\theta$$

moves steadily on the segment AB downward, as θ varies from 0 to π, where

$$A = x + i\sqrt{1 - x^2} \quad \text{and} \quad B = x - i\sqrt{1 - x^2}.$$

Applying this transformation to the Laplace-Mehler integral, we get

$$
\begin{aligned}
P_n(x) &= \frac{1}{\pi} \int_{AB} z^n \frac{d\theta}{dz} \, dz \\
&= \frac{i}{\pi} \int_{AB} \frac{z^n}{\sqrt{1 - x^2} \, \sin\theta} \, dz \\
&= \frac{i}{\pi} \int_{AB} \frac{z^n}{\sqrt{1 - 2xz + z^2}} \, dz,
\end{aligned}
$$

where the square root is determined in such a way that the real part is positive; namely, the function $\sqrt{w}$ maps the wedge on the full angle by exclusion of the negative real axis onto the right half plane. Hence, the function $\sqrt{1 - 2xz + z^2}$ is analytic on the region Ω by exclusion of the two vertical half lines defined by

$$
\ell^{\pm} = \left\{ z \in \mathbb{C} \;\middle|\; z = x \pm it, \; t \geq \sqrt{1 - x^2} \right\}.
$$

Now we can change the segment AB by the circular arc through the point $z = 1$ joining A and B. If $x = \cos\varphi$ for $0 < \varphi < \pi$, this arc is expressed as $z = e^{i\theta}$ for $\varphi \geq \theta \geq -\varphi$; hence

$$
P_n(\cos\varphi) = \frac{1}{\pi} \int_{-\varphi}^{\varphi} \frac{e^{i(n+1)\theta}}{\sqrt{1 - 2e^{i\theta} \cos\varphi + e^{2i\theta}}} \, d\theta.
$$

Since

$$
1 - 2e^{i\theta} \cos\varphi + e^{2i\theta} = 2(\cos\theta - \cos\varphi)e^{i\theta},
$$

we obtain

$$
P_n(\cos\varphi) = \frac{\sqrt{2}}{\pi} \int_0^{\varphi} \frac{\cos(n + 1/2)\theta}{\sqrt{\cos\theta - \cos\varphi}} \, d\theta. \qquad \square
$$

SOLUTION 14.7

The proof goes on the lines of the fourth proof of Szegö (1934). Since

$$
\Delta_n(x) = P_n^2(x) - P_{n+1}(x)P_{n-1}(x)
$$

is even, it suffices to consider the problem on the interval $[0, 1]$. Using the recursion formula satisfied by the Legendre polynomials, we get

$$
\Delta_n(x) = A^2(x) + B(x)P_{n-1}^2(x)
$$

where

$$A(x) = P_n(x) - \frac{2n+1}{2n+2} x P_{n-1}(x),$$

$$B(x) = \frac{n}{n+1} - \left(\frac{2n+1}{2n+2} x\right)^2.$$

Hence $\Delta_n(x)$ is clearly positive for

$$0 \le x < \frac{\sqrt{n(n+1)}}{n+1/2}.$$

Moreover, if $P_{n+1}(\xi) = 0$, then $P_n(\xi) \ne 0$; otherwise, we would have $P_0(\xi) = 0$ by the recursion formula, a contradiction; so, $\Delta_n(\xi) > 0$. We thus consider hereafter any point x in the interval

$$\left[\frac{\sqrt{n(n+1)}}{n+1/2}, 1\right)$$

satisfying $P_{n+1}(x) \ne 0$. Note that $\Delta_n(1) = 0$.

We next introduce the polynomial in y of degree $n+1$:

$$Q_{n+1}(y) = \sum_{k=0}^{n+1} \binom{n+1}{k} P_k(x) y^k.$$

It follows from the Laplace-Mehler integral that

$$Q_{n+1}(y) = \frac{1}{\pi} \int_0^{\pi} \left(1 + xy + iy\sqrt{1-x^2} \cos\theta\right)^{n+1} d\theta.$$

Solving the equation $\sigma^2 (1+xy)^2 + (\sigma y)^2 (1-x^2) = 1$ in σ, we get

$$\sigma = \frac{1}{\sqrt{1+2xy+y^2}};$$

therefore

$$Q_{n+1}(y) = \frac{1}{\sigma^{n+1}} P_{n+1}(\phi(y))$$

where

$$\phi(y) = \sigma(1+xy) = \frac{1+xy}{\sqrt{1+2xy+y^2}}.$$

The function $\phi(y)$ is strictly monotone increasing on $(-\infty, 0]$ and strictly monotone decreasing on $[0, \infty)$ satisfying $\phi(-\infty) = -x$, $\phi(0) = 1$ and $\phi(\infty) = x$. Since $P_{n+1}(x) \ne 0$, it follows that ϕ gives a one-to-one correspondence between the negative zeros of Q_{n+1} smaller than $\phi^{-1}(x) < 0$ and that of P_{n+1} lying on the interval

$(-x, x)$. Similarly ϕ gives a two-to-one correspondence between the zeros of Q_{n+1} greater than $\phi^{-1}(x)$ and that of P_{n+1} lying on the interval $(x, 1)$. Since the zeros of P_{n+1} locate symmetrically in the interval $(-1, 1)$ with respect to the origin, we see that $Q_{n+1}(y)$ has exactly $n + 1$ simple real roots, say $y_1 < y_2 < \cdots < y_{n+1}$. Therefore, if we put

$$s_1 = \sum_{k=1}^{n+1} y_k = -\binom{n+1}{1}\frac{P_n(x)}{P_{n+1}(x)}$$

and

$$s_2 = \sum_{1 \le j < k \le n+1} y_j y_k = \binom{n+1}{2}\frac{P_{n-1}(x)}{P_{n+1}(x)},$$

then

$$s_1^2 - 2s_2 = \sum_{k=1}^{n+1} y_k^2 = \frac{1}{n}\sum_{1 \le j < k \le n+1}\left(y_j^2 + y_k^2\right) > \frac{2}{n}s_2,$$

which implies $\Delta_n(x) > 0$. □

SOLUTION 14.8

Let $\phi(x, y) = G(x)$ be the generating function for the Legendre polynomials as defined in **SOLUTION 14.2**; that is,

$$\phi(x, y) = \frac{1}{\sqrt{1 - 2xy + x^2}} = \sum_{n=0}^{\infty} P_n(y)x^n$$

for $|y| \le 1$ and $|x| < 1$. For brevity we put $A = \phi(xz, y)$ and $B = \phi(x/z, y)$ where z is a complex variable on the unit circle $|z| = 1$. Using

$$x\frac{\partial\phi}{\partial x} = x(y - x)\phi^3 = \sum_{n=1}^{\infty} nP_n(y)x^n$$

it follows from the residue theorem that

$$\sum_{n=1}^{\infty} n(n + 1)P_n^2(y)x^{2n} = \frac{1}{2\pi i}\int_C \frac{\Phi_1(z)}{z}\,dz,$$

where

$$\Phi_1(z) = xz(y - xz)\left(1 + \frac{x}{z}\left(y - \frac{x}{z}\right)B^2\right)A^3B$$

$$= x(y - xz)(z - xy)A^3B^3$$

and C is the unit circle $|z| = 1$. Similarly we have

$$\sum_{n=1}^{\infty} n(n+1) P_{n+1}(y) P_{n-1}(y) x^{2n} = \frac{1}{2\pi i} \int_C \frac{\Phi_2(z)}{z} dz,$$

where

$$\Phi_2(z) = \frac{x}{z}(y - xz)\left(1 + \frac{x}{z}\left(y - \frac{x}{z}\right)B^2\right)A^3 B$$

$$= \frac{x}{z^2}(y - xz)(z - xy)A^3 B^3.$$

Therefore

$$\sum_{n=1}^{\infty} n(n+1)\Delta_n(y) x^{2n} = \frac{1}{2\pi i} \int_C \frac{\Phi_3(z)}{z} dz,$$

where

$$\Phi_3(z) = \Phi_1(z) - \Phi_2(z) = x(y - xz)(z - xy)\left(1 - \frac{1}{z^2}\right)A^3 B^3.$$

We now write $D(z) = x(y - xz)(z - xy)$. The transformation $w = 1/z$ maps the unit circle $|z| = 1$ onto $|w| = 1$ in the opposite direction and the integral

$$\int_C \frac{dz}{z}$$

is invariant under the transformation, as well as AB. Since $D(z)(1 - z^{-2})$ is transformed into

$$D\left(\frac{1}{z}\right)(1 - z^2) = -z^2 D\left(\frac{1}{z}\right)\left(1 - \frac{1}{z^2}\right),$$

we have

$$\int_C \frac{\Phi_3(z)}{z} dz = - \int_C z^2 D\left(\frac{1}{z}\right)\left(1 - \frac{1}{z^2}\right) A^3 B^3 \frac{dz}{z}.$$

Since

$$D(z) - z^2 D\left(\frac{1}{z}\right) = x^2(1 - y^2)(1 - z^2),$$

we obtain

$$\int_C \frac{\Phi_3(z)}{z} dz = -\frac{1}{2} x^2 (1 - y^2) \int_C \left(z - \frac{1}{z}\right)^2 A^3 B^3 \frac{dz}{z}.$$

On the other hand, it is easily seen that

$$\frac{\partial \phi}{\partial y} = x\phi^3 = \sum_{n=1}^{\infty} P_n'(y)x^n,$$

which implies

$$x^2 A^3 B^3 = \sum_{n,m \geq 1} P_n'(y)P_m'(y)x^{n+m}z^{n-m}.$$

We thus have

$$\sum_{n=1}^{\infty} n(n+1)\Delta_n(y)x^{2n} = (1-y^2)\sum_{n=1}^{\infty} E_n(y)x^{2n},$$

where

$$E_n(y) = (P_n'(y))^2 - P_{n+1}'(y)P_{n-1}'(y);$$

that is,

$$n(n+1)\Delta_n(y) = (1-y^2)E_n(y)$$

for any integer $n \geq 1$. Therefore $E_n(y) > 0$ for all $|y| \leq 1$ by virtue of Turán's inequality (**PROBLEM 14.7**), where the sign of equality is excluded because

$$E_n(\pm 1) = \frac{n(n+1)}{2}. \qquad \square$$

SOLUTION 14.9

Suppose, on the contrary, that the number of distinct real roots in the interval (a, b) of $p_n(x)$ is less than n. Then the number of roots of odd order is clearly less than n, which implies that there exists a non-zero polynomial $q(x)$ of degree less than n satisfying $p_n(x)q(x) \geq 0$ on $[a, b]$, contrary to the orthogonality:

$$\int_a^b p_n(x)q(x)\rho(x)\,dx = 0.$$

Let $\alpha < \beta$ be any consecutive real roots of $p_n(x)$ and ℓ be the number of distinct roots of $p_{n-1}(x)$ lying in the interval $[\alpha, \beta]$. If $\ell = 0$, then $p_{n-1}(x)$ has the constant sign on $[\alpha, \beta]$; so, it is geometrically obvious that the curves $cp_n(x)$ and $p_{n-1}(x)$ are tangent together at some point in $[\alpha, \beta]$ with some constant $c \neq 0$. Similarly if $\ell \geq 2$, then we can choose a suitable constant c' so that the curves $p_n(x)$ and $c'p_{n-1}(x)$ are tangent together at some point in $[\gamma, \delta] \subset [\alpha, \beta]$ where $\gamma < \delta$ are any two consecutive roots of $p_{n-1}(x)$. This is possible even if $\alpha = \gamma$, because we can take $c' = p_n'(\alpha)/p_{n-1}'(\alpha)$ in that case. Anyway we can choose some constant c such that

$$p_n(x) - cp_{n-1}(x) = (x - \xi)^2 q(x)$$

for some ξ and some polynomial $q(x)$ of degree $n - 2$. Therefore

$$\int_a^b \big(p_n(x) - cp_{n-1}(x)\big)\, q(x)\rho(x)\,dx = 0$$

by orthogonality; however this is a contradiction because the integral on the left-hand side is equal to

$$\int_a^b (x - \xi)^2 q^2(x)\rho(x)\,dx > 0.$$

Thus we have $\ell = 1$, as required. □

SOLUTION 14.10

We put

$$q_n(x) = \sum_{k=0}^n a_k x^k \quad \text{and} \quad q_n(x)\log(1 + x) = \sum_{m=0}^\infty c_m x^m.$$

Then we have

$$c_m = \sum_{k=0}^n (-1)^{m-k-1} \frac{a_k}{m - k} = 0$$

for $n < m \le 2n$. Since

$$0 = \sum_{k=0}^n (-1)^k \frac{a_k}{m - k} = \int_0^1 x^{m-1} \left(\sum_{k=0}^n (-1)^k a_k x^{-k}\right) dx$$

$$= \int_0^1 x^{m-n-1} \cdot x^n q_n\left(-\frac{1}{x}\right) dx,$$

it follows that $Q(x) = x^n q_n(-1/x)$ is a polynomial with $\deg Q \le n$ and is orthogonal to x^j for $0 \le j < n$. Such a polynomial of degree n with integer coefficients is given by

$$Q(x) = \frac{1}{n!}(x^n(1 - x)^n)^{(n)} = (-1)^n P_n(2x - 1).$$

Therefore, using $P_n(-x) = (-1)^n P_n(x)$ and adopting $(-1)^n q_n(x)$ instead of $q_n(x)$ we have a solution:

$$x^n P_n\left(\frac{2}{x} + 1\right).$$

□

Chapter 15

Chebyshev Polynomials

[Summary of Basic Points]

1. As is mentioned in the remark after SOLUTION 3.8, the polynomial $T_n(x)$ defined by the relation

$$T_n(\cos \theta) = \cos n\theta$$

is called the nth Chebyshev polynomial of the first kind, first appeared in Chebyshev (1854). We encounter these polynomials in various extremal problems as well as in best approximation problems. The Chebyshev polynomials form an orthogonal system over the interval $[-1, 1]$ with respect to the measure

$$d\mu = \frac{dx}{\sqrt{1 - x^2}}.$$

Obviously $T_n(x)$ is even or odd according to the parity of n.

2. The nth Chebyshev polynomial $T_n(x)$ is a polynomial with integer coefficients of degree n. The first eight polynomials are as follows:

$$T_0(x) = 1$$
$$T_1(x) = x$$
$$T_2(x) = 2x^2 - 1$$
$$T_3(x) = 4x^3 - 3x$$
$$T_4(x) = 8x^4 - 8x^2 + 1$$
$$T_5(x) = 16x^5 - 20x^3 + 5x$$
$$T_6(x) = 32x^6 - 48x^4 + 18x^2 - 1$$
$$T_7(x) = 64x^7 - 112x^5 + 56x^3 - 7x.$$

239

3. Let $\delta_{m,n}$ be the Kronecker delta. We have

$$\int_{-1}^{1} T_m(x) T_n(x) \, d\mu = \frac{\pi}{2} \delta_{m,n}$$

except for $(n, m) \neq (0, 0)$. If $n = m = 0$, then the corresponding integral is π.

4. $T_m(T_n(x)) = T_{mn}(x)$ and $2T_m(x) T_n(x) = T_{m+n}(x) + T_{|m-n|}(x)$ for all m and n.

5. [*Recursion Formula*]

$$T_{n+1}(x) = 2x T_n(x) - T_{n-1}(x).$$

6. [*Differential Equation*]

$$(1 - x^2) T_n''(x) - x T_n'(x) + n^2 T_n(x) = 0.$$

7. The nth Chebyshev polynomial of the second kind $U_n(x)$ is defined by

$$U_n(\cos \theta) = \frac{\sin(n + 1)\theta}{\sin \theta}.$$

It satisfies the same recursion formula as T_n, as an independent solution of the recursion formula. Obviously

$$U_n(x) = \frac{1}{n + 1} T_{n+1}'(x)$$

holds and it forms a system of orthogonal polynomials over $[-1, 1]$ with the weight function $\sqrt{1 - x^2}$.

PROBLEM 15.1

Show that

$$\frac{1 - xy}{1 - 2xy + x^2} = \sum_{n=0}^{\infty} T_n(y) x^n$$

for any $|x| < 1$ and $|y| \leq 1$.

Compare with **PROBLEM 14.2**.

PROBLEM 15.2

Show that

$$T_n(x) = \frac{1}{2}\left(\left(x + \sqrt{x^2 - 1}\right)^n + \left(x - \sqrt{x^2 - 1}\right)^n\right)$$

for any $|x| > 1$.

PROBLEM 15.3

Show that

$$T_n(x) = \sum_{k=0}^{[n/2]} (-1)^k \frac{n}{n-k} \binom{n-k}{k} 2^{n-2k-1} x^{n-2k}.$$

PROBLEM 15.4

Show the following *Rodrigues formula*:

$$T_n(x) = \frac{(-1)^n}{1 \cdot 3 \cdots (2n-1)} \sqrt{1 - x^2} \frac{d^n}{dx^n} (1 - x^2)^{n-1/2}.$$

PROBLEM 15.5

Show that

$$T_n^{(k)}(1) = \frac{n^2(n^2 - 1^2) \cdots (n^2 - (k-1)^2)}{1 \cdot 3 \cdot 5 \cdots (2k - 1)}$$

for every $1 \le k \le n$.

PROBLEM 15.6

Show that

$$\frac{\pi}{2} \sqrt{1 - x^2} = 1 - 2 \sum_{n=1}^{\infty} \frac{T_{2n}(x)}{4n^2 - 1}.$$

Note that to expand a given function $f(x)$ into the Chebyshev series on the interval $[-1, 1]$ is nothing but to expand $f(\cos \theta)$ into the cosine Fourier series on the interval $[-\pi, \pi]$.

PROBLEM 15.7

Show that the minimum of the integral

$$\int_0^1 |x^n + a_1 x^{n-1} + \cdots + a_n|\, dx$$

as $a_1, a_2, \ldots, a_n$ range over all real numbers, is equal to 4^{-n} and that the minimum is attained by the polynomial $U_n(2x-1)/4^n$.

Compare with **PROBLEM 14.4**.

PROBLEM 15.8

Let $Q(x)$ be any polynomial of degree $\leq n$ with real coefficients and let M be the maximum of $|Q(x)|$ on the interval $[-1, 1]$. Show then that

$$|Q(x)| \leq M\,|T_n(x)|$$

for any $|x| > 1$.

This is due to Chebyshev (1881).

PROBLEM 15.9

Let $Q(x)$ be any monic polynomial of degree n with real coefficients. Show then that

$$\max_{|x| \leq 1} |Q(x)| \geq \frac{1}{2^{n-1}}$$

and the equality occurs if and only if $Q(x) = 2^{1-n} T_n(x)$.

PROBLEM 15.10

We use the same notations as in **PROBLEM 1.10** for the interval $E = [-1, 1]$. Show that

$$\frac{1}{2^{n-1}} \leq \frac{M_{n+1}}{M_n} \leq \frac{n+1}{2^{n-1}}.$$

Deduce from this that the transfinite diameter of $[-1, 1]$ is equal to $1/2$.

Solutions for Chapter 15

SOLUTION 15.1

The proof is much easier than the Legendre case (**PROBLEM 14.2**). For brevity put $y = \cos\theta$. Let $G(y)$ be the generating function of the Chebyshev polynomials; that is,

$$G(y) = \sum_{n=0}^{\infty} T_n(\cos\theta)x^n = \sum_{n=0}^{\infty} (\cos n\theta)x^n.$$

Since the right-hand side is the real part of the geometric series $\sum_{n=0}^{\infty} (e^{i\theta}x)^n$, it follows that

$$G(y) = \mathrm{Re}\, \frac{1}{1 - e^{i\theta}x} = \frac{1 - x\cos\theta}{1 + x^2 - 2x\cos\theta}. \qquad \Box$$

SOLUTION 15.2

Put

$$f_n(x) = \frac{1}{2}\left(\left(x + \sqrt{x^2-1}\right)^n + \left(x - \sqrt{x^2-1}\right)^n\right)$$

and

$$g_n(x) = \frac{\sqrt{x^2-1}}{2}\left(\left(x + \sqrt{x^2-1}\right)^n - \left(x - \sqrt{x^2-1}\right)^n\right).$$

It is easily seen that

$$\begin{pmatrix} f_{n+1}(x) \\ g_{n+1}(x) \end{pmatrix} = \begin{pmatrix} x & 1 \\ x^2 - 1 & x \end{pmatrix}\begin{pmatrix} f_n(x) \\ g_n(x) \end{pmatrix},$$

from which we can get the recursion formula

$$f_{n+1}(x) = 2x f_n(x) - f_{n-1}(x).$$

Since $f_0(x) = 1$ and $f_1(x) = x$, we have $f_n(x) = T_n(x)$ for all n. $\qquad \Box$

SOLUTION 15.3

Putting $x = \cos\theta$ and expanding the right-hand side of

$$\cos n\theta = \text{Re}\,(\cos\theta + i\sin\theta)^n,$$

we have

$$T_n(x) = \text{Re}\sum_{k=0}^{n} \binom{n}{k}\cos^{n-k}\theta\,(i\sin\theta)^k$$

$$= \sum_{\ell=0}^{[n/2]} (-1)^\ell\binom{n}{2\ell}\cos^{n-2\ell}\theta\,(1-\cos^2\theta)^\ell.$$

Thus

$$T_n(x) = x^n - \binom{n}{2}x^{n-2}(1-x^2) + \binom{n}{4}x^{n-4}(1-x^2)^2 - \cdots,$$

which implies that

$$a_{n-2k} = (-1)^k\sum_{\ell=k}^{[n/2]} \binom{n}{2\ell}\binom{\ell}{k}$$

where a_{n-2k} is the coefficient of x^{n-2k} in the nth Chebyshev polynomial $T_n(x)$. Since a_{n-2k} can be written as

$$a_{n-2k} = \frac{(-1)^k}{k!}\,Q^{(k)}(1)$$

where

$$Q(x^2) = \sum_{\ell=0}^{[n/2]} \binom{n}{2\ell}x^{2\ell} = \frac{(1+x)^n + (1-x)^n}{2},$$

it follows from Cauchy's integral formula that

$$(-1)^k a_{n-2k} = \frac{1}{2\pi i}\int_{C_0} \frac{Q(1+z)}{z^{k+1}}\,dz$$

$$= \frac{1}{4\pi i}\int_{C_0} \frac{(1+\sqrt{1+z})^n + (1-\sqrt{1+z})^n}{z^{k+1}}\,dz,$$

where C_0 is a sufficiently small circle centered at the origin and the square root $\sqrt{1+z}$ is determined as the real part is positive. Since $1 - \sqrt{1+z} = O(|z|)$ in a neighborhood of the origin, we have

$$a_{n-2k} = \frac{(-1)^k}{4\pi i} \int\limits_{C_0} \frac{(1 + \sqrt{1+z})^n}{z^{k+1}} \, dz.$$

Therefore, substituting $w = \sqrt{1+z}$ we see that $(-1)^k a_{n-2k}$ is equal to

$$\frac{1}{2\pi i} \int\limits_{C_0} \frac{w(1+w)^n}{(w^2-1)^{k+1}} \, dw = \frac{1}{2\pi i} \int\limits_{C_1} \frac{w(1+w)^{n-k-1}}{(w-1)^{k+1}} \, dw$$

$$= \frac{1}{2\pi i} \int\limits_{C_2} \frac{(1+\zeta)(2+\zeta)^{n-k-1}}{\zeta^{k+1}} \, d\zeta,$$

where C_1 and C_2 are small circles centered at $w = 1$ and $\zeta = 0$ respectively. The last integral is clearly equal to

$$\binom{n-k-1}{k} 2^{n-2k-1} + \binom{n-k-1}{k-1} 2^{n-2k} = \frac{n}{n-k} \binom{n-k}{k} 2^{n-2k-1}. \qquad \square$$

SOLUTION 15.4

Put

$$Q(x) = \sqrt{1-x^2} \, \frac{d^n}{dx^n} (1-x^2)^{n-1/2}.$$

It follows from Leibniz' rule that

$$Q(x) = \sqrt{1-x^2} \sum_{k=0}^{n} \binom{n}{k} ((1+x)^{n-1/2})^{(n-k)} ((1-x)^{n-1/2})^{(k)}.$$

The right-hand side can be written as

$$\sum_{k=0}^{n} (-1)^k \binom{n}{k} \frac{(2n-1)\cdots(2k+1)}{2^{n-k}} (1+x)^k$$

$$\times \frac{(2n-1)\cdots(2n-2k+1)}{2^k} (1-x)^{n-k},$$

which implies that $Q(x)$ is a polynomial of degree $\leq n$. Indeed the coefficient of x^n in $Q(x)$ is equal to

$$\frac{(-1)^n}{2^n} \sum_{k=0}^{n} \binom{n}{k} (2n-1)\cdots(2k+1) \times (2n-1)\cdots(2n-2k+1) \neq 0.$$

The reader may evaluate this sum as

$$\frac{(-1)^n}{2} \cdot \frac{(2n)!}{n!},$$

using the expansion of

$$\frac{(1+x)^{2n} + (1-x)^{2n}}{2}.$$

On the other hand, integrating by parts for $k+1$ times, we get

$$\int_{-1}^{1} x^k Q(x)\,d\mu = (-1)^{k+1} \int_{-1}^{1} (x^k)^{(k+1)} \frac{d^{n-k-1}}{dx^{n-k-1}} (1-x^2)^{n-1/2}\,dx = 0$$

for all integers $0 \le k < n$. Since the degree of the polynomial

$$R(x) = T_n(x) - \frac{(-2)^n n!}{(2n)!} Q(x)$$

is less than n and satisfies

$$\int_{-1}^{1} x^k R(x)\,d\mu = 0$$

for all $0 \le k < n$, we have

$$\int_{-1}^{1} R^2(x)\,d\mu = 0;$$

hence $R(x)$ vanishes everywhere. In other words,

$$T_n(x) = \frac{(-2)^n n!}{(2n)!} Q(x) = \frac{(-1)^n}{1 \cdot 3 \cdots (2n-1)} Q(x). \qquad \square$$

SOLUTION 15.5

It follows from k times differentiation of the recursion formula that

$$\sum_{\ell=0}^{2} \binom{k}{\ell} (1-x^2)^{(\ell)} T_n^{(k+2-\ell)}(x) - \sum_{j=0}^{1} \binom{k}{\ell} x^{(\ell)} T_n^{(k+1-\ell)}(x) + n^2 T_n^{(k)}(x) = 0,$$

which implies that

$$T_n^{(k+1)}(1) = \frac{n^2 - k^2}{2k+1} T_n^{(k)}(1).$$

One can deduce the desired formula from this in view of $T_n(1) = 1$. $\qquad \square$

SOLUTION 15.6

Put

$$\phi(x) = \frac{\pi}{2}\sqrt{1-x^2} - 1 + 2 \sum_{n=1}^{\infty} \frac{T_{2n}(x)}{4n^2 - 1}.$$

Since $|T_n(x)| \leq 1$ for $-1 \leq x \leq 1$, the Chebyshev series on the right-hand side converges uniformly. Thus we can employ the termwise integration to obtain

$$\int_{-1}^{1} T_{2m}(x)\phi(x)\,d\mu = \frac{\pi}{2} \int_{-1}^{1} T_{2m}(x)\,dx - \int_{-1}^{1} T_{2m}(x)\,d\mu$$

$$+ 2\sum_{n=1}^{\infty} \frac{1}{4n^2 - 1} \int_{-1}^{1} T_{2m}(x)T_{2n}(x)\,d\mu$$

for all integers $m \geq 0$. By the substitution $x = \cos\theta$ the first integral on the right-hand side is transformed to

$$\frac{\pi}{2} \int_{0}^{\pi} \cos 2m\theta \sin\theta\,d\theta,$$

which is equal to $-\pi/(4m^2 - 1)$. By the orthogonality the second integral vanishes except for $m = 0$, in which it is equal to π. Similarly the third integral vanishes except for $n = m$ and in such a case it is equal to

$$\frac{\pi}{4m^2 - 1}$$

if $n = m \geq 1$. Hence ϕ is orthogonal to every even Chebyshev polynomial with respect to $d\mu$. But it is also orthogonal to every odd Chebyshev polynomial because ϕ is even. Thus $\phi(x)$ vanishes everywhere by the remark after **Problem 8.5**. $\quad\square$

SOLUTION 15.7

For a given polynomial

$$A(x) = x^n + a_1 x^{n-1} + \cdots + a_n$$

we define

$$B(x) = A\left(\frac{x+2}{4}\right) \tag{15.1}$$

so that

$$B(x) = \frac{x^n}{4^n} + a_1' x^{n-1} + \cdots + a_n'$$

with some real numbers $a_1', \ldots, a_n'$. Putting further

$$Q(x) = \int_{0}^{x} B(s)\,ds \tag{15.2}$$

we obtain

$$Q(x) = \frac{x^{n+1}}{4^n(n+1)} + a_1'' x^n + \cdots + a_n'' x$$

with some real numbers $a_1'', \ldots, a_n''$. Applying the same method as in the proof of PROBLEM 3.8 to $Q(x)$ with the same notations, we get

$$\frac{4(n+1)}{4^n(n+1)} = \left| \sum_{k=0}^{n+1} Q(\alpha_k) \right| \le \sum_{k=0}^{n} |Q(\alpha_k) - Q(\alpha_{k+1})| \qquad (15.3)$$

where

$$\alpha_k = 2\cos\frac{k\pi}{n+1}.$$

Therefore, by using (15.1) and (15.2) in (15.3), we have

$$\frac{1}{4^{n-1}} \le \sum_{k=0}^{n} \int_{\alpha_{k+1}}^{\alpha_k} |B(s)| \, ds = \int_{-2}^{2} |B(s)| \, ds = 4\int_0^1 |A(x)| \, dx.$$

The equality holds for

$$A_n(x) = \frac{1}{4^n(n+1)} T_{n+1}'(2x - 1)$$

where $T_m(x)$ is the mth Chebyshev polynomial of the first kind. Indeed we have

$$\int_0^1 |A_n(x)| \, dx = \frac{2}{4^{n+1}(n+1)} \int_{-1}^{1} |T_{n+1}'(s)| \, ds$$

$$= \frac{2}{4^{n+1}(n+1)} \int_0^{\pi} |T_{n+1}'(\cos\theta)| \sin\theta \, d\theta$$

$$= \frac{2}{4^{n+1}} \int_0^{\pi} |\sin(n+1)\theta| \, d\theta = \frac{1}{4^n}. \qquad \square$$

REMARK. Achieser (1956) stated on p. 88 that this inequality is due to Korkin and Zolotareff (1873) while Chebyshev (1859) already got it implicitly. However Cheney (1966) stated on p. 233 that Korkin and Zolotareff (1873) posed the problem and it was Stieltjes (1876) who actually solved it.

SOLUTION 15.8

Suppose, contrary to the conclusion, that $|Q(x_0)| > M |T_n(x_0)|$ for some point x_0 satisfying $|x_0| > 1$. Put $c = Q(x_0)/T_n(x_0)$ for brevity. Consider the polynomial

$$R(x) = cT_n(x) - Q(x)$$

of degree $\le n$. If we put

$$\alpha_k = \cos\frac{k\pi}{n}$$

for $0 \le k \le n$, then clearly $T_n(\alpha_k) = (-1)^k$; therefore,

$$\operatorname{sgn} R(\alpha_k) = (-1)^k \operatorname{sgn} c,$$

because $|c| > M$ and $|Q(\alpha_k)| \le M$. This implies that the polynomial R vanishes at least at n points in the interval $(-1, 1)$. But then, since $R(x_0) = 0$, R must vanish at least $n + 1$ points, a contradiction. $\qquad\square$

SOLUTION 15.9

Since

$$Q(\cos \theta) = \frac{\cos n\theta}{2^{n-1}},$$

it is easily seen that the absolute value of

$$Q(x) = \frac{T_n(x)}{2^{n-1}} = x^n + \cdots$$

attains its maximum 2^{1-n} at $y_k = \cos(k\pi/n)$ for each $0 \le k \le n$. Suppose now that there exists a monic polynomial $R(x) = x^n + \cdots$ satisfying

$$\max_{|x|\le 1} |R(x)| < \frac{1}{2^{n-1}}.$$

Then clearly we have $R(y_0) < Q(y_0), R(y_1) > Q(y_1), \ldots$ so that the polynomial $R(x) - Q(x)$ has at least one zero in each interval (y_{k+1}, y_k). Hence $R(x) - Q(x)$ has at least n zeros in the interval $(-1, 1)$, contrary to the fact that the degree of $R(x) - Q(x)$ is less than n.

Next let $U(x) = x^n + \cdots$ be any polynomial with real coefficients satisfying

$$\max_{|x|\le 1} |U(x)| = \frac{1}{2^{n-1}}.$$

Let m be the number of points x in $[-1, 1]$ satisfying $|U(x)| = 2^{1-n}$, which are denoted by $x_1 < \cdots < x_m$. We then have $m = n + 1$. To see this it suffices to show $m > n$, because $U'(x_k) = 0$ for $1 < k < m$. Suppose, on the contrary, that $m \le n$. For any two consecutive points x_i and x_{i+1} satisfying

$$\operatorname{sgn} U(x_i) \operatorname{sgn} U(x_{i+1}) = -1$$

there exists at least one zero ξ of $U(x)$ in the interval (x_i, x_{i+1}). We take only one zero point for each (x_i, x_{i+1}) and name them $\xi_1 < \cdots < \xi_M$. Clearly

$$M \le m - 1 \le n - 1.$$

Now we consider the polynomial

$$V(x) = c(x - \xi_1) \cdots (x - \xi_M)$$

where $c = \pm 1$ is chosen so that the sign of V on the interval (ξ_i, ξ_{i+1}) coincides with sgn $U(x_k)$ for some $x_k \in (\xi_i, \xi_{i+1})$. In each interval (ξ_i, ξ_{i+1}) where V is positive, U may take a negative value somewhere in this interval. If so, the local minimum of U on (ξ_i, ξ_{i+1}) is certainly greater than -2^{1-n}. This means that the local maximum of the absolute value of $U(x) - \epsilon V(x)$ on this interval is less than 2^{1-n} for any sufficiently small $\epsilon > 0$. Moreover this argument is valid for the intervals $(-1, \xi_1)$ and $(\xi_M, 1)$. Therefore

$$\max_{|x| \le 1} |U(x) - \epsilon V(x)| < \frac{1}{2^{n-1}}.$$

This is contrary to the previous result, because the degree of V is less than n. Hence we have $m = n + 1$. Since

$$\max_{|x| \le 1} \frac{|Q(x) + U(x)|}{2} \le \frac{1}{2} \max_{|x| \le 1} |Q(x)| + \frac{1}{2} \max_{|x| \le 1} |U(x)|$$

$$= \frac{1}{2^{n-1}},$$

there exist $n + 1$ points $w_1 < \cdots < w_{n+1}$ in the interval $[-1, 1]$ satisfying

$$|Q(w_k) + U(w_k)| = \frac{1}{2^{n-2}}$$

by applying the same argument as above to the polynomial $(Q(x) + U(x))/2$. Thus we obtain

$$Q(w_k) = U(w_k) = \pm \frac{1}{2^{n-1}}$$

for all $1 \le k \le n + 1$, which implies that U coincides with Q. $\square$

| SOLUTION 15.10 |

Let $\xi_1, \ldots, \xi_n$ be the points in $[-1, 1]$ at which $|V(x_1, \ldots, x_n)|$ attains its maximum M_n and let ξ_0 be the point in $[-1, 1]$ at which the absolute value of the polynomial

$$\phi(x) = (x - \xi_1) \cdots (x - \xi_n)$$

attains its maximum. Then

$$|\phi(\xi_0)| = \frac{|V(\xi_0, \xi_1, \ldots, \xi_n)|}{|V(\xi_1, \ldots, \xi_n)|} \le \frac{M_{n+1}}{M_n};$$

hence it follows from **PROBLEM 15.9** that

$$\frac{1}{2^{n-1}} \le \max_{|x| \le 1} |\phi(x)| = |\phi(\xi_0)| \le \frac{M_{n+1}}{M_n},$$

because the leading coefficient of $\phi(x)$ is unity.

On the other hand, let

$$Q(x) = \frac{1}{2^{n-1}} T_n(x) = x^n + \cdots$$

and $(\eta_1, \ldots, \eta_{n+1})$ be the point at which $|V(x_1, \ldots, x_{n+1})|$ attains its maximum M_{n+1}. Then

$$V(\eta_1, \ldots, \eta_{n+1}) = \begin{vmatrix} 1 & \eta_1 & \cdots & \eta_1^{n-1} & \eta_1^n \\ 1 & \eta_2 & \cdots & \eta_2^{n-1} & \eta_2^n \\ \vdots & \vdots & \ddots & \vdots & \vdots \\ 1 & \eta_n & \cdots & \eta_n^{n-1} & \eta_n^n \\ 1 & \eta_{n+1} & \cdots & \eta_{n+1}^{n-1} & \eta_{n+1}^n \end{vmatrix} = \begin{vmatrix} 1 & \eta_1 & \cdots & \eta_1^{n-1} & Q(\eta_1) \\ 1 & \eta_2 & \cdots & \eta_2^{n-1} & Q(\eta_2) \\ \vdots & \vdots & \ddots & \vdots & \vdots \\ 1 & \eta_n & \cdots & \eta_n^{n-1} & Q(\eta_n) \\ 1 & \eta_{n+1} & \cdots & \eta_{n+1}^{n-1} & Q(\eta_{n+1}) \end{vmatrix}.$$

By the cofactor expansion along the last column we get

$$M_{n+1} \le |Q(\eta_1)| \cdot |V(\eta_2, \ldots, \eta_{n+1})| + \cdots + |Q(\eta_{n+1})| \cdot |V(\eta_1, \ldots, \eta_n)|$$
$$\le \frac{n+1}{2^{n-1}} M_n.$$

The above inequalities hold even for $n = 1$ if we define $M_1 = 1$. Multiplying these inequalities

$$\frac{1}{2^{k-1}} \le \frac{M_{k+1}}{M_k} \le \frac{k+1}{2^{k-1}},$$

from $k = 1$ to $n - 1$, we obtain

$$\frac{1}{2^{(n-1)(n-2)/2}} \le M_n \le \frac{n!}{2^{(n-1)(n-2)/2}},$$

which implies that $M_n^{2/(n(n-1))}$ converges to $1/2$ as $n \to \infty$. $\qquad\square$

Chapter 16

Gamma Function

[Summary of Basic Points]

1. The function $\Gamma(s)$ defined by the improper integral

$$\Gamma(s) = \int_0^\infty x^{s-1} e^{-x}\, dx$$

convergent for $s > 0$ is called the *Gamma function* and satisfies the functional relation

$$\Gamma(s + 1) = s\Gamma(s).$$

The Gamma function is also called the second Eulerian integral. It follows that $\Gamma(n + 1) = n!$ for any integer $n \geq 1$. The reader should notice the shift of the argument in this formula.

2. L. Euler (1707–1783) introduced the interpolation formula (**PROBLEM 16.1**) in a correspondence to C. Goldbach (1690–1764) in 1729 as a generalization of the factorial for the case s is rational. We owe to A. M. Legendre (1752–1833) its formulation in that form as well as the consideration for any $s > 0$, who also introduced the notation $\Gamma(s)$ and gave the names for two types of Euler's integrals.

3. It is known that

$$\Gamma\left(\frac{1}{2}\right) = \sqrt{\pi}.$$

4. The Gamma function is closely related to Euler's constant γ through

$$\Gamma'(1) = -\gamma.$$

253

5. The function defined by the improper integral

$$B(s,t) = \int_0^1 x^{s-1}(1-x)^{t-1}\,dx$$

for $s, t > 0$ is called the Beta function or the first Eulerian integral. The name 'Beta' was introduced for the first time in Binet(1839). The reason why this is the first is that Euler started his derivation with this integral. It is known that

$$B(s,t) = \frac{\Gamma(s)\Gamma(t)}{\Gamma(s+t)}.$$

For the proof see the former part of **Solution 16.3**. This yields $\Gamma(1/2) = \sqrt{\pi}$ when $s = t = 1/2$.

For various topics about the Gamma function involving Hadamard's factorial function, as well as Euler's experimental derivation, see Davis(1959). There are also good elementary expositions about the Gamma function; for example, see Barnes(1899), Jensen(1916), Gronwall(1918), etc.

Problem 16.1

For any $s > 0$ show that

$$\Gamma(s) = \lim_{n\to\infty} \frac{n^s n!}{s(s+1)\cdots(s+n)}.$$

This formula is due to Euler.

Problem 16.2

Show that the Gamma function $\Gamma(s)$ is logarithmically convex as well as convex. Show moreover that the Beta function $B(s,t)$ is logarithmically convex as well as convex with respect to s and t.

Recall that a positive function $f(x)$ on I is said to be logarithmically convex if $\log f(x)$ is convex on I. See Chapter 9.

PROBLEM 16.3

Let $\triangle_{n-1}$ be the $(n-1)$-dimensional simplex defined by $x_1, x_2, \ldots, x_{n-1} \geq 0$ and $x_1 + \cdots + x_{n-1} \leq 1$ for each integer $n \geq 2$. For $s_1 > 0, \ldots, s_n > 0$ show that the integral

$$\int \cdots \int_{\triangle_{n-1}} x_1^{s_1-1} \cdots x_{n-1}^{s_{n-1}-1} (1 - x_1 - \cdots - x_{n-1})^{s_n-1} \, dx_1 \cdots dx_{n-1}$$

is equal to

$$\frac{\Gamma(s_1) \cdots \Gamma(s_n)}{\Gamma(s_1 + \cdots + s_n)}.$$

PROBLEM 16.4

For any non-zero polynomial $P(x; z_0, z_1, \ldots, z_m)$ with $m + 2$ variables show that the Gamma function $\Gamma(x)$ does not satisfy the differential equation

$$P(x; y, y', y'', \ldots, y^{(m)}) = 0.$$

This was first shown by Hölder(1887) and Moore(1897) gave another proof. Barnes(1899), Ostrowski(1919) and Hausdorff(1925) also gave simpler and shorter proofs. Ostrowski(1925) corrected a mistake in his earlier paper that was pointed out by Hausdorff and gave a further shorter proof. Ostrowski and Moore used the functional relation

$$f(x + 1) = xf(x)$$

satisfied by the Gamma function $\Gamma(x)$, while Hölder, Barnes and Hausdorff used

$$\psi(x + 1) = \psi(x) + \frac{1}{x}$$

satisfied by the digamma function $\psi(x) = \Gamma'(x)/\Gamma(x)$.

PROBLEM 16.5

Suppose that $f(x) \in \mathscr{C}(0, \infty)$ is positive, logarithmically convex and satisfies the functional equation

$$f(x + 1) = xf(x)$$

with $f(1) = 1$. Then show that $f(x) = \Gamma(x)$.

This is known as the Bohr-Mollerup theorem. See Bohr and Mollerup(1922). Later their proof was simplified by Artin(1964).

PROBLEM 16.6

Show that

$$\frac{1}{\Gamma(s)} = s e^{\gamma s} \prod_{n=1}^{\infty} \left(1 + \frac{s}{n}\right) e^{-s/n}$$

where γ is Euler's constant.

This is known as Weierstrass' canonical product of order 1. This is valid for any complex number s because the convergence is uniform on compact sets in the whole complex plane. This was first found by Schlömilch (1844) but it was Weierstrass (1856) who established as the product theorem in the theory of functions, in which he used the notation $1/Fc(s)$ for the gamma function.

PROBLEM 16.7

Show that

$$\int_{x}^{x+1} \log \Gamma(s)\, ds = x(\log x - 1) + \frac{1}{2} \log(2\pi)$$

for any $x > 0$.

This is known as Raabe's integral, first shown by Raabe (1843) for any positive integer and subsequently (1844) for any positive real number x.

PROBLEM 16.8

Show that

$$\Gamma(s)\Gamma(1 - s) = \frac{\pi}{\sin \pi s}$$

for any $0 < s < 1$.

This is known as Euler's reflection formula, which also produces $\Gamma(1/2) = \sqrt{\pi}$.

PROBLEM 16.9

Using the Bohr-Mollerup theorem show that

$$\log \Gamma(s) = \int_{0}^{\infty} \left(s - 1 - \frac{1 - e^{-(s-1)x}}{1 - e^{-x}}\right) \frac{e^{-x}}{x}\, dx$$

for any $s > 0$.

This formula is due to Malmstén (1847). Cauchy (1841) had obtained this when s is a positive integer.

PROBLEM 16.10 ─────────────────────────────

Using Malmstén's formula in the previous problem show that

$$\log \Gamma(s) = \left(s - \frac{1}{2}\right) \log s - s + \frac{1}{2}\log(2\pi) + \omega(s),$$

where

$$\omega(s) = \int_0^\infty \left(\frac{1}{2} - \frac{1}{x} + \frac{1}{e^x - 1}\right) \frac{e^{-sx}}{x} \, dx$$

for any s > 0.

This is known as Binet's first formula for the log Gamma function. The function $\omega(s)$ is the Laplace transform of

$$\frac{1}{x}\left(\frac{1}{2} - \frac{1}{x} + \frac{1}{e^x - 1}\right),$$

which is a monotone decreasing function in $\mathscr{C}[0, \infty)$. Note that $\omega(s)$ gives the error term for Stirling's approximation (see the remark after **SOLUTION 16.7**). Binet (1839) also found another integral expression for $\omega(s)$, known as the second formula, as follows:

$$\omega(s) = 2 \int_0^\infty \frac{\arctan(x/s)}{e^{2\pi x} - 1} \, dx.$$

PROBLEM 16.11 ─────────────────────────────

Using the Fourier series of $\log \dfrac{\Gamma(s)}{\Gamma(1-s)}$ *show that*

$$\log \Gamma(s) + \frac{1}{2} \log \frac{\sin \pi s}{\pi} + (\gamma + \log 2\pi)\left(s - \frac{1}{2}\right) = \sum_{n=2}^\infty \frac{\log n}{n\pi} \sin 2n\pi s$$

for 0 < s < 1, where γ is Euler's constant.

This is known as Kummer's series (1847), in which he used the following integral representation for Euler's constant:

$$\gamma = \int_0^\infty \left(e^{-x} - \frac{1}{1+x}\right) \frac{dx}{x}.$$

Note that Kummer's series at $s = 3/4$ yields

$$\frac{\log 3}{3} - \frac{\log 5}{5} + \frac{\log 7}{7} - \cdots = \pi\left(\frac{\gamma - \log \pi}{4} + \log \Gamma\left(\frac{3}{4}\right)\right),$$

while Hardy (1912) obtained

$$\frac{\log 2}{2} - \frac{\log 3}{3} + \frac{\log 4}{4} - \cdots = \frac{1}{2}\log^2 2 - \gamma \log 2.$$

Solutions for Chapter 16

SOLUTION 16.1

For the function

$$\Phi_n(s) = \frac{n^s n!}{s(s+1)\cdots(s+n)}$$

we have by the partial fraction expansion

$$\Phi_n(s) = n^s \sum_{k=0}^{n} (-1)^k \binom{n}{k} \frac{1}{s+k}$$

$$= n^s \int_0^1 x^{s-1}(1-x)^n \, dx$$

$$= \int_0^n t^{s-1}\left(1-\frac{t}{n}\right)^n \, dt.$$

Let $\epsilon > 0$ be any number satisfying $\epsilon(s+2) < 1$. The last integral can be written as $J_1 + J_2$, where J_1 and J_2 are the integrals over the interval $[0, n^\epsilon]$ and $[n^\epsilon, n]$ respectively. For $0 \le t \le n^\epsilon$ we have

$$n\log\left(1-\frac{t}{n}\right) = -t + O(n^{\epsilon-1});$$

therefore

$$J_1 = \int_0^{n^\epsilon} t^{s-1}e^{-t}\, dt + O(n^{\epsilon(s+2)-1})$$

as $n \to \infty$.

For $n^\epsilon \le t \le n$ there exists a positive constant c satisfying

$$n\log\left(1-\frac{t}{n}\right) \le n\log(1-n^{\epsilon-1}) \le -cn^\epsilon;$$

hence $J_2 = O(n^s \exp(-cn^\epsilon))$, which converges to 0 as $n \to \infty$. Therefore $\Phi_n(s)$ converges to $\Gamma(s)$ as $n \to \infty$. $\qquad\square$

SOLUTION 16.2

Note that we can differentiate in s repeatedly under the integral sign in the second Eulerian integral, because

$$\int_0^\infty (\log x)^k x^{s-1} e^{-x}\, dx$$

converges uniformly on any compact sets in s in the region $s > 0$ for any integer $k \geq 1$. Thus the convexity of $\Gamma(s)$ follows from

$$\int_0^\infty (\log x)^2 x^{s-1} e^{-x}\, dx > 0$$

and the logarithmic convexity follows from the fact that the quadratic function in σ:

$$\int_0^\infty (\sigma + \log x)^2 x^{s-1} e^{-x}\, dx$$

is positive for any $\sigma \in \mathbb{R}$, because this implies that $\Gamma'^2(s) < \Gamma(s)\Gamma''(s)$.

Similar argument can be applied to the first Eulerian integral for the convexity and the logarithmic convexity with respect to s and t. $\qquad\square$

REMARK. Since

$$\int_0^1 (\sigma \log x + \log(1 - x))^2\, x^{s-1}(1 - x)^{t-1}\, dx > 0$$

for any $\sigma \in \mathbb{R}$ we have

$$\left(\frac{\partial^2}{\partial s \partial t} B(s,t)\right)^2 < \frac{\partial^2}{\partial s^2} B(s,t) \frac{\partial^2}{\partial t^2} B(s,t).$$

In other words, the Beta function has the positive Hessian.

SOLUTION 16.3

We first treat the case $n = 2$. For a fixed number $t > 0$ put

$$\Psi(s) = \frac{\Gamma(s + t)}{\Gamma(t)} B(s,t)$$

for $s > 0$. Obviously $\Psi(s) > 0$ and $\Psi(1) = 1$. Moreover we have

$$\Psi(s + 1) = (s + t)\frac{\Gamma(s + t)}{\Gamma(t)} B(s + 1, t) = s\Psi(s),$$

because it follows from integration by parts that

$$B(s + 1, t) = \frac{s}{t} B(s, t + 1) = \frac{s}{t} B(s, t) - \frac{s}{t} B(s + 1, t).$$

As is shown in **SOLUTION 16.2**, both $\Gamma(s+t)$ and $B(s, t)$ are logarithmically convex with respect to s. This means that $\Psi(s)$ is logarithmically convex and it follows from **PROBLEM 16.5** that $\Psi(s) = \Gamma(s)$. The formula for $n = 2$ is thus proved. This method is due to Artin (1964).

The general case will be shown by induction on n. Suppose that the formula holds true for n. For an arbitrarily fixed $(x_1, \ldots, x_{n-1})$ in the $(n - 1)$-dimensional simplex $\triangle_{n-1}$ we consider the substitution

$$x_n = (1 - x_1 - x_2 - \cdots - x_{n-1}) t$$

for $0 < t < 1$. Then the integral

$$\int \cdots \int_{\triangle_n} x_1^{s_1-1} \cdots x_n^{s_n-1} (1 - x_1 - \cdots - x_n)^{s_{n+1}-1} \, dx_1 \cdots dx_n$$

is transformed to

$$\int \cdots \int_{\triangle_{n-1}} x_1^{s_1-1} \cdots x_{n-1}^{s_{n-1}-1} (1 - x_1 - \cdots - x_{n-1})^{s_n+s_{n+1}-1} \, dx_1 \cdots dx_{n-1}$$

$$\times \int_0^1 t^{s_n-1} (1 - t)^{s_{n+1}-1} \, dt$$

$$= \frac{\Gamma(s_1) \cdots \Gamma(s_{n-1}) \Gamma(s_n + s_{n+1})}{\Gamma(s_1 + \cdots + s_n + s_{n+1})} \cdot \frac{\Gamma(s_n) \Gamma(s_{n+1})}{\Gamma(s_n + s_{n+1})},$$

which shows the formula for $n + 1$. $\qquad\square$

SOLUTION 16.4

The proof is based on Ostrowski (1925). For any term of P of the form

$$A(x) z_0^{n_0} z_1^{n_1} \cdots z_m^{n_m},$$

where $A(x)$ is a polynomial only in x, we assign the index $(n_0, n_1, \ldots, n_m)$ and introduce a lexicographical order; namely, we say that $(n_0, n_1, \ldots, n_m)$ is higher than $(n'_0, n'_1, \ldots, n'_m)$ if $n_m = n'_m, \ldots, n_{j+1} = n'_{j+1}$ and $n_j > n'_j$ for some $0 \le j \le m$. Clearly the indices form a totally ordered set. Note that we do not distinguish $(n_0, n_1, \ldots, n_m)$ from $(n_0, n_1, \ldots, n_m, 0, \ldots, 0)$.

For any polynomial $P(x; z_0, z_1, \ldots, z_m)$ satisfying

$$P(x; \Gamma, \Gamma', \Gamma'', \ldots, \Gamma^{(m)}) = 0 \qquad (16.1)$$

if exists, we assign the highest index $(n_0, n_1, \ldots, n_m)$ among the terms of the form discussed above, which is denoted by ind P. We then pick up a polynomial P^* having the lowest ind P among the all polynomials P satisfying (16.1). Let

$$A^*(x) z_0^{v_0} z_1^{v_1} \cdots z_m^{v_m}$$

be the corresponding highest term with ind $P^* = (v_0, v_1, \ldots, v_m)$. We can also assume that $\deg A^*$ is the smallest degree among such polynomials and that the leading coefficient is unity.

Let P be any polynomial satisfying (16.1) with ind $P = \text{ind } P^*$ and let $A(x)$ be the coefficient of the highest term of P. Put $A(x) = q(x) A^*(x) + r(x)$ with $\deg r < \deg A^*$. If $r \neq 0$, the coefficient of the highest term of $P - q(x) P^*$ would be $r(x)$, contrary to the choice of A^*; hence $A(x) = q(x) A^*(x)$. If $P - q(x) P^* \neq 0$, then the highest index of $P - q(x) P^*$ would be certainly lower than ind P^*, contrary to the choice of P^*. We thus have $P = q(x) P^*$.

Substituting $\Gamma(x + 1) = x \Gamma(x)$ in the differential equation (16.1) with $P = P^*$, we obtain a new equation

$$0 = P^*(x + 1; x\Gamma, x\Gamma' + \Gamma, \ldots, x\Gamma^{(m)} + m\Gamma^{(m-1)})$$
$$= Q(x; \Gamma, \Gamma', \Gamma'', \ldots, \Gamma^{(m)}), \quad \text{say.}$$

The highest term of Q certainly comes from the expansion of

$$A^*(x + 1)(xz_0)^{v_0}(xz_1 + z_0)^{v_1} \cdots (xz_m + mz_{m-1})^{v_m};$$

therefore, ind $Q = \text{ind } P^*$ and the coefficient of the highest term of Q becomes $x^N A^*(x + 1)$ with $N = v_0 + v_1 + \cdots + v_m$. It follows from the above argument that

$$B(x) = \frac{x^N A^*(x + 1)}{A^*(x)}$$

is a polynomial of degree N and $Q = B(x) P^*$; hence

$$P^*(x + 1; xz_0, xz_1 + z_0, \ldots, xz_m + mz_{m-1}) = B(x) P^*(x; z_0, z_1, \ldots, z_m).$$

If $B(\alpha) = 0$ for some $\alpha \neq 0$, then $P^*(\alpha + 1; w_0, w_1, \ldots, w_m) = 0$. Let

$$M = 1 + \max\left(\deg_{z_0} P^*, \ldots, \deg_{z_m} P^*\right)$$

where $\deg_{z_k}$ means the degree with respect to z_k. Since an index $(n_0, n_1, \ldots, n_m)$ is higher than another index $(n'_0, n'_1, \ldots, n'_m)$ if and only if

$$n_0 + n_1 M + \cdots + n_m M^m > n'_0 + n'_1 M + \cdots + n'_m M^m,$$

it follows from

$$P^*(\alpha + 1; t, t^M, \ldots, t^{M^m}) = 0$$

that the coefficient of every term of P^* vanishes at $\alpha + 1$, contrary to the choice of A^*. Therefore $B(x) = x^N$ and

$$P^*(x + 1; xz_0, xz_1 + z_0, \ldots, xz_m + mz_{m-1}) = x^N P^*(x; z_0, z_1, \ldots, z_m). \qquad (16.2)$$

Putting $x = 0$ in (16.2) we get

$$P^*(1; 0, z_0, \ldots, mz_{m-1}) = 0,$$

which implies that $R(1; w_1, \ldots, w_m) = 0$, where

$$R(x; w_1, \ldots, w_m) = P^*(x; 0, w_1, \ldots, w_m);$$

hence R has a factor $x - 1$ by the same argument as above. Note that $R \neq 0$; otherwise P^* would have a factor z_0, contrary to the choice of P^*. Similarly putting $x = 1$ and $z_0 = 0$ in (16.2) we get

$$P^*(2; 0, z_1, z_2 + z_1, \ldots, z_m + mz_{m-1}) = 0,$$

which implies $R(2; w_1, \ldots, w_m) = 0$; so, R has a factor $x - 2$. Repeating this argument we see that R has a factor $x - k$ for all integer $k \geq 1$, which is a contradiction. $\qquad \square$

SOLUTION 16.5

By the functional equation it is easily seen that $f(n) = (n - 1)!$ for all integers $n \geq 1$. For any fixed $x \in (0, 1)$ it follows from **PROBLEM 9.4** that

$$\log f(x + n + 1) \leq (1 - x) \log f(n + 1) + x \log f(n + 2)$$
$$= (1 - x) \log n! + x \log(n + 1)!;$$

namely, $f(x + n + 1) \leq (n + 1)^x n!$. Similarly it follows from

$$\log f(n + 1) \leq \frac{x}{1 + x} \log f(n) + \frac{1}{1 + x} \log f(x + n + 1)$$

that $n^x n! \leq f(x + n + 1)$. Hence, using $f(x + n + 1) = (x + n) \cdots (x + 1) x f(x)$,

$$\Phi_n(x) \leq f(x) \leq \left(1 + \frac{1}{n}\right)^x \Phi_n(x),$$

where

$$\Phi_n(x) = \frac{n^x n!}{x(x + 1) \cdots (x + n)}.$$

Since $\Phi_n(x)$ converges to $\Gamma(x)$ as $n \to \infty$ by **PROBLEM 16.1**, we obtain $f(x) = \Gamma(x)$, as required. □

SOLUTION 16.6

It follows from **PROBLEM 16.1** that

$$\frac{1}{\Gamma(s)} = \lim_{n\to\infty} \frac{s(s+1)\cdots(s+n)}{n^s n!}$$

$$= s \lim_{n\to\infty} \frac{(s+1)\cdots(s+n)}{(n+1)^s n!}.$$

The sequence on n inside of the last limit sign can be written in the form

$$\prod_{k=1}^{n}\left(1 + \frac{s}{k}\right)\exp\left(s\log\frac{k}{k+1}\right),$$

which is equal to the product of

$$\prod_{k=1}^{n}\left(1 + \frac{s}{k}\right)e^{-s/k}$$

and

$$\exp\left(s\left(1 + \frac{1}{2} + \cdots + \frac{1}{n} - \log(n+1)\right)\right).$$

Obviously the last expression converges to $e^{\gamma s}$ where γ is Euler's constant. □

SOLUTION 16.7

Put

$$f(x) = \int_{x}^{x+1} \log\Gamma(s)\,ds$$

for $x > 0$. Differentiation yields

$$f'(x) = \log\frac{\Gamma(x+1)}{\Gamma(x)} = \log x;$$

hence

$$f(x) = x(\log x - 1) + c$$

for some constant c. Note that $c = f(1) + 1$. To determine the value of c we use the formula in **PROBLEM 16.1** in the form

$$\log\Gamma(s) = -\log s + \lim_{n\to\infty}\left(s\log n + \log\frac{1}{s+1} + \cdots + \log\frac{n}{s+n}\right),$$

which can be written as

$$\log \Gamma(s+1) = \sum_{k=1}^{\infty} \left(s \log \frac{k+1}{k} - \log \frac{s+k}{k} \right),$$

and the series converges uniformly in $s \in [0, 1]$ in view of

$$s \log \frac{k+1}{k} - \log \frac{s+k}{k} = O\left(\frac{1}{k^2}\right).$$

Hence we may integrate the above expression termwisely to obtain

$$
\begin{aligned}
f(1) &= \int_0^1 \log \Gamma(s+1)\,ds \\
&= \lim_{n \to \infty} \left(\frac{1}{2} \log n - (n+1) \log(n+1) + n + \log n! \right) \\
&= -1 + \lim_{n \to \infty} A_n,
\end{aligned}
$$

where

$$A_n = n + \log n! - \left(n + \frac{1}{2} \right) \log n.$$

Hence $c = \lim_{n \to \infty} A_n$, and so c is also the limit of the sequence

$$2A_n - A_{2n} = \log \frac{n!^2}{(2n)!} + \left(2n + \frac{1}{2} \right) \log 2 - \frac{1}{2} \log n,$$

which is equal to the logarithm of

$$\sqrt{2}\left(1 + \frac{1}{2n} \right) \frac{\sqrt{n}\, n!}{\dfrac{1}{2}\left(\dfrac{1}{2} + 1 \right) \cdots \left(\dfrac{1}{2} + n \right)}.$$

This converges to $\sqrt{2}\,\Gamma(1/2) = \sqrt{2\pi}$ as $n \to \infty$ by **Problem 16.1**. Hence $c = \log \sqrt{2\pi}$. $\square$

Remark. The above proof shows that the factorial $n!$ is asymptotically

$$\sqrt{2\pi n}\left(\frac{n}{e} \right)^n$$

as $n \to \infty$, known as *Stirling's approximation*. J. Stirling (1692–1770) gave this asymptotic formula for $n!$ in Methodus Differentialis (1730).

SOLUTION 16.8

It follows from **PROBLEM 16.1** that

$$\frac{1}{\Gamma(s)\Gamma(1-s)} = \lim_{n\to\infty} \frac{s(s+1)\cdots(s+n)}{n^s n!} \cdot \frac{(1-s)(2-s)\cdots(n+1-s)}{n^{1-s} n!}$$

$$= s \lim_{n\to\infty} \left(1 + \frac{1-s}{n}\right) \prod_{k=1}^{n}\left(1 - \frac{s^2}{k^2}\right),$$

which is equal to

$$s \prod_{n=1}^{\infty}\left(1 - \frac{s^2}{n^2}\right) = \frac{\sin \pi s}{\pi}$$

by **PROBLEM 2.5**. □

REMARK. This can be used to evaluate Raabe's integral (**PROBLEM 16.7**). For, it follows that

$$\int_0^1 \log \Gamma(s)\,ds = \frac{1}{2}\left(\int_0^1 \log \Gamma(s)\,ds + \int_0^1 \log \Gamma(1-s)\,ds\right)$$

$$= \frac{1}{2}\int_0^1 \log \frac{\pi}{\sin \pi s}\,ds.$$

Then it is not hard to see that the last expression is equal to $\log \sqrt{2\pi}$, which is the value of $c = f(1) + 1$.

SOLUTION 16.9

Let $\phi(s)$ denote the integral on the right-hand side of the equality to be shown. For any fixed closed subinterval $[a, b]$ in $(0, \infty)$ we have

$$\left(s - 1 - \frac{1 - e^{-(s-1)x}}{1 - e^{-x}}\right)\frac{e^{-x}}{x} = \frac{(s-1)(s-3)}{2} + O(x)$$

as $x \to 0+$, where the constant in O-symbol can be taken uniformly in $s \in [a, b]$. Similarly we have

$$\left(s - 1 - \frac{1 - e^{-(s-1)x}}{1 - e^{-x}}\right)\frac{e^{-x}}{x} = \begin{cases} (s-2)\dfrac{e^{-x}}{x} + o(e^{-\delta x}) & (s > 1) \\[2mm] 0 & (s = 1) \\[2mm] \dfrac{e^{-sx}}{x} + o(e^{-x}) & (0 < s < 1) \end{cases}$$

where $\delta = \max(2, s)$ and the constant in the estimates can be taken uniformly in $s \in [a, b]$. This implies that $\phi \in \mathscr{C}(0, \infty)$. For any $s, t > 0$, we have

$$\frac{\phi(s) + \phi(t)}{2} = \int_0^\infty \left(\frac{s+t}{2} - 1 - \frac{1 - (e^{-sx} + e^{-tx})e^x/2}{1 - e^{-x}} \right) \frac{e^{-x}}{x} \, dx$$

$$\geq \phi\left(\frac{s+t}{2}\right),$$

because e^{-sx} is convex with respect to s and $1 - e^{-x} > 0$ for any $x > 0$. Therefore ϕ is convex on $(0, \infty)$. Since $\phi(1) = 0$ obviously, it follows from the Bohr-Mollerup theorem that $\phi(s) = \log \Gamma(s)$ if $\phi(s + 1) - \phi(s) = \log s$ for any $s > 0$. However this is easily verified from

$$\phi(s + 1) - \phi(s) = \int_0^\infty \frac{e^{-x} - e^{-sx}}{x} \, dx,$$

which can be seen to be $\log s$ by integrating

$$\int_0^\infty e^{-sx} \, dx = \frac{1}{s}$$

with respect to s.   □

SOLUTION 16.10

It follows from Malmstén's formula that

$$\log \Gamma(s + y) = \int_0^\infty \left(s + y - 1 - \frac{1 - e^{-(s+y-1)x}}{1 - e^{-x}} \right) \frac{e^{-x}}{x} \, dx$$

for any $0 \leq y \leq 1$. Integrating with respect to y we get

$$\int_s^{s+1} \log \Gamma(x) \, dx = \int_0^\infty \left(s - \frac{1}{2} + \frac{e^{-(s-1)x}}{x} - \frac{1}{1 - e^{-x}} \right) \frac{e^{-x}}{x} \, dx.$$

By Raabe's integral (**PROBLEM 16.7**) the left-hand side is equal to

$$s \log s - s + \frac{1}{2} \log(2\pi).$$

Hence, subtracting

$$\frac{1}{2} \log s = \frac{1}{2} \int_0^\infty \frac{e^{-x} - e^{-sx}}{x} \, dx$$

we see that

$$\left(s - \frac{1}{2}\right) \log s - s + \frac{1}{2} \log(2\pi)$$

$$= \int_0^\infty \left(s - 1 + \left(\frac{1}{2} + \frac{1}{x}\right) e^{-(s-1)x} - \frac{1}{1 - e^{-x}}\right) \frac{e^{-x}}{x} \, dx.$$

By Malmstén's formula again, we conclude that

$$\omega(s) = \int_0^\infty \left(-\frac{1}{2} - \frac{1}{x} + \frac{1}{1 - e^{-x}}\right) \frac{e^{-sx}}{x} \, dx,$$

as asserted. $\square$

SOLUTION 16.11

Since the function $\log \Gamma(s)$ is asymptotic to $-\log s$ as $s \to 0+$, the improper integral

$$\int_0^1 \log \frac{\Gamma(s)}{\Gamma(1 - s)} \, ds$$

converges absolutely. Hence it follows from **PROBLEM 7.13** that

$$\log \frac{\Gamma(s)}{\Gamma(1 - s)} = \frac{a_0}{2} + \sum_{n=1}^\infty (a_n \cos 2n\pi x + b_n \sin 2n\pi x)$$

for any $0 < s < 1$, where a_n, b_n are the Fourier coefficients over $[0, 1]$. Substituting $\sigma = 1 - s$ we get

$$a_n = 2 \int_0^1 \log \frac{\Gamma(1 - \sigma)}{\Gamma(\sigma)} \cos 2n\pi\sigma \, d\sigma = -a_n;$$

hence $a_n = 0$ for all $n \geq 0$. To evaluate b_n we use the formula in **PROBLEM 16.1**:

$$\log \frac{\Gamma(s)}{\Gamma(1 - s)} = \log \frac{1 - s}{s} + \lim_{m \to \infty} A_m(s),$$

where

$$A_m(s) = (2s - 1) \log m + \sum_{k=1}^m \log \frac{k + 1 - s}{k + s}.$$

We show that the limit on the right-hand side converges uniformly in $s \in [0, 1]$. To this end, note first that, for any integer $p > q \geq 1$ we have

$$\left| \sum_{k=q+1}^p \frac{1}{k + s} - \log \frac{p}{q} \right| < \frac{1}{q + 1}$$

for any $0 \leq s \leq 1$; therefore

$$|A_p(s) - A_q(s)| = \left| (2s-1) \log \frac{p}{q} + \sum_{k=q+1}^{p} \log\left(1 + \frac{1-2s}{k+s} \right) \right|$$

$$\leq \sum_{k=q+1}^{p} \left| \log\left(1 + \frac{1-2s}{k+s} \right) + \frac{2s-1}{k+s} \right| + \frac{1}{q+1}$$

$$< \frac{2}{q}$$

by the inequality $|\log(1 + x) - x| \leq x^2$ for $|x| < 1/2$. Thus we can interchange the order of limit and integration to obtain

$$b_n = 2 \lim_{m \to \infty} \int_0^1 \widetilde{A}_m(s) \sin 2n\pi s \, ds,$$

where

$$\widetilde{A}_m(s) = (2s-1) \log m + \sum_{k=0}^{m} \log \frac{k+1-s}{k+s}.$$

Since

$$\int_0^1 (2s-1) \sin 2n\pi s \, ds = -\frac{1}{n\pi}$$

and

$$\sum_{k=0}^{m-1} \int_0^1 \log \frac{k+1-s}{k+s} \sin 2n\pi s \, ds = -2 \int_0^m \log t \sin 2n\pi t \, dt,$$

we obtain

$$b_n = -2 \lim_{m \to \infty} \left(\frac{\log m}{n\pi} + 2 \int_0^m \log t \sin 2n\pi t \, dt \right).$$

Here we ignored the last term corresponding to $k = m$ in $\widetilde{A}_m(s)$, because it converges to 0 uniformly in s. It follows from integration by parts that

$$\int_0^m \log t \sin 2n\pi t \, dt = \frac{1 - \cos 2n\pi t}{2n\pi} \log t \Big|_{0+}^{m} - \frac{1}{2n\pi} \int_0^m \frac{1 - \cos 2n\pi t}{t} \, dt$$

$$= -\frac{1}{2n\pi} \int_0^1 \frac{1 - \cos 2mn\pi x}{x} \, dx.$$

Hence

$$b_n = \frac{2(\log n + C)}{n\pi},$$

where

$$C = \lim_{N \to \infty} \left(\int_0^1 \frac{1 - \cos 2N\pi x}{x} \, dx - \log N \right)$$

is a constant independent of n. Then it is easily seen that

$$C = \lim_{N \to \infty} \left(\int_0^{2N\pi} \frac{1 - \cos s}{s} \, ds - \log N \right)$$

$$= \int_0^1 \frac{1 - \cos s}{s} \, ds - \int_1^\infty \frac{\cos s}{s} \, ds + \log 2\pi,$$

which is equal to $\gamma + \log 2\pi$ by **Problem 6.10**. We thus have

$$\log \frac{\Gamma(s)}{\Gamma(1 - s)} = \frac{2}{\pi} \sum_{n=1}^\infty \frac{\log n + \gamma + \log 2\pi}{n} \sin 2n\pi x,$$

which implies Kummer's series by virtue of **Problem 7.12** and **Problem 16.8**. □

Remark. Kummer's series converges uniformly on the interval $[\delta, 1 - \delta]$ for each $\delta > 0$ by Dirichlet's test.

❖ ❋ ❖

Chapter 17

Prime Number Theorem

[Summary of Basic Points]

1. Let $\pi(x)$ be the number of primes not exceeding x. The prime number theorem states that

$$\lim_{x \to \infty} \frac{\pi(x) \log x}{x} = 1,$$

in other words, $\pi(x)$ is asymptotic to $x/\log x$ as $x \to \infty$, which is one of the most celebrated results in mathematics. The first major step toward the prime number theorem was made by Chebyshev (1852), who showed that

$$A \frac{x}{\log x} \leq \pi(x) \leq A' \frac{x}{\log x}$$

for sufficiently large x where

$$A = \log\left(\frac{2^{1/2} 3^{1/3} 5^{1/5}}{30^{1/30}}\right) = 0.92129... \quad \text{and} \quad A' = \frac{6}{5}A = 1.10555....$$

2. Chebyshev introduced therein two important functions

$$\theta(x) = \sum_{p \leq x} \log p$$

and

$$\psi(x) = \sum_{p^k \leq x} \log p = \sum_{p \leq x} \left[\frac{\log x}{\log p}\right] \log p$$

for $x \geq 0$, where p runs over all primes not exceeding x. Note that $\psi(n)$ is the least common multiple of $1, 2, ..., n$.

3. The function $\psi(x)$ can be written as

$$\psi(x) = \sum_{n \leq x} \Lambda(n),$$

where

$$\Lambda(n) = \begin{cases} \log p & (n \text{ is a power of a prime } p) \\ 0 & (\text{otherwise}) \end{cases},$$

known as the von Mangoldt function, was introduced in 1895. We put $\Lambda(x) = \Lambda([x])$ for $x \geq 1$ for convenience.

4. The Möbius function $\mu(n)$, introduced by Möbius (1832), is defined by

$$\mu(n) = \begin{cases} 1 & (n = 1) \\ (-1)^k & (n \text{ is a product of } k \text{ distinct primes}) \\ 0 & (\text{otherwise}) \end{cases}.$$

It is easily seen that

$$\sum_{d\mid n} \mu(d) = \begin{cases} 1 & (n = 1) \\ 0 & (n \geq 2) \end{cases} \tag{17.1}$$

where d runs over all divisors of n.

5. Let f and g be any functions defined on all divisors of n. The *Möbius inversion formula* states that if $g(m) = \sum_{d\mid m} f(d)$ for all divisors m of n, then

$$f(n) = \sum_{d\mid n} \mu(d) g\left(\frac{n}{d}\right).$$

As an application of the inversion formula we have

$$\Lambda(n) = -\sum_{d\mid n} \mu(d) \log d, \tag{17.2}$$

as the inversion of

$$\sum_{d\mid n} \Lambda(d) = \sum_{p^k\mid n} \log p = \log \prod_{p^k\mid n} p^k = \log n. \tag{17.3}$$

The prime number theorem was first established by Hadamard (1896) and by de la Vallée Poussin (1896) independently, following Riemann's program in 1859. Their arguments are based on the absence of zeros of $\zeta(z)$ on the vertical line $\text{Re } z = 1$ and also on the existence of some zero-free region of $\zeta(z)$ in the critical strip $0 \leq \text{Re } z \leq 1$. Wiener's Tauberian theory on Fourier analysis implies the equivalence of the prime number theorem and the absence of zeros of $\zeta(z)$ on Re $z = 1$. Later Newman (1980) found a simple analytic proof of the prime number theorem. See also Korevaar (1982) and Zagier (1997).

Selberg (1949) and Erdös (1949) succeeded in giving elementary proofs of the prime number theorem, in the sense that they treated only finite sums, not infinite sums or integrals. However elementary proofs do not necessarily mean simple proofs.

Fine historical perspectives of the prime number theorem will be found in Levinson (1969), Goldstein (1973), Diamond (1982), Bateman and Diamond (1996).

Below we shall give a chain of simple problems towards a proof of the prime number theorem in the spirit of Erdös and Selberg, culminating in **PROBLEM 17.13**. All the solutions are assumed to be understandable with academic ability of the calculus level.

◆ ◞◠ ❋ ◞◠ ◆

PROBLEM 17.1

Show that

$$\sum_{n \le x} \psi\left(\frac{x}{n}\right) = \log [x]!$$

for any $x \ge 1$, where n runs over positive integers not exceeding x.

This is due to Chebyshev (1852). Using Stirling's approximation and (17.4) we see that

$$\frac{1}{N} \sum_{n=1}^{N} \left(\psi\left(\frac{N}{n}\right) - \frac{N}{n} \right) = \frac{\log N!}{N} - \sum_{n=1}^{N} \frac{1}{n}$$

$$= -1 - \gamma + O\left(\frac{\log N}{N}\right)$$

as $N \to \infty$. The left-hand side may suggest the improper integral

$$\int_0^1 \left(\psi\left(\frac{1}{x}\right) - \frac{1}{x} \right) dx,$$

however it seems to be hard to justify this derivation in brief discussion. Instead, it is easily seen that the integral over $[\epsilon, 1]$ is bounded for $\epsilon > 0$, which derives

$$\liminf_{x \to \infty} \frac{\psi(x)}{x} \le 1 \le \limsup_{x \to \infty} \frac{\psi(x)}{x}.$$

PROBLEM 17.2

Let $f(x)$ be any function defined on $[1, \infty)$ and put

$$g(x) = (\log x) \sum_{n \le x} f\left(\frac{x}{n}\right).$$

Show then that

$$\sum_{n \le x} \mu(n) g\left(\frac{x}{n}\right) = f(x) \log x + \sum_{n \le x} \Lambda(n) f\left(\frac{x}{n}\right).$$

This is due to Tatuzawa and Iseki (1951).

PROBLEM 17.3

Apply the formula in **PROBLEM 17.2** to $f(x) = \log x - 1$ to deduce that

$$\sum_{n \le x} \Lambda(n) \log \frac{x}{n} = O(x)$$

as $x \to \infty$.

PROBLEM 17.4

Apply the formula in **PROBLEM 17.2** to $f(x) = \psi(x) - x + \gamma + 1$ to deduce that

$$\psi(x) \log x + \sum_{n \le x} \Lambda(n) \psi\left(\frac{x}{n}\right) = 2x \log x + O(x)$$

as $x \to \infty$, where γ is Euler's constant.

This is also due to Tatuzawa and Iseki (1951), which can be shown to be equivalent to

$$\theta(x) \log x + \sum_{p \le x} \theta\left(\frac{x}{p}\right) \log p = 2x \log x + O(x),$$

as $x \to \infty$, known as Selberg's inequality (1949).

PROBLEM 17.5

Show that

$$\pi(x) = \frac{\theta(x)}{\log x} + \int_2^x \frac{\theta(t)}{t \log^2 t} \, dt$$

for any $x \ge 2$. Deduce from this that the prime number theorem is equivalent to $\psi(x) \sim x$ as $x \to \infty$.

PROBLEM 17.6

Show that

$$U(x)\log x + \int_1^x \Lambda(t) U\left(\frac{x}{t}\right) dt = O(x),$$

as $x \to \infty$, where

$$U(x) = \int_1^x \frac{\psi(t) - t}{t} dt.$$

Levinson (1966) introduced the function $U(x)$ to return the prime number theorem to $U(x) = o(x)$ as $x \to \infty$. To see this, for any $0 < \epsilon < 1$ let $x_\epsilon > 0$ be a number satisfying

$$(1 - \epsilon)x < \int_1^x \frac{\psi(t)}{t} dt < (1 + \epsilon)x$$

for any $x > x_\epsilon$. Put $y = (1 + \sqrt{\epsilon})x$ for brevity. Since $\psi(x)$ is a monotone increasing function, we obtain

$$(y - x)\frac{\psi(x)}{y} \leq \int_x^y \frac{\psi(t)}{t} dt$$

$$= \int_1^y \frac{\psi(t)}{t} dt - \int_1^x \frac{\psi(t)}{t} dt$$

$$< (1 + \epsilon)y - (1 - \epsilon)x;$$

that is,

$$\psi(x) < (1 + \sqrt{\epsilon})^3 x$$

and therefore

$$\limsup_{x \to \infty} \frac{\psi(x)}{x} \leq 1.$$

By a similar way we have

$$\frac{1 - 2\sqrt{\epsilon} - \epsilon}{1 + \sqrt{\epsilon}} < \frac{\psi(y)}{y},$$

which implies that

$$\liminf_{x \to \infty} \frac{\psi(x)}{x} \geq 1.$$

Thus we have shown the prime number theorem by **PROBLEM 17.5**. So the remainder of this chapter is devoted to show that $U(x) = o(x)$.

Our method is substantially based on Wright (1954) and Levinson (1969), however we replace finite sums by integrals as often as possible.

PROBLEM 17.7

Using **PROBLEMS 17.3** *and* **17.4** *show that*

$$\iint\limits_{\Delta_x} (\Lambda(st) + \Lambda(s)\Lambda(t))\, ds dt = 2x \log x + O(x)$$

as $x \to \infty$.

PROBLEM 17.8

Iterate the estimate given in the previous problem to show that

$$U(x) \log^2 x + \iint\limits_{\Delta_x} (\Lambda(st) - \Lambda(s)\Lambda(t)) U\left(\frac{x}{st}\right) ds dt = O(x \log x)$$

as $x \to \infty$, *where* $\Delta_x = \{(s,t) \mid st \le x, s \ge 1, t \ge 1\}$.

PROBLEM 17.9

For $x > 1$ *let* $f \in \mathscr{C}[1, x]$ *and* g *be a Lipschitz function with constant* L *satisfying* $g(1) = 0$ *defined on* $[1, x]$. *Show then that*

$$\left| \int_1^x f(t) g\left(\frac{x}{t}\right) dt \right| \le L x \int_1^x \left| \int_1^t f(s)\, ds \right| \frac{dt}{t^2}.$$

If $g(x) \in \mathscr{C}^1[1, x]$, then $|g'(t)| \le L$ and it follows from integration by parts that

$$\int_1^x f(t) g\left(\frac{x}{t}\right) dt = F(t) g\left(\frac{x}{t}\right)\Big|_{t=1}^{t=x} + x \int_1^x F(t) g'\left(\frac{x}{t}\right) \frac{dt}{t^2},$$

which implies immediately the above inequality, where

$$F(t) = \int_1^t f(s)\, ds.$$

This observation would be a good hint.

PROBLEM 17.10

Using **PROBLEMS 17.8** *and* **17.9** *prove that*

$$|U(x)| \log^2 x \le 2 \iint\limits_{\Delta_x} \left| U\left(\frac{x}{st}\right) \right| ds dt + C x \log x$$

for some constant $C > 0$.

PROBLEM 17.11

Put $V(x) = e^{-x}U(e^x)$ for $x \geq 0$. Show that

$$\limsup_{x\to\infty} |V(x)| = \limsup_{x\to\infty} \frac{1}{x} \int_0^x |V(s)| \, ds.$$

PROBLEM 17.12

Let $f(x)$ be a bounded Lipschitz function with a constant L defined on the interval $10, \infty)$. Put

$$\alpha = \limsup_{x\to\infty} |f(x)| \quad and \quad \beta = \limsup_{x\to\infty} \frac{1}{x} \int_0^x |f(s)| \, ds.$$

Suppose further that

$$\delta = \limsup_{x\to\infty} \left| \int_0^x f(s) \, ds \right|$$

is finite. Show then that

$$\beta(\alpha^2 + 2\delta L) \leq 2\alpha\delta L.$$

PROBLEM 17.13

Deduce the prime number theorem from the above facts.

◆ ❊ ◆

Solutions for Chapter 17

SOLUTION 17.1

It follows from the definition of ψ that

$$S_0 = \sum_{n \le x} \psi\left(\frac{x}{n}\right) = \sum_{mn \le x} \Lambda(m),$$

where m and n run over all positive integers satisfying $mn \le x$. We now rearrange such pairs (m, n) according to $k = mn$; namely,

$$S_0 = \sum_{k \le x} \sum_{mn = k} \Lambda(m).$$

Since m runs over all divisors of k, it follows from (17.3) that

$$S_0 = \sum_{k \le x} \sum_{d \mid k} \Lambda(d) = \sum_{k \le x} \log k = \log[x]!. \qquad \square$$

SOLUTION 17.2

We have

$$S_1 = \sum_{n \le x} \mu(n) g\left(\frac{x}{n}\right) = \sum_{mn \le x} \mu(n) f\left(\frac{x}{mn}\right) \log \frac{x}{n}$$

and rearrange pairs (m, n) in the same manner as in SOLUTION 17.1; so,

$$S_1 = \sum_{k \le x} f\left(\frac{x}{k}\right) \sum_{d \mid k} \mu(d) \log \frac{x}{d}$$

$$= (\log x) \sum_{k \le x} f\left(\frac{x}{k}\right) \sum_{d \mid k} \mu(d) - \sum_{k \le x} f\left(\frac{x}{k}\right) \sum_{d \mid k} \mu(d) \log d.$$

The first term of the last expression is $f(x) \log x$ by (17.1) and in the second term we can substitute $\sum_{d \mid k} \mu(d) \log d$ by $-\Lambda(k)$ in view of (17.2). $\qquad \square$

SOLUTION 17.3

It follows from Stirling's approximation that

$$\sum_{n\le x} f\left(\frac{x}{n}\right) = \sum_{n\le x} \log\frac{x}{n} - [x] = [x]\log x - [x] - \log[x]!$$

is of order $O(\log x)$, and thus $g(x) = O(\log^2 x)$ as $x \to \infty$. Hence

$$\left|\sum_{n\le x} \mu(n)g\left(\frac{x}{n}\right)\right| \le K\sum_{n\le x} \log^2\left(\frac{x}{n}\right)$$

$$\le K\log^2 x + K\int_1^x \log^2\left(\frac{x}{t}\right) dt$$

$$< K\log^2 x + Kx\int_1^\infty \left(\frac{\log t}{t}\right)^2 dt = O(x)$$

for some constant $K > 0$. Therefore we have by **PROBLEM 17.2**

$$\sum_{n\le x} \Lambda(n)\log\frac{x}{n} = -\log^2 x + \log x + O(x) = O(x)$$

as $x \to \infty$. $\qquad\square$

SOLUTION 17.4

It follows from **PROBLEM 17.1** and from Stirling's approximation that

$$S_0 = \sum_{n\le x} \psi\left(\frac{x}{n}\right) = \log[x]!$$

$$= x\log x - x + O(\log x)$$

as $x \to \infty$. We next have

$$I = \int_1^\infty \frac{\{x\}}{x^2}\, dx = \sum_{k=1}^{n-1} \int_k^{k+1} \frac{x-k}{x^2}\, dx + O\left(\frac{1}{n}\right)$$

$$= \log n - \sum_{k=2}^n \frac{1}{k} + O\left(\frac{1}{n}\right)$$

as $n \to \infty$, where $\{x\}$ denotes the fractional part of x. This implies that $I = 1 - \gamma$ and therefore

$$\sum_{n\le x} \frac{1}{n} = \log x + \gamma + O\left(\frac{1}{x}\right). \tag{17.4}$$

Hence, applying the above estimates to $f(x) = \psi(x) - x + \gamma + 1$ we have

$$\sum_{n \leq x} f\left(\frac{x}{n}\right) = S_0 - \sum_{n \leq x} \frac{x}{n} + (\gamma + 1) \sum_{n \leq x} 1$$

$$= \log [x]! - x\left(\log x + \gamma + O\left(\frac{1}{x}\right)\right) + (\gamma + 1)x + O(1)$$

$$\doteqdot O(\log x).$$

Hence by the same way as the above proof one has

$$\sum_{n \leq x} \mu(n) g\left(\frac{x}{n}\right) = O(x)$$

as $x \to \infty$. Therefore it follows from **PROBLEM 17.2** that

$$\phi(x) = \psi(x) \log x + \sum_{n \leq x} \Lambda(n) \psi\left(\frac{x}{n}\right)$$

$$= x \log x + x \sum_{n \leq x} \frac{\Lambda(n)}{n} + O(x)$$

as $x \to \infty$, because $\psi(x) = O(x)$ by the remark after this solution.

On the other hand, since

$$S_0 = \sum_{mn \leq x} \Lambda(n) = \sum_{n \leq x} \left[\frac{x}{n}\right] \Lambda(n),$$

we have

$$\left|S_0 - x \sum_{n \leq x} \frac{\Lambda(n)}{n}\right| = \sum_{n \leq x} \left\{\frac{x}{n}\right\} \Lambda(n) \leq \psi(x) = O(x).$$

Thus we conclude that

$$\phi(x) = 2x \log x + O(x)$$

as $x \to \infty$. Observe that we have shown that

$$\sum_{n \leq x} \frac{\Lambda(n)}{n} = \log x + O(1) \tag{17.5}$$

in the proof. $\square$

REMARK. We will show the estimate $\psi(x) = O(x)$. The central binomial coefficient

$$\binom{2m}{m} = 2\frac{(2m-1)(2m-2)\cdots(m+1)}{(m-1)!}$$

is a multiple of the product of all primes lying in $[m, 2m)$, which is clearly less than $(1 + 1)^{2m} = 2^{2m}$. Hence

$$\theta(2m) - \theta(m - 1) = \sum_{m \le p < 2m} \log p < 2m \log 2$$

holds for every $m \ge 1$. For any $x > 0$ put $k = [x/2] + 1$. Then

$$\theta(x) - \theta\left(\frac{x}{2}\right) \le \theta(2k) - \theta(k - 1) < 2k \log 2 \le (x + 2) \log 2;$$

so, we have

$$\theta(x) = \sum_{n=0}^{N-1} \left(\theta\left(\frac{x}{2^n}\right) - \theta\left(\frac{x}{2^{n+1}}\right)\right) < 2x \log 2 + O(\log x) = O(x)$$

as $x \to \infty$, where $N = [\log x / \log 2]$. We thus have

$$\psi(x) = \theta(x) + \theta(x^{1/2}) + \theta(x^{1/3}) + \cdots + \theta(x^{1/(N+1)})$$
$$\le \theta(x) + N\theta(x^{1/2}) = O(x),$$

as required.

SOLUTION 17.5

Let $m \ge 2$ be any integer. Since

$$\theta(m) - \theta(m - 1) = \begin{cases} \log m & (m \text{ is a prime}) \\ 0 & (\text{otherwise}) \end{cases},$$

we have

$$\pi(n) = \sum_{m=2}^{n} \frac{\theta(m) - \theta(m - 1)}{\log m}$$

$$= \frac{\theta(n)}{\log n} + \sum_{m=2}^{n-1} \theta(m) \left(\frac{1}{\log m} - \frac{1}{\log(m + 1)}\right).$$

Noticing next that

$$\frac{1}{\log m} - \frac{1}{\log(m + 1)} = \int_m^{m+1} \frac{dt}{t \log^2 t}$$

we get

$$\pi(n) = \frac{\theta(n)}{\log n} + \int_2^n \frac{\theta(t)}{t \log^2 t} dt,$$

because $\theta(t) = \theta(m)$ for $m \le t < m + 1$. Moreover we can replace n by a real variable $x \ge 2$ in view of

$$\int_{[x]}^{x} \frac{\theta(t)}{t \log^2 t} \, dt = \theta([x]) \left(\frac{1}{\log [x]} - \frac{1}{\log x} \right).$$

Since there exists a constant $C > 0$ satisfying $\theta(x) \le Cx$ by the remark after **Solution 17.4**, we have

$$\int_2^x \frac{\theta(t)}{t \log^2 t} \, dt \le C \int_2^x \frac{dt}{\log^2 t}.$$

Then it follows from integrating by parts that

$$\int_2^x \frac{dt}{\log^2 t} = \frac{t}{\log^2 t} \Big|_2^x + 2 \int_2^x \frac{dt}{\log^3 t}$$

$$\le \frac{x}{\log^2 x} + C' + \frac{1}{2} \int_2^x \frac{dt}{\log^2 t}$$

for some constant $C' > 0$, because $2/\log^3 t \le 1/(2 \log^2 t)$ for any $t \ge e^4$. Therefore

$$\int_2^x \frac{\theta(t)}{t \log^2 t} \, dt = O\left(\frac{x}{\log^2 x} \right)$$

and

$$\frac{\pi(x) \log x}{x} - \frac{\theta(x)}{x} = O\left(\frac{1}{\log x} \right)$$

as $x \to \infty$, which implies that the prime number theorem is equivalent to $\theta(x) \sim x$. This completes the proof, because

$$\psi(x) - \theta(x) = O(\sqrt{x} \log x)$$

as is already seen in the remark after **Solution 17.4**. $\square$

Solution 17.6

Put $R(x) = \psi(x) - x$ for $x \ge 1$. It follows from (17.5) and **Problem 17.4** that

$$R(x) \log x + \sum_{n \le x} \Lambda(n) R\left(\frac{x}{n} \right) = O(x)$$

as $x \to \infty$. Replacing x by t, dividing by t and integrating from 1 to x with respect to t we see that

$$\int_1^x \frac{R(t)}{t} \log t \, dt + \int_1^x \sum_{n \le t} \Lambda(n) R\left(\frac{t}{n} \right) \frac{dt}{t} = O(x). \tag{17.6}$$

We first note that the function $U(x)$ is a Lipschitz function. To see this, using $|R(t)| \le Kt$ for some constant $K > 0$ we get

$$|U(x) - U(y)| = \left| \int_x^y \frac{R(t)}{t} \, dt \right| \le K|x - y|.$$

In particular, we see that $U(x) = O(x)$ as $x \to \infty$. Regardless of the apparent, it is not easy to replace this $O(x)$ by $o(x)$.

By integration by parts the first integral in (17.6) becomes

$$U(x) \log x - \int_1^x \frac{U(t)}{t} \, dt = U(x) \log x + O(x).$$

To deal with the second integral in (17.6) let $\widetilde{R}(x)$ and $\widetilde{U}(x)$ be the zero extensions of $R(x)$ and $U(x)$ respectively. Then the second integral can be written as

$$\int_1^x \sum_{n=1}^\infty \Lambda(n) \widetilde{R}\left(\frac{t}{n}\right) \frac{dt}{t} = \sum_{n=1}^\infty \Lambda(n) \int_{1/n}^{x/n} \frac{\widetilde{R}(s)}{s} \, ds$$

$$= \sum_{n=1}^\infty \Lambda(n) \widetilde{U}\left(\frac{x}{n}\right);$$

therefore the second integral differs from $\int_1^x \Lambda(t) U(x/t) \, dt$ by at most

$$\sum_{n=1}^\infty \Lambda(n) \int_n^{n+1} \left| \widetilde{U}\left(\frac{x}{n}\right) - \widetilde{U}\left(\frac{x}{t}\right) \right| dt = O(x),$$

because $\widetilde{U}(x)$ is also a Lipschitz function and

$$\sum_{n=1}^\infty \frac{\Lambda(n)}{n^2} < \infty.$$

Note that the above sums are actually finite. $\square$

SOLUTION 17.7

By partial integration we have

$$\int_1^x \Lambda(v) \log \frac{x}{v} \, dv = \int_1^x \left(\int_1^v \Lambda(u) \, du \right) \frac{dv}{v} = O(x)$$

as $x \to \infty$, which is an integral version of **PROBLEM 17.3**. Hence

$$\iint_{\Delta_x} \Lambda(st) \, ds \, dt = \int_1^x \Lambda(v) \log v \, dv$$

$$= (\log x) \int_1^x \Lambda(v) \, dv + O(x)$$

$$= \psi(x) \log x + O(x).$$

We now estimate the difference

$$\iint_{\Delta_x} \Lambda(s) \Lambda(t) \, ds \, dt - \sum_{mn \le x} \Lambda(m) \Lambda(n) = I + 2J,$$

where

$$I = \left(\int_1^{\sqrt{x}} \Lambda(x) \, dx \right)^2 - \left(\sum_{n \le \sqrt{x}} \Lambda(n) \right)^2,$$

$$J = \iint_{\substack{\sqrt{x} \le s \le x \\ st \le x}} \Lambda(s) \Lambda(t) \, ds \, dt - \sum_{\substack{\sqrt{x} \le n \le x \\ mn \le x}} \Lambda(m) \Lambda(n).$$

It is easily seen that

$$|I| \le \left(\psi(\sqrt{x}) + O(\log x) \right)^2 + \psi^2(\sqrt{x}) = O(x)$$

and

$$|J| \le \sum_{\sqrt{x} \le n \le x} \Lambda(n) \left| \iint_{\Delta_n} \Lambda(t) \, ds \, dt - \sum_{m \le x/n} \Lambda(m) \right| + O(\sqrt{x} \log x)$$

where $\Delta_n = \{ (s,t) \mid n \le s < n+1, 1 \le t \le x/s \}$. Since the length of the interval $[x/(n+1), x/n]$ is less than 1, it contains at most one integer; so $\Lambda(t) \le \log t$ on the interval and

$$\left| \iint_{\Delta_n} \Lambda(t) \, ds \, dt - \sum_{m \le x/n} \Lambda(m) \right| \le \log \frac{x}{n}.$$

Therefore we have $I + 2J = O(x)$ by **PROBLEM 17.3**. Since

$$\sum_{mn \le x} \Lambda(m) \Lambda(n) = \sum_{n \le x} \Lambda(n) \psi\left(\frac{x}{n} \right),$$

it follows from **PROBLEM 17.4** that

$$\iint_{\Delta_x} \left(\Lambda(st) + \Lambda(s) \Lambda(t) \right) ds \, dt = 2x \log x + O(x) \tag{17.7}$$

as $x \to \infty$. $\qquad\qquad\qquad\qquad\qquad\qquad\qquad\qquad\qquad\qquad\qquad\qquad$ $\square$

SOLUTION 17.8

Replacing x by x/s in the estimate given in **PROBLEM 17.6**, multiplying by $\Lambda(s)$ and integrating from 1 to x with respect to s we get

$$(\log x) \int_1^x \Lambda(s) U\left(\frac{x}{s}\right) ds - \int_1^x \Lambda(s)(\log s) U\left(\frac{x}{s}\right) ds$$
$$+ \iint\limits_{\Delta_x} \Lambda(s)\Lambda(t) U\left(\frac{x}{st}\right) ds\,dt = O(x \log x)$$

as $x \to \infty$, because

$$\int_1^x \frac{\Lambda(s)}{s} ds = \sum_{n \le x} \frac{\Lambda(n)}{n} + O(1) = \log x + O(1)$$

by (17.5). Since

$$\int_1^x \Lambda(s)(\log s) U\left(\frac{x}{s}\right) ds = \iint\limits_{\Delta_x} \Lambda(st) U\left(\frac{x}{st}\right) ds\,dt,$$

the estimate in question follows immediately from **PROBLEM 17.6**. $\qquad$ $\square$

SOLUTION 17.9

Let $\epsilon > 0$ be any number. By the uniform continuity of g, there exists a $\delta \in (0, 1)$ such that $|g(s) - g(t)| < \epsilon$ whenever $s, t \in [1, x]$ satisfy $|s - t| < \delta$. Instead of the Lipschitz function $g(t)$ we now consider

$$G(t) = \frac{1}{\delta} \int_{t-\delta}^t \tilde{g}(s)\,ds$$

for $t \ge 1$, where $\tilde{g}(s)$ is the zero extension of $g(s)$. Obviously $G(t)$ is continuously differentiable and satisfies $G(1) = 0$. Since $\tilde{g}$ is also a Lipschitz function with the same constant L, we have

$$|G'(t)| = \frac{|\tilde{g}(t) - \tilde{g}(t - \delta)|}{\delta} \le L.$$

Moreover, since $G(t) = \tilde{g}(\xi_t)$ for some $\xi_t \in (t - \delta, t)$, the function G is very close to g in the sense that

$$|G(t) - g(t)| = |\tilde{g}(\xi_t) - g(t)| < \epsilon$$

for any $t \in [1, x]$. Hence

$$\left| \int_1^x f(t) g\left(\frac{x}{t}\right) dt \right| \le \epsilon \int_1^x |f(t)| \, dt + \left| \int_1^x f(t) G\left(\frac{x}{t}\right) dt \right|.$$

By integration by parts we obtain

$$\int_1^x f(t) G\left(\frac{x}{t}\right) dt = \left(\int_1^t f(s) \, ds \right) G\left(\frac{x}{t}\right) \Big|_{t=1}^{t=x}$$

$$+ x \int_1^x \left(\int_1^t f(s) \, ds \right) G'\left(\frac{x}{t}\right) \frac{dt}{t^2}$$

and therefore, using $|G'(t)| \le L$,

$$\left| \int_1^x f(t) g\left(\frac{x}{t}\right) dt \right| \le \epsilon \int_1^x |f(t)| \, dt + Lx \int_1^x \left| \int_1^t f(s) \, ds \right| \frac{dt}{t^2}.$$

This completes the proof, because ϵ is arbitrary. $\qquad\square$

SOLUTION 17.10

We consider the continuous function

$$\Phi(v) = \int_1^v \left(\Lambda(v) + \Lambda(u) \Lambda\left(\frac{v}{u}\right) - 2 \right) \frac{du}{u}.$$

Then it follows from (17.7) that

$$\int_1^x \Phi(v) \, dv = \iint_{\varDelta_x} (\Lambda(st) + \Lambda(s) \Lambda(t) - 2) \, ds \, dt = O(x),$$

because

$$\iint_{\varDelta_x} ds \, dt = \int_1^x \left(\frac{x}{s} - 1 \right) ds = x \log x - x.$$

We then apply the estimate in **PROBLEM 17.9** to $f(x) = \Phi(x)$ and a Lipschitz function $g(x) = |U(x)|$ so that

$$\iint_{\varDelta_x} (\Lambda(st) + \Lambda(s) \Lambda(t) - 2) \left| U\left(\frac{x}{st}\right) \right| ds \, dt = O(x \log x) \qquad (17.8)$$

as $x \to \infty$. Thus, by **PROBLEM 17.8** and (17.8) we have

$$|U(x)| \log^2 x \leq \iint_{\Delta_x} (\Lambda(st) + \Lambda(s)\Lambda(t)) \left| U\left(\frac{x}{st}\right) \right| ds\,dt + Cx \log x$$

$$\leq 2 \iint_{\Delta_x} \left| U\left(\frac{x}{st}\right) \right| ds\,dt + C'x \log x$$

for some constants $C, C' > 0$. $\qquad\qquad\square$

SOLUTION 17.11

Write $x = e^u$ and make the change of variables $x/s = e^v$, $x/(st) = e^w$ in the integral in

$$|U(x)| \log^2 x \leq 2 \iint_{\Delta_x} \left| U\left(\frac{x}{st}\right) \right| ds\,dt + Cx \log x.$$

Then the triangular region $\{(v,w) \mid 0 \leq w \leq v \leq u\}$ in vw-plane is mapped to Δ_x and the Jacobian is $e^{u-w} > 0$. Hence, putting $V(u) = e^{-u} U(e^u)$ and dividing by $u^2 e^u$ we have

$$|V(u)| \leq \frac{2}{u^2} \int_0^u \int_0^v |V(w)|\, dw\,dv + \frac{C}{u}.$$

Put

$$\alpha = \limsup_{u \to \infty} |V(u)| \quad \text{and} \quad \beta = \limsup_{v \to \infty} \frac{1}{v} \int_0^v |V(w)|\, dw.$$

For any $\epsilon > 0$ there exists a $v_\epsilon > 0$ such that

$$\int_0^v |V(w)|\, dw < (\beta + \epsilon)v$$

for any $v > v_\epsilon$. We then have, using the above estimate,

$$\alpha \leq \limsup_{u \to \infty} \frac{2}{u^2} \left(C_\epsilon u + \frac{\beta + \epsilon}{2} u^2 \right) = \beta + \epsilon$$

where

$$C_\epsilon = \int_0^{v_\epsilon} |V(w)|\, dw.$$

Hence we have $\alpha \leq \beta$. Since the reverse inequality $\alpha \geq \beta$ is obvious, this completes the proof. $\qquad\square$

SOLUTION 17.12

Without loss of generality, we can assume that $\beta > 0$. We first note that the function $f(x)$ must have arbitrarily large zeros. For otherwise, the sign of $f(x)$ does not change on the interval $[x_0, \infty)$ for some x_0 and we can find a divergent sequence $x_0 < x_1 < x_2 < \cdots$ satisfying

$$\int_0^{x_n} |f(s)|\, ds > \frac{\beta}{2} x_n$$

for all $n \geq 1$. Then

$$\left| \int_0^{x_n} f(s)\, ds \right| \geq \int_{x_1}^{x_n} |f(s)|\, ds - \int_0^{x_1} |f(s)|\, ds$$

$$\geq \frac{\beta}{2} x_n - 2 \int_0^{x_1} |f(s)|\, ds,$$

which diverges to $+\infty$, contrary to the assumption that δ is finite.

It follows from the definition of superior limits that, for any $\epsilon > 0$ there is a point $x_\epsilon > 0$ satisfying

$$|f(x)| < \alpha + \epsilon \quad \text{and} \quad \left| \int_0^x f(s)\, ds \right| < \delta + \epsilon$$

for any $x \geq x_\epsilon$. For brevity put

$$\kappa_\epsilon = \frac{8(\delta + \epsilon)L}{(\alpha + \epsilon)^2 + 2(\delta + \epsilon)L} \quad \text{and} \quad \mu_\epsilon = 1 + \frac{2(\delta + \epsilon)L}{(\alpha + \epsilon)^2}.$$

Note that $\mu_\epsilon \geq \kappa_\epsilon$. Then let $a < b$ be any two consecutive zeros of f with $x_\epsilon \leq a$. We distinguish three cases as follows, according to $\sigma = (b - a)L/(\alpha + \epsilon)$.

Case (i) $\sigma < \kappa_\epsilon$.

Since the Lipschitz condition implies that the graph of $|f(x)|$ on the interval $[a, b]$ is contained in the equilateral triangular region with base $[a, b]$ and in height $(b - a)L/2$, we have clearly

$$\mathscr{M}_{[a,b]}(|f|) \leq \frac{(b - a)L}{4} < \frac{\kappa_\epsilon(\alpha + \epsilon)}{4}.$$

Case (ii) $\kappa_\epsilon \leq \sigma \leq \mu_\epsilon$.

By the same reason the graph of $|f(x)|$ is contained in the same triangular region as above. However in this case, since $|f(x)| \leq \alpha + \epsilon$, we can replace it by a smaller trapezoidal region in height $\alpha + \epsilon$ with the same base. Hence

$$\mathcal{M}_{[a,b]}(|f|) \leq \frac{(\alpha + \epsilon)^2}{(b-a)L} + (\alpha + \epsilon)\left(1 - \frac{2(\alpha + \epsilon)}{(b-a)L}\right)$$

$$= (\alpha + \epsilon)\left(1 - \frac{1}{\sigma}\right) \leq (\alpha + \epsilon)\left(1 - \frac{1}{\mu_\epsilon}\right).$$

Case (iii) $\sigma > \mu_\epsilon$.

Since

$$\int_a^b |f(s)|\, ds = \left|\int_a^b f(s)\, ds\right|$$

$$\leq \left|\int_0^a f(s)\, ds\right| + \left|\int_0^b f(s)\, ds\right| < 2(\delta + \epsilon),$$

we have

$$\mathcal{M}_{[a,b]}(|f|) < \frac{2(\delta + \epsilon)}{b-a} < \frac{2(\delta + \epsilon)L}{\mu_\epsilon(\alpha + \epsilon)}.$$

Consequently in each case we have

$$\mathcal{M}_{[a,b]}(|f|) < \frac{2(\alpha + \epsilon)(\delta + \epsilon)L}{(\alpha + \epsilon)^2 + 2(\delta + \epsilon)L}.$$

For any $x > x_\epsilon$ let a^* and b^* be the smallest and the largest zeros of $f(x)$ in the interval $[x_\epsilon, x]$ respectively. Since the sign of $f(x)$ on the interval $(b^*, x]$ does not change, it follows that

$$\int_0^x |f(s)|\, ds \leq \int_0^{a^*} |f(s)|\, ds + \sum_{a<b} \int_a^b |f(s)|\, ds + 2(\delta + \epsilon);$$

hence

$$\beta \leq \frac{2(\alpha + \epsilon)(\delta + \epsilon)L}{(\alpha + \epsilon)^2 + 2(\delta + \epsilon)L}.$$

Letting $\epsilon \to 0+$ we get the desired inequality. $\qquad\qquad\square$

SOLUTION 17.13

If the function $V(x)$ defined in **PROBLEM 17.11** satisfies all the conditions stated in **PROBLEM 17.12**, then

$$\alpha(\alpha^2 + 2\delta L) \leq 2\alpha\delta L,$$

which implies $\alpha = 0$; namely, we would have $U(x) = o(x)$ as $x \to \infty$. Therefore for the proof of the prime number theorem it suffices to show the following two properties:

(i) $V(x)$ is a Lipschitz function.

(ii) $\displaystyle\int_0^x V(s)\,ds$ is bounded.

We first show that $V(x)$ is a Lipschitz function. For any $0 \le x < y$ we have

$$\begin{aligned}
|V(y) - V(x)| &= \left| e^{-y}U(e^y) - e^{-x}U(e^x) \right| \\
&\le e^{-y}|U(e^y) - U(e^x)| + (e^{-x} - e^{-y})|U(e^x)| \\
&\le e^{-y}\int_{e^x}^{e^y} \frac{|\psi(t) - t|}{t}\,dt + (e^{-x} - e^{-y})|U(e^x)| \\
&\le C(1 - e^{x-y})
\end{aligned}$$

for some constant $C > 0$. Thus $V(x)$ is a Lipschitz function in view of

$$1 - e^{x-y} < y - x.$$

Secondly we show the boundedness of the integral $\displaystyle\int_0^x V(s)\,ds$. As is already seen in **Solution 17.8**,

$$\int_1^x \frac{\Lambda(s) - 1}{s}\,ds = O(1)$$

as $x \to \infty$. This integral is equal to

$$\frac{1}{x}\int_1^x (\Lambda(s) - 1)\,ds + \int_1^x \frac{1}{s^2}\int_1^s (\Lambda(t) - 1)\,dt\,ds$$

by integration by parts. Since

$$\int_1^s (\Lambda(t) - 1)\,dt = \psi(s) - s + O(\log s)$$

as $s \to \infty$, the integral

$$\int_1^x \frac{\psi(s) - s}{s^2}\,ds$$

is also bounded. Again it follows by integration by parts that

$$\frac{U(x)}{x} + \int_1^x \frac{U(s)}{s^2}\,ds$$

is bounded, and finally we get the boundedness of the integral

$$\int_1^x \frac{U(s)}{s^2}\, ds = \int_0^{\log x} V(s)\, ds,$$

as required. This completes the proof of the prime number theorem. □

◆　❈　◆

Chapter 18

Bernoulli Numbers

[Summary of Basic Points]

1. The nth *Bernoulli number* B_n is defined as the coefficient of z^n in the Taylor series of the function $z/(e^z - 1)$ about $z = 0$:

$$\frac{z}{e^z - 1} = \sum_{n=0}^{\infty} \frac{B_n}{n!} z^n. \tag{18.1}$$

Note that the radius of convergence of the series is 2π. Multiplying $e^z - 1$ to (18.1) and comparing the coefficients of z^{n+1} in both sides we get the recursion:

$$\sum_{k=0}^{n} \binom{n+1}{k} B_k = \begin{cases} 1 & (n = 0) \\ 0 & (n \geq 1) \end{cases}. \tag{18.2}$$

2. The Bernoulli numbers were introduced by Jakob Bernoulli(1654–1705) in the book 'Ars Conjectandi' published in 1713, eight years after his death. The first eleven non-zero B_n's are:

$$B_0 = 1, \quad B_1 = -\frac{1}{2}, \quad B_2 = \frac{1}{6}, \quad B_4 = -\frac{1}{30},$$

$$B_6 = \frac{1}{42}, \quad B_8 = -\frac{1}{30}, \quad B_{10} = \frac{5}{66}, \quad B_{12} = -\frac{691}{2730},$$

$$B_{14} = \frac{7}{6}, \quad B_{16} = -\frac{3617}{510}, \quad B_{18} = \frac{43867}{798}.$$

Each B_n is rational and $B_{2n+1} = 0$ for all $n \geq 1$. Notice that the numbering and the signs of the Bernoulli numbers are defined differently from the first edition.

3. The nth Bernoulli polynomial $B_n(x)$ is defined by the generating function:

$$\frac{ze^{xz}}{e^z - 1} = \sum_{n=0}^{\infty} B_n(x) \frac{z^n}{n!}. \tag{18.3}$$

It follows easily that $B_n(0) = B_n$, $B_n(1-x) = (-1)^n B_n(x)$ and $B_n'(x) = nB_{n-1}(x)$ for all $n \geq 0$. It also follows easily from (18.1) and (18.3) that

$$B_n(x) = \sum_{k=0}^{n} \binom{n}{k} B_k x^{n-k}.$$

The first eight polynomials are as follows:

$$B_0(x) = 1$$

$$B_1(x) = x - \frac{1}{2}$$

$$B_2(x) = x^2 - x + \frac{1}{6}$$

$$B_3(x) = x^3 - \frac{3}{2}x^2 + \frac{x}{2}$$

$$B_4(x) = x^4 - 2x^3 + x^2 - \frac{1}{30}$$

$$B_5(x) = x^5 - \frac{5}{2}x^4 + \frac{5}{3}x^3 - \frac{x}{6}$$

$$B_6(x) = x^6 - 3x^5 + \frac{5}{2}x^4 - \frac{1}{2}x^2 + \frac{1}{42}$$

$$B_7(x) = x^7 - \frac{7}{2}x^6 + \frac{7}{2}x^5 - \frac{7}{6}x^3 + \frac{x}{6}.$$

The reader may notice that $P_1(2x-1) = 2B_1(x)$ and $P_2(2x-1) = 6B_2(x)$ where $P_n(x)$ is the nth Legendre polynomial.

4. The Bernoulli polynomials have a connection with the sum of mth powers of positive integers. Since $B_n(x)$ satisfies the difference equation

$$f(x+1) - f(x) = nx^{n-1},$$

we have

$$\sum_{k=1}^{n-1} k^m = \frac{B_{m+1}(n) - B_{m+1}(0)}{m+1}$$

$$= \frac{1}{m+1} \sum_{k=0}^{m} \binom{m+1}{k} B_k n^{m-k+1} \qquad (18.4)$$

when $m \geq 1$.

PROBLEM 18.1

Show that

$$\frac{1}{\sin x} = \frac{1}{x} + \sum_{n=1}^{\infty} (-1)^{n-1} \frac{2^{2n} - 2}{(2n)!} B_{2n} x^{2n-1} \quad for \quad 0 < |x| < \pi.$$

PROBLEM 18.2

Show that

$$B_n = (-1)^n n! \begin{vmatrix} \dfrac{1}{2!} & 1 & 0 & \cdots & 0 & 0 \\[2mm] \dfrac{1}{3!} & \dfrac{1}{2!} & 1 & \cdots & 0 & 0 \\[2mm] \dfrac{1}{4!} & \dfrac{1}{3!} & \dfrac{1}{2!} & \cdots & 0 & 0 \\[2mm] \vdots & \vdots & \vdots & \ddots & \vdots & \vdots \\[2mm] \dfrac{1}{n!} & \dfrac{1}{(n-1)!} & \dfrac{1}{(n-2)!} & \cdots & \dfrac{1}{2!} & 1 \\[2mm] \dfrac{1}{(n+1)!} & \dfrac{1}{n!} & \dfrac{1}{(n-1)!} & \cdots & \dfrac{1}{3!} & \dfrac{1}{2!} \end{vmatrix} \quad for \quad n \geq 1.$$

This is referred as 'Laplace's formula' in Lyusternik and Yanpol'skii (1965).

PROBLEM 18.3

Let A/B be the irreducible fraction of

$$1 + \frac{1}{2^m} + \frac{1}{3^m} + \cdots + \frac{1}{(p-1)^m}$$

where $m \geq 1$ is an integer and $p > m+1$ is a prime. Show that $A \equiv 0 \pmod{p}$.

PROBLEM 18.4

Show that

$$\int_0^1 B_n(x) e^{2\pi i m x}\, dx = \begin{cases} 1 & (m = n = 0) \\[2mm] 0 & (m = 0, n \geq 1) \\[2mm] (-1)^{n-1} \dfrac{n!}{(2\pi i m)^n} & (m \in \mathbb{Z}, m \neq 0) \end{cases}.$$

PROBLEM 18.5 ──────────────────────────────

Show that

$$B_{2n} \equiv -\sum_{d}^{*} \frac{1}{d+1} \pmod 1$$

for all integers $n \geq 1$, where d runs over all divisors of $2n$ such that $d + 1$ is a prime. (So, $d = 1$ and $d = 2$ always contribute the sum.)

For example,

$$B_2 = 1 - \frac{1}{2} - \frac{1}{3},$$

$$B_4 = 1 - \frac{1}{2} - \frac{1}{3} - \frac{1}{5},$$

$$B_6 = 1 - \frac{1}{2} - \frac{1}{3} - \frac{1}{7},$$

$$B_8 = 1 - \frac{1}{2} - \frac{1}{3} - \frac{1}{5},$$

$$B_{10} = 1 - \frac{1}{2} - \frac{1}{3} - \frac{1}{11},$$

$$B_{12} = 1 - \frac{1}{2} - \frac{1}{3} - \frac{1}{5} - \frac{1}{7} - \frac{1}{13},$$

$$B_{14} = 2 - \frac{1}{2} - \frac{1}{3}.$$

This is known as the von Staudt-Clausen theorem found by von Staudt (1840) and by Clausen (1840) independently. The latter was published in 'Astronomische Nachrichten', the oldest astronomical journal founded in 1821 by H. C. Schumacher (one of the friends and astronomical collaborators of Gauss), as a brief announcement of the result without proof. See also von Staudt (1845). Later Schwering (1899) gave another proof using

$$\frac{1}{x+1} + \frac{1}{2(x+1)(x+2)} + \cdots + \frac{(n-1)!}{n(x+1)(x+2)\cdots(x+n)} + \cdots$$

$$= \frac{B_0}{x} + \frac{B_1}{x^2} + \frac{B_2}{x^3} + \frac{B_4}{x^5} + \frac{B_6}{x^7} + \cdots$$

and Wilson's theorem, which gave rise to the proof by Kluyver (1900) using Fermat's little theorem. Rado (1934) also gave another proof using (18.4) and Fermat's little theorem. The von Staudt-Clausen theorem is used to define the p-adic interpolation of the Riemann zeta function. See Chapter II in Koblitz (1977).

Clausen was a lover of numbers. He calculated correctly the 248 decimals of π in 1847 (Delahaye (1997)) and factored the 6th Fermat number $2^{64} + 1$ to

$$274, 177 \times 67, 280, 421, 310, 721$$

in 1854 (Schönbeck (2004)).

PROBLEM 18.6

Generalize the method employed in **SOLUTION 10.3** *to obtain*

$$\zeta(2n) = (-1)^{n-1} \frac{2^{2n-1}}{(2n)!} B_{2n} \pi^{2n}. \tag{18.5}$$

Notice that this implies

$$\operatorname{sgn} B_{2n} = (-1)^{n-1} \quad \text{and} \quad |B_{2n}| \sim \frac{2(2n)!}{(2\pi)^{2n}}$$

as $n \to \infty$. This proof is due to Apostol (1973). Williams (1971) gave a similar proof but used complex function theory in addition. Skau and Selmer (1971) also discussed a similar method to obtain the rationality of $\zeta(2n)/\pi^{2n}$. See also Hovstad (1972). Chen (1975) gave a proof along the same lines. Robbins (1999) gave another proof for Chen's recursive formula, using the Fourier expansion of x^{2n} on the interval $[-\pi, \pi]$. Ji and Chen (2000) proved inductively using the Fourier expansion of some quadratic function and iterated integrations.

On the other hand, Kuo (1949) obtained a rather complicated formula which represents $\zeta(2n)$ as a sum of $\zeta(2), \ldots, \zeta(2[n/2])$ and their products, while the proof requires the Fourier series appeared in **PROBLEM 7.13** and Parseval's theorem. Kuo's formula is equivalent to a formula for the Bernoulli numbers, for which Carlitz (1961) gave a simpler and slightly general formula using the Bernoulli polynomials.

Stark (1972) considered even moments of the Dirichlet kernel

$$D_n(\theta) = \frac{\sin(n + 1/2)\theta}{2 \sin(\theta/2)} = \frac{1}{2} + \sum_{k=1}^{n} \cos k\theta,$$

as a generalization of his earlier work in 1969, in which he had used the Fejér kernel. See the comment after **PROBLEM 10.4**. He also pointed out that the method employed in **SOLUTION 10.7** is implicitly based upon the de la Vallée Poussin kernel. Again Stark (1974) used the Fejér kernel to give a simpler proof for the recursive formula obtained in 1972.

Berndt (1975) gave two elementary proofs, the first is based on the calculation of the Fourier coefficients of the Bernoulli polynomials, and the second is based on the partial fraction expansion

$$\pi^2 \operatorname{cosec}^2 \pi x = \sum_{k=-\infty}^{\infty} \frac{1}{(k + x)^2},$$

whose elementary proof was given by Neville (1951).

Osler (2004) used the product formula of sines presented in **PROBLEM 2.5**. Tsumura (2004) evaluated $\zeta(2m)$ in a different way.

PROBLEM 18.7 ─────────────────────────────────

Show that

$$\left(n + \frac{1}{2}\right)\zeta(2n) = \sum_{k=1}^{n-1} \zeta(2k)\zeta(2n-2k) \tag{18.6}$$

for any integer $n \geq 2$ without using the Bernoulli numbers.

This is due to Williams (1953), which enables us to determine the value of $\zeta(2n)$ from $\zeta(2)$. Two somewhat complicated recursive formulae satisfied by $\zeta(2k)$ were already given in Titchmarsh (1926). Estermann (1947) gave a recursive formula

$$(2^{2n} - 1)\zeta(2n) = \sum_{k=1}^{n-1} a_{k,n}\,\zeta(2k)\,\zeta(2n-2k)$$

where $a_{1,n} = 2^{2n} - 10$ and $a_{k,n} = -2(2k-1)(2^{2n-2k}-1)$ for $k \geq 2$. Their formulae are obtained only by rearranging absolutely convergent series.

In order to deduce (18.5) from (18.6), it suffices to show that

$$(2n + 1)B_{2n} = -\sum_{k=1}^{n-1}\binom{2n}{2k}B_{2k}B_{2n-2k}$$

for all $n \geq 2$. This formula is given by von Staudt (1845), in which he uses the notations:

$$\overset{n}{A} = \frac{B_n}{n!} \quad \text{and} \quad \overset{n}{B} = (-1)^{n-1}B_{2n}.$$

See also Underwood (1928). The direct proof from the definition of Bernoulli numbers can be found in Berndt (1975), as follows. Put

$$f(z) = \frac{z}{e^z - 1} = 1 - \frac{z}{2} + g(z)$$

for brevity. The coefficient of z^{2n} for $n \geq 2$ in the Taylor series of $(z f(z))'$ about $z = 0$ is

$$\frac{2n + 1}{(2n)!}B_{2n}.$$

On the other hand,

$$(z f(z))' = (2 - z)f(z) - f^2(z) = \left(1 - \frac{z}{2}\right)^2 - g^2(z)$$

and the coefficient of z^{2n} for $n \geq 2$ on the right-hand side is equal to

$$-\sum_{k=1}^{n-1} \frac{B_{2k}B_{2n-2k}}{(2k)!(2n-2k)!},$$

which implies the desired recursive formula.

PROBLEM 18.8

To generalize the method in **SOLUTION 10.12** *consider the 2n-dimensional improper integral:*

$$\int\!\!\cdots\!\!\int\limits_{(0,1)^{2n}} \frac{dx_1 dx_2 \cdots dx_{2n}}{1 - (x_1 x_2 \cdots x_{2n})^2}.$$

Carry out the transformation

$$x_1 = \frac{\sin\theta_1}{\cos\theta_2}, \quad \ldots, \quad x_{2n-1} = \frac{\sin\theta_{2n-1}}{\cos\theta_{2n}}, \quad x_{2n} = \frac{\sin\theta_{2n}}{\cos\theta_1}$$

to show that

$$\zeta(2n) = (-1)^{n-1} \frac{2^{2n-1}}{(2n)!} B_{2n} \pi^{2n}.$$

This is due to Beukers, Calabi and Kolk (1993). See also the remark given after **PROBLEM 10.12**.

PROBLEM 18.9

Let $\Phi(z) = \sum\limits_{n=0}^{\infty} a_n z^n$ be a series whose radius of convergence is greater than $(2\pi)^{-1}$. Show that

$$\lim_{N\to\infty} \sum_{1 \le |n| \le N} \frac{1}{(2\pi n)^2} \Phi\left(\frac{i}{2\pi n}\right) = \sum_{n=0}^{\infty} a_n \frac{B_{n+2}}{(n+2)!}.$$

Note that the limit is not affected by any odd term of $\Phi(z)$. For example, if $\Phi(z) = z^{2n-2}$, then we have the formula (18.5). As another example, we take

$$\Phi(z) = \frac{1}{1 - 2\pi s z}$$

for any $|s| < 1$. Then, using (18.1) we have

$$\sum_{n=1}^{\infty} \frac{1}{n^2 + s^2} = \frac{1}{2s^2} \left(\frac{2\pi s}{e^{2\pi s} - 1} - 1 + \pi s \right),$$

which holds for every $s \in \mathbb{C}$ except for $\pm i, \pm 2i, \ldots$ by analytic continuation. In particular, putting $s = 1, 1 + i$ we get

$$\sum_{n=1}^{\infty} \frac{1}{n^2 + 1} = \sum_{n=1}^{\infty} \frac{4}{n^4 + 4} = \frac{\pi}{2} \coth \pi - \frac{1}{2}$$

respectively, where 'coth' is the hyperbolic cotangent function defined by

$$\coth x = \frac{\cosh x}{\sinh x} = \frac{e^x + e^{-x}}{e^x - e^{-x}}.$$

Solutions for Chapter 18

SOLUTION 18.1

We have

$$\frac{1}{\sin x} = \frac{2ie^{ix}}{e^{2ix} - 1} = \frac{2i}{e^{ix} - 1} - \frac{2i}{e^{2ix} - 1}$$

$$= 2i \sum_{n=0}^{\infty} \frac{B_n}{n!}(ix)^{n-1} - 2i \sum_{n=0}^{\infty} \frac{B_n}{n!}(2ix)^{n-1}$$

for $0 < |x| < \pi$. Hence

$$\frac{1}{\sin x} = \frac{1}{x} + \sum_{n=1}^{\infty} (2i^n - (2i)^n) \frac{B_n}{n!} x^{n-1}$$

$$= \frac{1}{x} + \sum_{n=1}^{\infty} (-1)^{n-1} \frac{2^{2n} - 2}{(2n)!} B_{2n} x^{2n-1}.$$ □

SOLUTION 18.2

Put $C_k = B_k/k!$. From (18.2) we have

$$\sum_{k=0}^{n} \frac{C_k}{(n+1-k)!} = \begin{cases} 1 & (n = 0) \\ 0 & (n \geq 1) \end{cases},$$

which can be written as

$$\begin{pmatrix} 1 & 0 & 0 & \cdots & 0 & 0 \\ \frac{1}{2!} & 1 & 0 & \cdots & 0 & 0 \\ \frac{1}{3!} & \frac{1}{2!} & 1 & \cdots & 0 & 0 \\ \vdots & \vdots & \vdots & \ddots & \vdots & \vdots \\ \frac{1}{n!} & \frac{1}{(n-1)!} & \frac{1}{(n-2)!} & \cdots & 1 & 0 \\ \frac{1}{(n+1)!} & \frac{1}{n!} & \frac{1}{(n-1)!} & \cdots & \frac{1}{2!} & 1 \end{pmatrix} \begin{pmatrix} C_0 \\ C_1 \\ C_2 \\ \vdots \\ C_{n-1} \\ C_n \end{pmatrix} = \begin{pmatrix} 1 \\ 0 \\ 0 \\ \vdots \\ 0 \\ 0 \end{pmatrix}.$$

Since the $(n + 1) \times (n + 1)$ matrix in the left-hand side is lower triangular, its determinant is 1. Hence it follows from Cramer's rule that

$$
C_n = \begin{vmatrix}
1 & 0 & \cdots & 0 & 1 \\
\dfrac{1}{2!} & 1 & \cdots & 0 & 0 \\
\vdots & \vdots & \ddots & \vdots & \vdots \\
\dfrac{1}{n!} & \dfrac{1}{(n-1)!} & \cdots & 1 & 0 \\
\dfrac{1}{(n+1)!} & \dfrac{1}{n!} & \cdots & \dfrac{1}{3!} & 0
\end{vmatrix}.
$$

Laplace expansion along the last column yields the desired formula. $\qquad\square$

SOLUTION 18.3

Let $1 \le a_k < p$ be the first digit in the canonical p-adic expansion of $1/k \in \mathbb{Z}_p$ for each $1 \le k < p$. Since the correspondence $x \mapsto x^{-1}$ is an automorphism on $(\mathbb{Z}/p\mathbb{Z})^\times$, we have

$$
\{a_1, a_2, ..., a_{p-1}\} = \{1, 2, ..., p - 1\}
$$

as sets. Let b_k be the first digit in the canonical p-adic expansion of $1/k^m \in \mathbb{Z}_p$. Then $b_k \equiv a_k^m \pmod{p}$ and hence

$$
b_1 + b_2 + \cdots + b_{p-1} \equiv 1^m + 2^m + \cdots + (p-1)^m = S_m(p) \pmod{p}
$$

where

$$
S_m(p) = \frac{1}{m+1} \sum_{k=0}^{m} \binom{m+1}{k} B_k p^{m-k+1}
$$

from (18.4). Let $|x|_p$ be the p-adic norm of $x \in \mathbb{Q}_p$. Since $p > m + 1$, we get

$$
|S_m(p)|_p \le \max_{0 \le k \le m} \frac{|B_k|_p}{p^{m-k+1}}.
$$

Now it follows from the recursion formula (18.2) that $(n + 1)!B_n \in \mathbb{Z}$ for every $n \ge 0$. Hence we have

$$
1 \ge |(k + 1)!B_k|_p = |(k + 1)!|_p \cdot |B_k|_p = |B_k|_p
$$

for $1 \le k \le m$; therefore $|S_m(p)|_p < 1$ and $S_m(p) \in p\mathbb{Z}_p$; so, $|A/B|_p < 1$. $\qquad\square$

SOLUTION 18.4

For $m \in \mathbb{Z}, n \geq 0$ put

$$c_{m,n} = \int_0^1 B_n(x) e^{2\pi imx} \, dx.$$

Applying Cauchy's integral formula to (18.3) we have

$$B_n(x) = \frac{n!}{2\pi i} \int_{C_r} \frac{e^{xz}}{z^n(e^z - 1)} \, dz$$

where C_r is the circle centered at $z = 0$ with radius $r < 2\pi$. Thus

$$\max_{0 \leq x \leq 1} |B_n(x)| \leq e^r \frac{n!}{r^{n-1}} \max_{|z|=r} \frac{1}{|e^z - 1|} = O\left(\frac{n!}{r^n}\right)$$

as $n \to \infty$. This means that the series (18.3) converges uniformly on $[0, 1]$ as the series of functions in x for each fixed $|z| < 2\pi$. Hence, integrating by parts we obtain

$$\sum_{n=0}^{\infty} \frac{c_{m,n}}{n!} z^n = \frac{z}{e^z - 1} \int_0^1 e^{x(z+2\pi im)} \, dx$$

$$= \frac{z}{e^z - 1} \cdot \frac{e^z - 1}{z + 2\pi im} = \frac{z}{z + 2\pi im}.$$

When $m \neq 0$, the right-hand side can be expanded as

$$\sum_{n=1}^{\infty} (-1)^{n-1} \left(\frac{z}{2\pi im}\right)^n,$$

which completes the proof.   □

SOLUTION 18.5

This is based on Kluyver's proof. Expanding $z = \log(1 + e^z - 1)$ in $e^z - 1$ in a neighborhood of $z = 0$ and dividing by $e^z - 1$, we get

$$f(z) = \frac{z}{e^z - 1} = 1 - \frac{e^z - 1}{2} + \frac{(e^z - 1)^2}{3} - \cdots.$$

Hence

$$B_{2n} = f^{(2n)}(0) = -\frac{c_{1,2n}}{2} + \frac{c_{2,2n}}{3} - \frac{c_{3,2n}}{4} + \cdots + \frac{c_{2n,2n}}{2n + 1},$$

where

$$c_{k,m} = \left((e^z - 1)^k\right)^{(m)}\bigg|_{z=0}.$$

The basic fact is that $c_{k,m}$ is always divisible by $k!$. We will show this by induction on $k \geq 1$. The first case $k = 1$ is trivial. Suppose that there exists $k \in \mathbb{N}$ such that mth derivative of $(e^z - 1)^k$ at $z = 0$ is divisible by $k!$ for all $m \geq 0$. We see that

$$((e^z - 1)^{k+1})^{(m)}\Big|_{z=0} = (k + 1)((e^z - 1)^k e^z)^{(m-1)}\Big|_{z=0}$$

$$= (k + 1) \sum_{\ell=0}^{m-1} \binom{m-1}{\ell}((e^z - 1)^k)^{(\ell)}\Big|_{z=0}$$

and the right-hand side is divisible by $(k + 1)k!$ from the assumption, which completes the induction step.

If $k + 1 = ab$ is a composite number greater than 4, then $a + b \leq k$ and

$$c_{k,2n} = \left((e^z - 1)^a (e^z - 1)^b (e^z - 1)^{k-a-b}\right)^{(2n)}\Big|_{z=0}$$

is divisible by $a!b!(k - a - b)!$; hence $c_{k,2n}/ab$ is an integer. If $k + 1 = 4$, then

$$c_{3,2n} = 3 - 3 \cdot 2^{2n} + 3^{2n} \equiv 0 \pmod{4}$$

so that $c_{3,2n}/4$ is an integer. If $k + 1 = p$ is an odd prime, then $\ell^k \equiv 1 \pmod{p}$ for $1 \leq \ell < p$ by Fermat's little theorem. Since

$$c_{k,m} = \sum_{\ell=1}^{k} (-1)^{k-\ell} \binom{k}{\ell} \ell^m,$$

it follows that $c_{k,m} \pmod{p}$ is periodic in m with period k. Furthermore we know that $c_{k,1} = c_{k,2} = \cdots = c_{k,k-1} = 0$ and

$$c_{k,k} \equiv \sum_{\ell=1}^{k} (-1)^{k-\ell} \binom{k}{\ell} \equiv -1 \pmod{p}.$$

Therefore $c_{p-1,2n} \equiv -1 \pmod{p}$ if and only if $2n$ is divisible by $p - 1$; otherwise $c_{p-1,2n} \equiv 0 \pmod{p}$. This holds also for $p = 2$, because $c_{1,2n} = 1$. $\square$

REMARK. It can be easily seen that $c_{m,m} = m!$ for all $m \geq 1$. Thus it follows from Wilson's theorem that $c_{p-1,p-1} \equiv -1 \pmod{p}$.

| SOLUTION 18.6 |

Since $\cot^2 \theta < \theta^{-2} < 1 + \cot^2 \theta$ for $0 < \theta < \pi/2$, we have

$$\cot^{2n} \theta < \frac{1}{\theta^{2n}} < (1 + \cot^2 \theta)^n.$$

As is already seen in **SOLUTION 10.3**, the m roots of the polynomial

$$\varphi(x) = \sum_{k=0}^{m} (-1)^k \binom{2m+1}{2k+1} x^{m-k}$$

of degree m are

$$x_k = \cot^2 \frac{k\pi}{2m+1}$$

for $1 \le k \le m$. Putting $s_n(m) = x_1^n + \cdots + x_m^n$ and using the inequalities mentioned above we get

$$s_n(m) < \frac{(2m+1)^{2n}}{\pi^{2n}} \sum_{k=1}^{m} \frac{1}{k^{2n}} < \sum_{k=1}^{m} (1 + x_k)^n.$$

Observe that the right-hand side is expanded as

$$s_n(m) + \binom{n}{1} s_{n-1}(m) + \cdots. \tag{18.7}$$

Thus, if $s_n(m)$ is asymptotic to $c_n m^{2n}$ with some constant c_n as $m \to \infty$, then (18.7) is asymptotic to the same one and we can conclude that

$$\zeta(2n) = c_n \left(\frac{\pi}{2} \right)^{2n}.$$

Since $s_n(m)$ is a symmetric function in $x_1, \ldots, x_m$, it can be expressed as a sum of elementary symmetric functions. Indeed, by Newton's identities we have

$$\binom{2m+1}{1} s_n(m) - \binom{2m+1}{3} s_{n-1}(m) + \cdots$$
$$+ (-1)^{n-1} \binom{2m+1}{2n-1} s_1(m) + (-1)^n \binom{2m+1}{2n+1} n = 0$$

if $n \le m$. From this formula we will show that

$$s_n(m) \sim c_n m^{2n} \quad (m \to \infty) \quad \text{with} \quad c_n = (-1)^{n-1} \frac{2^{4n-1}}{(2n)!} B_{2n}$$

by induction on n. Obviously it is true when $n = 1$ in view of $c_1 = 2/3 = 4B_2$. Suppose now that it is true up to n. Then by Newton's identities

$$\lim_{m \to \infty} \frac{s_{n+1}(m)}{m^{2n+2}}$$

exists and is equal to

$$\frac{2^2}{3!} c_n - \frac{2^4}{5!} c_{n-1} + \cdots - (-1)^n \frac{2^{2n}}{(2n+1)!} c_1 + (-1)^n \frac{(n+1)2^{2n+2}}{(2n+3)!}.$$

By the assumption this is written as

$$(-1)^{n-1}2^{4n+4}\left(\sum_{k=1}^{n}\frac{2^{-(2n-2k+3)}}{(2n-2k+3)!}\cdot\frac{B_{2k}}{(2k)!}-\frac{n+1}{2^{2n+2}(2n+3)!}\right).$$

The expression in the parentheses added to

$$\frac{1}{2(2n+2)!}B_{2n+2}$$

is exactly equal to the coefficient of z^{2n+3} in the Taylor series of $ze^{z/2}/(e^z-1)$ at $z=0$. However this function is even and the coefficient of z^{2n+3} vanishes. Hence we obtain

$$\lim_{m\to\infty}\frac{s_{n+1}(m)}{m^{2n+2}}=(-1)^n\frac{2^{4n+3}}{(2n+2)!}B_{2n+2},$$

which completes the induction. $\qquad\square$

SOLUTION 18.7

We start with

$$\sum_{k=1}^{n-1}\zeta(2k)\zeta(2n-2k)=\lim_{m\to\infty}S_m,$$

where

$$S_m=\sum_{k=1}^{n-1}\sum_{j=1}^{m}\sum_{\ell=1}^{m}\frac{1}{j^{2k}\ell^{2n-2k}}$$

$$=(n-1)\sum_{j=1}^{m}\frac{1}{j^{2n}}+\sum_{1\le j\ne\ell\le m}\frac{\ell^{-2n+2}-j^{-2n+2}}{j^2-\ell^2}.$$

Let T_m denote the second sum on the right-hand side. Since

$$\sum_{1\le j\ne\ell\le m}\frac{-j^{-2n+2}}{j^2-\ell^2}=\sum_{1\le j\ne\ell\le m}\frac{-\ell^{-2n+2}}{\ell^2-j^2},$$

it follows that

$$T_m=2\sum_{1\le j\ne\ell\le m}\frac{1}{\ell^{2n-2}(j^2-\ell^2)}=2\sum_{\ell=1}^{m}\frac{c_\ell}{\ell^{2n-2}},$$

where

$$c_\ell=\sum_{j}\frac{1}{j^2-\ell^2}$$

and the j in the summation runs through $[1, m]$ except for ℓ. We then have

$$c_\ell = \frac{1}{2\ell} \sum_j \left(\frac{1}{j-\ell} - \frac{1}{j+\ell} \right) = \frac{3}{4\ell^2} - \frac{1}{2\ell} \sum_{j=m-\ell+1}^{m+\ell} \frac{1}{j},$$

so that

$$T_m = \frac{3}{2} \sum_{\ell=1}^{m} \frac{1}{\ell^{2n}} + O\left(\sum_{\ell=1}^{m} \frac{1}{\ell^{2n-1}} \sum_{j=m-\ell+1}^{m+\ell} \frac{1}{j} \right).$$

Note that the cancellation is similar to that in **SOLUTION 10.1**. Since

$$\sum_{j=m-\ell+1}^{m+\ell} \frac{1}{j} < \frac{2\ell}{m-\ell+1},$$

we see that the error term to T_m is estimated above by

$$\sum_{\ell=1}^{m} \frac{2\ell}{\ell^{2n-1}(m-\ell+1)} \le 2 \sum_{\ell=1}^{m} \frac{1}{\ell(m-\ell+1)}$$

$$= \frac{2}{m+1} \sum_{\ell=1}^{m} \left(\frac{1}{\ell} + \frac{1}{m-\ell+1} \right) = O\left(\frac{\log m}{m} \right)$$

as $m \to \infty$. Therefore

$$S_m = \left(n + \frac{1}{2} \right) \sum_{j=1}^{m} \frac{1}{j^{2n}} + O\left(\frac{\log m}{m} \right),$$

which completes the proof. $\square$

SOLUTION 18.8

Let I_n be the improper integral in the problem. It follows from termwise integration that

$$I_n = \sum_{k=0}^{\infty} \frac{1}{(2k+1)^{2n}} = \left(1 - \frac{1}{2^{2n}} \right) \zeta(2n).$$

Under the transformation

$$x_1 = \frac{\sin\theta_1}{\cos\theta_2}, \quad \ldots, \quad x_{2n-1} = \frac{\sin\theta_{2n-1}}{\cos\theta_{2n}}, \quad x_{2n} = \frac{\sin\theta_{2n}}{\cos\theta_1},$$

the hypercube $\square = (0, 1)^{2n}$ is the image of the polytope $\triangle$ defined by

$$\theta_1 + \theta_2 < \frac{\pi}{2}, \quad \ldots, \quad \theta_{2n-1} + \theta_{2n} < \frac{\pi}{2}, \quad \theta_{2n} + \theta_1 < \frac{\pi}{2}.$$

To see this, first observe that $0 < x_1, \ldots, x_{2n} < 1$ for any $(\theta_1, \ldots, \theta_{2n}) \in \Delta$. Conversely, for any $(x_1, \ldots, x_{2n}) \in \square$,

$$\phi(t) = x_{2n}^2 (1 - x_1^2 (1 - \cdots (1 - x_{2n-1}^2 (1 - t)) \cdots))$$

is a linear function on t and there exists a unique t^* in the open interval $(0, 1)$ satisfying $\phi(t) = t$, because $\phi(0), \phi(1) \in (0, 1)$. Using the point t^* we can determine $\theta_{2n}, \ldots, \theta_1$ in the interval $(0, \pi/2)$ in this order by the formulae

$$\sin \theta_{2n} = \sqrt{t^*}, \quad \sin \theta_{2n-1} = x_{2n-1} \cos \theta_{2n}, \quad \ldots, \quad \sin \theta_1 = x_1 \cos \theta_2.$$

The polytope Δ is thus mapped onto the hypercube $\square$ diffeomorphically, because its Jacobian

$$\begin{vmatrix} u_{1,2} & 0 & 0 & \cdots & 0 & v_{2n,1} \\ v_{1,2} & u_{2,3} & 0 & \cdots & 0 & 0 \\ 0 & v_{2,3} & u_{3,4} & \cdots & 0 & 0 \\ \vdots & \vdots & \vdots & \ddots & \vdots & \vdots \\ 0 & 0 & 0 & \cdots & u_{2n-1,2n} & 0 \\ 0 & 0 & 0 & \cdots & v_{2n-1,2n} & u_{2n,1} \end{vmatrix}$$

where

$$u_{j,k} = \frac{\cos \theta_j}{\cos \theta_k} \quad \text{and} \quad v_{j,k} = \tan \theta_k \frac{\sin \theta_j}{\cos \theta_k},$$

is equal to

$$u_{1,2} u_{2,3} \cdots u_{2n,1} - v_{1,2} v_{2,3} \cdots v_{2n,1} = 1 - (x_1 \cdots x_{2n})^2 > 0.$$

Notice that this is the denominator of the integrand considered.

Let Δ^* be the image of Δ by the linear contraction in the ratio of $2/\pi$; so,

$$\Delta^* = \left\{ (s_1, \ldots, s_{2n}) \in \mathbb{R}^{2n} \mid s_1, \ldots, s_{2n} > 0, s_1 + s_2 < 1, \ldots, s_{2n} + s_1 < 1 \right\}.$$

Thus we have

$$\zeta(2n) = \frac{2^{2n}}{2^{2n} - 1} \int \cdots \int_\Delta d\theta_1 \cdots d\theta_{2n}$$

$$= \frac{\pi^{2n}}{2^{2n} - 1} |\Delta^*|_{2n},$$

where $|X|_m$ denotes the m-dimensional volume of a set X. For each $1 \leq j \leq 2n$ we put

$$\Omega_j = \left\{ (s_1, \ldots, s_{2n}) \subset \Delta^* \mid s_j < s_k \text{ for any } k \neq j \right\}.$$

Obviously Ω_j's are disjoint and congruent mutually; so,

$$\overline{\Delta^*} = \bigcup_{j=1}^{2n} \overline{\Omega_j}$$

where $\overline{X}$ denotes the closure of a set X and hence

$$|\Delta^*|_{2n} = 2n\,|\Omega_{2n}|_{2n}.$$

Now let Σ be the subset of $\overline{\Omega_{2n}}$ defined by $s_{2n} = 0$; namely, Σ is the set of points $(s_1, ..., s_{2n-1}, 0)$ in $\mathbb{R}^{2n}$ satisfying $s_1, ..., s_{2n-1} \geq 0$, $s_1 + s_2 \leq 1, ..., s_{2n-2} + s_{2n-1} \leq 1$ and $s_{2n-1} \leq 1$. The set $\overline{\Omega_{2n}}$ has a pyramid-like shape with the base Σ and

$$\left(\frac{1}{2}, \frac{1}{2}, ..., \frac{1}{2}\right)$$

is one of vertices of $\overline{\Omega_{2n}}$, because every point in $\overline{\Omega_{2n}}$ can be uniquely expressed as

$$\left((1-t)s_1 + \frac{t}{2}, ..., (1-t)s_{2n-1} + \frac{t}{2}, \frac{t}{2}\right)$$

with some $(s_1, ..., s_{2n-1}, 0) \in \Sigma$ and $t \in [0, 1]$. This gives the transformation from $\Sigma \times [0, 1]$ onto $\overline{\Omega_{2n}}$ with the Jacobian

$$\begin{vmatrix} 1-t & 0 & \cdots & 0 & 0 \\ 0 & 1-t & \cdots & 0 & 0 \\ \vdots & \vdots & \ddots & \vdots & \vdots \\ 0 & 0 & \cdots & 1-t & 0 \\ \frac{1}{2} - s_1 & \frac{1}{2} - s_2 & \cdots & \frac{1}{2} - s_{2n-1} & \frac{1}{2} \end{vmatrix} = \frac{1}{2}(1-t)^{2n-1} > 0.$$

Hence we have

$$|\Omega_{2n}|_{2n} = \frac{1}{2} \int\cdots\int_{\Sigma} \int_0^1 (1-t)^{2n-1}\, dt\, ds_1 \cdots ds_{2n-1}$$

$$= \frac{1}{4n} \int\cdots\int_{\Sigma} ds_1 \cdots ds_{2n-1};$$

therefore

$$\zeta(2n) = \frac{\pi^{2n}}{2(2^{2n} - 1)}\,|\Sigma|_{2n-1}.$$

Our problem is thus reduced to the evaluation of the $(2n-1)$-dimensional volume of Σ. It follows from the definition of Σ that

$$|\Sigma|_{2n-1} = \int_0^1 \left(\int_0^{1-s_1} \cdots \left(\int_0^{1-s_{2n-2}} ds_{2n-1} \right) \cdots ds_2 \right) ds_1.$$

We now define a sequence of polynomials $p_m(x)$ inductively by

$$p_{m+1}(x) = \int_0^{1-x} p_m(t)\,dt$$

with the initial condition $p_0(x) = 1$. We see that $|\Sigma|_{2n-1} = p_{2n-1}(0)$ and that $\{p_m(x)\}_{m \geq 0}$ is bounded on $[0, 1]$. Then the generating function

$$F(x, \theta) = \sum_{m=0}^{\infty} p_m(x) \theta^m$$

defined for $|\theta| < 1$ satisfies the second-order differential equation

$$\frac{\partial^2 F}{\partial x^2} + \theta^2 F = 0$$

and its general solution is

$$F = a(\theta) \cos \theta x + b(\theta) \sin \theta x.$$

The functions $a(\theta)$ and $b(\theta)$ are determined by the initial conditions $F(1, \theta) = 1$ and $\partial F/\partial x(0, \theta) = -\theta$; thus,

$$F(x, \theta) = \frac{1 + \sin \theta}{\cos \theta} \cos \theta x - \sin \theta x.$$

Substituting $x = 0$ we obtain

$$\sum_{m=0}^{\infty} p_m(0) \theta^m = \sec \theta + \tan \theta,$$

where $\sec \theta$ is the reciprocal of $\cos \theta$. Since $\sec \theta$ is an even function, $p_{2n-1}(0)$ is determined only by the Taylor series of $\tan \theta$ about $\theta = 0$. Indeed,

$$\tan \theta = -\frac{4i}{e^{4\theta i} - 1} + \frac{2i}{e^{2\theta i} - 1} - i$$

$$= -i + \frac{1}{\theta} \sum_{n=0}^{\infty} ((2\theta i)^n - (4\theta i)^n) \frac{B_n}{n!}$$

$$= \sum_{n=1}^{\infty} (-1)^{n-1} \frac{2^{2n}(2^{2n} - 1)}{(2n)!} B_{2n} \theta^{2n-1},$$

which implies that

$$p_{2n-1}(0) = (-1)^{n-1} \frac{2^{2n}(2^{2n} - 1)}{(2n)!} B_{2n}.$$

This completes the proof. $\square$

REMARK. The method used here to evaluate the volume of the polytope Σ can be found in Macdonald and Nelsen (1979). Also see Elkies (2003).

SOLUTION 18.9

Let $\rho > (2\pi)^{-1}$ be the radius of convergence of the power series $\Phi(z)$. It then follows from (18.1) that

$$\frac{\Phi(z)}{e^{1/z} - 1} = \sum_{m,n \geq 0} \frac{a_m B_n}{n!} z^{m-n+1}$$

holds on the annular domain $\{(2\pi)^{-1} < |z| < \rho\}$. Hence

$$I = \frac{1}{2\pi i} \int_C \frac{\Phi(z)}{e^{1/z} - 1} dz = \sum_{m=0}^{\infty} \frac{a_m B_{m+2}}{(m+2)!},$$

where C is a circle centered at $z = 0$ and oriented counterclockwise with radius $r \in ((2\pi)^{-1}, \rho)$. Since the function $f(z) = \Phi(z)/(e^{1/z} - 1)$ is holomorphic in the disk $\{|z| < \rho\}$ except for $z = 0$ and $1/(2n\pi i)$, $n \in \mathbb{Z}$, it follows that

$$I = \frac{1}{2\pi i} \int_{C_N} \frac{\Phi(z)}{e^{1/z} - 1} dz + \sum_{1 \leq |n| \leq N} \operatorname*{Res}_{z=i/(2n\pi)} f(z),$$

where C_N is a circle centered at $z = 0$ and oriented counterclockwise with radius $1/((2N + 1)\pi)$.

Observe now that there exists a positive constant c_0, independent of N, satisfying $|e^{1/z} - 1| \geq c_0$ on all circles C_N. To see this, putting $w = 1/z = (2N + 1)\pi e^{i\theta}$ and $1 - e^{-\pi} = \sin \theta^*$ ($0 < \theta^* < \pi/2$) we distinguish two cases, as follows.

Case (i) $\cos((2N + 1)\pi \sin \theta) \leq \cos \theta^*$.

Putting $X = e^{(2N+1)\pi \cos \theta}$ we have

$$\begin{aligned} |e^w - 1|^2 &= X^2 + 1 - 2X \cos((2N + 1)\pi \sin \theta) \\ &\geq X^2 + 1 - 2X \cos \theta^* \\ &= (X - \cos \theta^*)^2 + \sin^2 \theta^* \geq \sin^2 \theta^*. \end{aligned}$$

Case (ii) $\cos((2N + 1)\pi \sin\theta) > \cos\theta^*$.

Since $\sin\theta^* > |\sin((2N + 1)\pi \sin\theta)|$, we can write $(2N + 1)\pi \sin\theta = 2k\pi + \epsilon$ for some $|\epsilon| < \theta^*$ and $k \in \mathbb{Z}$. Hence

$$\left| \sin\theta - \frac{2k}{2N + 1} \right| < \frac{|\epsilon|}{(2N + 1)\pi}$$
$$< \frac{\theta^*}{(2N + 1)\pi} < \frac{1}{2(2N + 1)};$$

that is, $2|k| \leq 2N + 1 + 1/2$, so we have $|k| \leq N$ and

$$|(2N + 1)\pi \cos\theta|^2 = (2N + 1)^2\pi^2 - (2k\pi + \epsilon)^2$$
$$\geq (2N + 2|k| + 1)\pi^2 - 4|k|\pi\theta^* - \theta^{*2}$$
$$> \pi^2.$$

Thus $|e^w - 1| \geq |X - 1| \geq 1 - e^{-\pi}$.

Therefore we can take $c_0 = 1 - e^{-\pi}$ and

$$\left| \int_{C_N} \frac{\Phi(z)}{e^{1/z} - 1} dz \right| \leq \frac{1}{c_0} \max_{|z| \leq 1/(2\pi)} |\Phi(z)| \frac{2}{2N + 1} \to 0$$

as $N \to \infty$. Hence

$$I = \lim_{N \to \infty} \sum_{1 \leq |n| \leq N} \operatorname*{Res}_{z = i/(2n\pi)} f(z)$$

$$= \lim_{N \to \infty} \sum_{1 \leq |n| \leq N} \frac{1}{(2\pi n)^2} \Phi\left(\frac{i}{2\pi n} \right). \qquad \square$$

✦ ✸ ✦

Chapter 19

Metric Spaces

[Summary of Basic Points]

1. For a nonempty set X and a function $d : X \times X \to \mathbb{R}$ we say that (X, d) is a metric space if the following axioms are fulfilled:

   (i) $d(x, y) \geq 0$ for any $x, y \in X$ and the equality holds if and only if $x = y$.

   (ii) $d(x, y) = d(y, x)$ for any $x, y \in X$.

   (iii) $d(x, y) \leq d(x, z) + d(z, y)$ for any $x, y, z \in X$.

   The last inequality is called *triangle inequality*. We sometimes say simply that X is a metric space when it is clear what metric is used.

2. A subset of a metric space is compact if and only if it is sequentially compact.

3. Let f be a mapping defined on a metric space X into itself. We define $f^n(x) = f(f^{n-1}(x))$ for all $n \geq 1$ with $f^0(x) = x$, which is called the nth iterate of f. For any $x \in X$ the subset

$$\{x, \ f(x), \ f^2(x), \ ..., \ f^n(x), \ ...\}$$

   is called the orbit of f starting from x and denoted by $\mathrm{Orb}(x)$. We sometimes say that an orbit starting from x converges to y if $d(f^n(x), y)$ converges to 0 as $n \to \infty$.

4. A point $x_0 \in X$ is said to be a fixed point of f if $f(x_0) = x_0$; or, equivalently, if $\mathrm{Orb}(x_0) = \{x_0\}$.

5. A mapping $f : X \to X$ is said to be a *contraction* provided that there exists a constant $0 \leq \lambda < 1$ satisfying

$$d(f(x), f(y)) \leq \lambda d(x, y)$$

   for any $x, y \in X$. So any contraction is automatically continuous. Every contraction f on a complete metric space has a unique fixed point and any orbit converges to this fixed point. This is known as the 'contraction mapping principle'. .

6. A metric function d is said to be an *ultrametric* or a *non-Archimedean metric* if d satisfies the following stronger inequality than (iii):

(iv) $d(x, y) \le \max(d(x, z), d(z, y))$ for any $x, y, z \in X$,

which is called 'strong triangle inequality'. For example, the discrete metric

$$\varrho(x, y) = \begin{cases} 1 & (x \ne y) \\ 0 & (x = y) \end{cases}$$

is an ultrametric. Equipped with the discrete metric ϱ any set can be regarded as an ultrametric space.

7. For $r > 0$ and a point a in a metric space (X, d) the set

$$B_{<r}(a) = \{ x \in X \mid d(a, x) < r \}$$

is called an open ball centered at a of radius r. Similarly we say that $B_{\le r}(a) = \{ x \in X \mid d(a, x) \le r \}$ is a closed ball. In an ultrametric space every open ball is closed and every closed ball is open.

8. For a linear space E and a function $\|\cdot\| : E \to \mathbb{R}$ we say that $(E, \|\cdot\|)$ is a normed linear space if the following axioms are fulfilled:

(v) $\|x\| \ge 0$ for any $x \in E$ and the equality holds if and only if $x = 0$.
(vi) $\|\alpha x\| = |\alpha| \|x\|$ for any $x \in E$ and scalar α.
(vii) $\|x + y\| \le \|x\| + \|y\|$ for any $x, y \in E$.

Remark that (vi) depends on a choice of an absolute value $|\alpha|$ on the field of scalars. A normed linear space E induces a metric $d(x, y) = \|x - y\|$.

A typical example of normed linear space is the Euclidean space $\mathbb{R}^n$ with the *Euclidean norm*:

$$\|x\| = \sqrt{x_1^2 + x_2^2 + \cdots + x_n^2}$$

where $x = (x_1, \ldots, x_n)$.

9. If a norm $\|\cdot\|$ satisfies the following stronger inequality than (vii):

(viii) $\|x + y\| \le \max(\|x\|, \|y\|)$ for any $x, y \in E$,

then it is called a *non-Archimedean norm*. A non-Archimedean norm $\|\cdot\|$ induces an ultrametric $d(x, y) = \|x - y\|$.

10. For a field $\mathbb{K}$ a function $|\cdot| : \mathbb{K} \to \mathbb{R}$ is said to be an *absolute value* if

(ix) $|x| \geq 0$ for any $x \in \mathbb{K}$ and the equality holds if and only if $x = 0$.

(x) $|xy| = |x||y|$ for any $x, y \in \mathbb{K}$.

(xi) $|x + y| \leq |x| + |y|$ for any $x, y \in \mathbb{K}$.

An absolute value is sometimes called a valuation or a norm. We say that $|\cdot|$ is a non-Archimedean absolute value if

(xii) $|x + y| \leq \max(|x|, |y|)$ for any $x, y \in \mathbb{K}$.

PROBLEM 19.1

Let (X, d) be an ultrametric space. Show that

$$d(x, y) = \max(d(x, z), d(z, y))$$

if $d(x, z) \neq d(z, y)$.

This means that every triangle in an ultrametric space is an isosceles triangle having sides of lengths α, α and β with $\alpha \geq \beta$. In a non-Archimedean norm space we have

$$\|x \pm y\| = \max(\|x\|, \|y\|)$$

if $\|x\| \neq \|y\|$; or, equivalently,

$$\|x_0 + x_1 + \cdots + x_n\| = \|x_0\|$$

if $\|x_1\|, \|x_2\|, \ldots, \|x_n\| < \|x_0\|$. To this property Katok (2007) gave a turn of phrase 'the strongest wins'.

PROBLEM 19.2

Let (X, d) be a compact ultrametric space. Show that a subset E in X is open and compact if and only if E is a finite union of disjoint open balls.

PROBLEM 19.3

Let (X, d) be a compact metric space and let $f : X \to X$ be an isometry; that is,

$$d(f(x), f(y)) = d(x, y)$$

for any $x, y \in X$. Show that f is bijective. (So, f is a homeomorphism.)

┌─ **PROBLEM 19.4** ───┐

Let (X, d) be a compact metric space. Suppose that $f : X \to X$ satisfies

$$d(f(x), f(y)) \geq d(x, y)$$

for any $x, y \in X$. Show that f is a bijective isometry.

└───┘

┌─ **PROBLEM 19.5** ───┐

Let (X, d) be a compact metric space. Suppose that $f : X \to X$ is surjective and satisfies

$$d(f(x), f(y)) \leq d(x, y)$$

for any $x, y \in X$. Show that f is an isometry.

└───┘

┌─ **PROBLEM 19.6** ───┐

Let (X, d) be a compact metric space. Suppose that $f : X \to X$ satisfies

$$d(f(x), f(y)) < d(x, y)$$

for any $x \neq y \in X$. Show that f has a unique fixed point in X and every orbit of f converges to the fixed point.

└───┘

┌─ **PROBLEM 19.7** ───┐

Let (X, d) be a complete metric space and $\omega(t)$ be a right-continuous function defined on $[0, \infty)$ with $\omega(0) = 0$ and $0 \leq \omega(t) < t$ for any $t > 0$. Suppose that $f : X \to X$ satisfies

$$d(f(x), f(y)) \leq \omega(d(x, y))$$

for any $x, y \in X$. Show that f has a unique fixed point in X and every orbit of f converges to the fixed point.

└───┘

Note that we do not assume that $\omega(t)$ is monotonously increasing.

┌─ **PROBLEM 19.8** ───┐

Let p_n/q_n be the irreducible fraction of

$$\frac{2}{1} + \frac{2^2}{2} + \frac{2^3}{3} + \cdots + \frac{2^n}{n}.$$

Show that $p_n \equiv 0 \pmod{2^{\sigma(n)}}$, where $\sigma(n) = n + 1 - [\log(n + 1)/\log 2]$.

└───┘

Solutions for Chapter 19

If both $d(x, z)$ and $d(x, y)$ are less than $d(z, y)$, then

$$d(z, y) \leq \max(d(z, x), d(x, y)) < d(z, y),$$

a contradiction. □

Let E be an open and compact subset in X. Since

$$\bigcup_{x \in E} B_{<1/n}(x)$$

is an open covering of E for each integer $n \geq 1$, it follows from the compactness of E that there exists a finite set of points $\{x_1^{(n)}, \ldots, x_{m_n}^{(n)}\}$ in E satisfying

$$E \subset \bigcup_{k=1}^{m_n} B_{<1/n}\left(x_k^{(n)}\right).$$

Suppose now that

$$E \subsetneqq \bigcup_{k=1}^{m_n} B_{<1/n}\left(x_k^{(n)}\right)$$

for every $n \geq 1$. For each $n \geq 1$ there exists a point y_n in the set $B_{<1/n}(x_n) \setminus E$ where x_n is some point among $x_k^{(n)}, 1 \leq k \leq m_n$. Since $\{x_n\}$ is a sequence in E there exists a subsequence $\{n_j\}$ such that x_{n_j} converges to some $x^* \in E$ as $j \to \infty$. Then y_{n_j} converges also to the same point x^*, because $d(x_{n_j}, y_{n_j}) \leq 1/n_j$.

On the other hand, since E is open, there exists an open ball $B_{<r}(x^*)$ contained in E. Then we have $d(x^*, y_n) \geq r$ for all n, because $y_n \notin E$. This contradiction implies that there exists an integer N such that

$$E = \bigcup_{k=1}^{m_N} B_{<1/N}\left(x_k^{(N)}\right). \tag{19.1}$$

Suppose that $B_{<\rho}(a) \cap B_{<\rho}(b) \neq \emptyset$. Then for some $z \in B_{<\rho}(a) \cap B_{<\rho}(b)$ we have

$$d(x, b) \leq \max(d(x, a), d(a, z), d(z, b)) < \rho$$

for any $x \in B_{<\rho}(a)$; hence $B_{<\rho}(a) \subset B_{<\rho}(b)$. Similarly we have $B_{<\rho}(a) \supset B_{<\rho}(b)$; so $B_{<\rho}(a) = B_{<\rho}(b)$. Hence the finite union (19.1) can be taken as a disjoint union. This completes the proof. $\Box$

SOLUTION 19.3

For any $x_0 \in X$ put $x_n = f^n(x_0)$ for $n \geq 1$. We denote by $E(x_0)$ the closure of $\text{Orb}(x_1)$. Note that this orbit does not start from x_0. Since X is compact, there exists a subsequence $\{n_j\}$ such that $\{x_{n_j}\}$ converges as $j \to \infty$; so, for any $\epsilon > 0$, there exist two integers $n > m \geq 1$ satisfying $d(x_m, x_n) < \epsilon$. Then we have

$$d(x_0, x_{n-m}) = d(x_m, x_n) < \epsilon,$$

which means that $x_0 \in E(x_0)$ because ϵ is arbitrary. Since a continuous image of compact set is compact, it follows that $E(x_0) \subset \overline{f(X)} = f(X)$; hence $x_0 \in f(X)$, so $X \subset f(X)$, as required. $\Box$

SOLUTION 19.4

Suppose, on the contrary, that there exist two points $x_0, y_0 \in X$ with

$$\delta = d(f(x_0), f(y_0)) - d(x_0, y_0) > 0.$$

We write $x_n = f^n(x_0)$ and $y_n = f^n(y_0)$ for $n \geq 1$. Since X is compact, there exists a subsequence $\{n_j\}$ such that both $\{x_{n_j}\}$ and $\{y_{n_j}\}$ converge as $j \to \infty$. Thus there exist two integers $n > m \geq 1$ satisfying

$$d(x_m, x_n) < \frac{\delta}{2} \quad \text{and} \quad d(y_m, y_n) < \frac{\delta}{2},$$

which imply that

$$d(x_0, x_{n-m}) \leq d(x_m, x_n) < \frac{\delta}{2} \quad \text{and} \quad d(y_0, y_{n-m}) \leq d(y_m, y_n) < \frac{\delta}{2}$$

respectively. However we have

$$\begin{aligned}
d(x_{n-m}, y_{n-m}) &\geq d(x_1, y_1) \\
&= \delta + d(x_0, y_0) \\
&\geq \delta + d(x_{n-m}, y_{n-m}) - d(x_0, x_{n-m}) - d(y_0, y_{n-m}) \\
&> d(x_{n-m}, y_{n-m}),
\end{aligned}$$

a contradiction. Hence f is an isometry and f is bijective by **PROBLEM 19.3**. $\Box$

SOLUTION 19.5

Suppose, on the contrary, that there exist two points $x_0, y_0 \in X$ with

$$\delta = d(x_0, y_0) - d(f(x_0, f(y_0)) > 0.$$

Since f is surjective, the inverse image $f^{-1}(\{x_0\})$ is not empty; so there exists at least one point $x_{-1} \in X$ with $f(x_{-1}) = x_0$. Repeating this process we can construct a sequence $\{x_{-n}\}_{n \geq 0}$ in X satisfying $f(x_{-n}) = x_{-n+1}$ for all $n \geq 1$. We can get a similar sequence $\{y_{-n}\}$ in X satisfying $f(y_{-n}) = y_{-n+1}$ for all $n \geq 1$. Since X is compact, there exists a subsequence $\{n_j\}$ such that both $\{x_{-n_j}\}$ and $\{y_{-n_j}\}$ converge as $j \to \infty$; so, there exist two integers $n > m \geq 1$ satisfying

$$d(x_{-m}, x_{-n}) < \frac{\delta}{2} \quad \text{and} \quad d(y_{-m}, y_{-n}) < \frac{\delta}{2},$$

which imply that

$$\frac{\delta}{2} > d(x_{-m}, x_{-n}) \geq d(x_0, x_{m-n}) \quad \text{and} \quad \frac{\delta}{2} > d(y_{-m}, y_{-n}) \geq d(y_0, y_{m-n})$$

respectively. However we have

$$d(x_0, y_0) = \delta + d(x_1, y_1)$$
$$\geq \delta + d(x_{m-n}, y_{m-n})$$
$$\geq \delta + d(x_0, y_0) - d(x_0, x_{m-n}) - d(y_0, y_{m-n})$$
$$> d(x_0, y_0),$$

a contradiction. Hence f is an isometry. $\qquad\square$

SOLUTION 19.6

For any $x_0 \in X$ put $x_n = f^n(x_0)$ for $n \geq 1$. Put

$$E = \bigcap_{n=0}^{\infty} \overline{\text{Orb}(x_n)},$$

which is not empty by Cantor's intersection theorem. This set is known as the omega-limit set of x_0. For any $x \in E$ there exists a subsequence $\{n_j\}$ such that $\{x_{n_j}\}$ converges to x as $j \to \infty$. Since X is compact, without loss of generality, we can assume that $\{x_{n_j-1}\}$ also converge to some $z \in E$ as $j \to \infty$ respectively. By the continuity of f we have $f(z) = x$; thus $x \in f(E)$ and so, $E \subset f(E)$.

Next, for any $x \in f(E)$ there exists at least one point $y \in E$ with $x = f(y)$. Let $\{m_j\}$ be a subsequence such that $\{x_{m_j}\}$ converges to y as $j \to \infty$. Since $\{x_{m_j+1}\}$

converges to $f(y) = x$, we have $x \in E$ and so, $f(E) \subset E$. Therefore $E = f(E)$.

If E consists of more than one point, then its diameter

$$\kappa = \sup_{x,y \in E} d(x,y)$$

is finite and positive. Since E is compact, the supremum is certainly attained at some points u, v in E. Let a, b be distinct two points in E satisfying $u = f(a)$ and $v = f(b)$. However

$$\kappa = d(u, v) = d(f(a), f(b)) < d(a, b) \le \kappa,$$

a contradiction. Thus the set E exactly consists of one point, say $E = \{x^*\}$, which is obviously a fixed point of f. Moreover the orbit starting from x_0 converges to x^*. If f has two fixed points $x^* \ne x^{**} \in X$, then we would have

$$d(x^*, x^{**}) = d(f(x^*), f(x^{**})) < d(x^*, x^{**}),$$

a contradiction; hence x^* is a unique fixed point of f and every orbit converges to the fixed point x^*. $\square$

SOLUTION 19.7

For any $x_0 \in X$ put $x_n = f^n(x_0)$ for $n \ge 1$. Since $\omega(t) \le t$ holds for any $t \ge 0$, we have

$$d(x_n, x_{n+1}) = d(f(x_{n-1}), f(x_n)) \le \omega(d(x_{n-1}, x_n)) \le d(x_{n-1}, x_n). \tag{19.2}$$

Thus $\{d(x_{n-1}, x_n)\}$ is a monotone decreasing sequence bounded below. Let $\alpha \ge 0$ be the limit of this sequence. Letting $n \to \infty$ in (19.2) we have $\alpha \le \omega(\alpha)$, because ω is right-continuous; hence we get $\alpha = 0$.

We next show that $\{x_n\}$ is a Cauchy sequence in X. Suppose, on the contrary, that $\{x_n\}$ is not a Cauchy sequence. Then there exist a positive constant ϵ_0 and two subsequences $m_j < n_j$ satisfying

$$d(x_{m_j}, x_{n_j}) \ge \epsilon_0 \tag{19.3}$$

for all j. Now we can assume that n_j is the least integer satisfying (19.3); namely,

$$d(x_{m_j}, x_{n_j-1}) < \epsilon_0 \le d(x_{m_j}, x_{n_j}).$$

Put $\xi_j = d(x_{m_j-1}, x_{n_j-1})$ for brevity. Then

$$\xi_j \le d(x_{m_j-1}, x_{m_j}) + d(x_{m_j}, x_{n_j-1}) < d(x_{m_j-1}, x_{m_j}) + \epsilon_0.$$

Moreover, since $\epsilon_0 \le d(x_{m_j}, x_{n_j}) \le \omega(\xi_j) \le \xi_j$, it follows that the sequence ξ_j

converges to ϵ_0 from above as $j \to \infty$. Hence, by the right-continuity of ω we have $\epsilon_0 \le \omega(\epsilon_0)$, a contradiction. Thus $\{x_n\}$ is a Cauchy sequence.

Let x^* be the limit of $\{x_n\}$. Since

$$
\begin{aligned}
d(x^*, f(x^*)) &\le d(x^*, x_{n+1}) + d(x_{n+1}, f(x^*)) \\
&\le d(x^*, x_{n+1}) + \omega(d(x_n, x^*)) \\
&\le d(x^*, x_{n+1}) + d(x_n, x^*) \to 0
\end{aligned}
$$

as $n \to \infty$, x^* is a fixed point of f and the orbit starting from x_0 converges to this fixed point x^*. If f has two fixed points $x^* \ne x^{**}$ in X, then we would have

$$
d(x^*, x^{**}) = d(f(x^*), f(x^{**})) \le \omega(d(x^*, x^{**})) < d(x^*, x^{**}),
$$

a contradiction. Therefore every orbit converges to the fixed point x^*.   □

SOLUTION 19.8

This problem is discussed in Katok (2007) and we will give some remark later. We denote by ord_p the p-adic ordinal, which is also called the p-adic valuation. Put $c_n = p_n/q_n$ for brevity. The p-adic logarithm $\log_p$ on the field $\mathbf{Q}_p$ is defined by the series

$$
\log_p(1 + x) = \sum_{n=1}^{\infty} (-1)^{n-1} \frac{x^n}{n},
$$

converging in $\{x \in \mathbf{Q}_p \mid |x|_p < 1\} = p\mathbf{Z}_p$, and satisfies $\log_p(xy) = \log_p x + \log_p y$. It is known that $p = 2$ is a unique prime such that the p-adic function $\log_p(1 + x)$ is *not* injective on $p\mathbf{Z}_p$. In particular, we have $\log_2(-1) = 0$ and hence

$$
\begin{aligned}
\operatorname{ord}_2(c_n) = \operatorname{ord}_2\left(\frac{2^{n+1}}{n+1} + \frac{2^{n+2}}{n+2} + \cdots \right) \\
\ge \min_{k>n} \tau(k), \quad\quad\quad (19.4)
\end{aligned}
$$

where $\tau(k) = k - \operatorname{ord}_2(k)$. Since $\operatorname{ord}_2(k) \le [\log k / \log 2]$ and $\{k - [\log k / \log 2]\}$ is a monotone increasing sequence for $k \ge 2$, we have

$$
\operatorname{ord}_2(c_n) \ge n + 1 - \left[\frac{\log(n+1)}{\log 2} \right]. \quad\quad\quad □
$$

REMARK. If the minimum is attained at exactly one $k = k_n$ in (19.4), it follows from 'the strongest wins' property stated after **PROBLEM 19.1** that

$$
\operatorname{ord}_2(c_n) = k_n - \operatorname{ord}_2(k_n).
$$

For example, let ℓ and m be integers satisfying $\tau(k) < \ell$ for $1 \le k \le m$. Then it is easily seen that $\tau(2^\ell - k) > \tau(2^\ell) = 2^\ell - \ell$ for $1 \le k \le m$ and $\tau(2^\ell + k) > \tau(2^\ell)$ for all $k \ge 1$; hence we conclude that

$$\mathrm{ord}_2(c_n) = 2^\ell - \ell$$

for $n = 2^\ell - m, \ldots, 2^\ell - 1$.

On the other hand, since $\tau(2^\ell + k) \ge \tau(2^\ell + 1) = \tau(2^\ell + 2) = 2^\ell + 1$ for all $k \ge 3$ and $\ell \ge 2$, we have

$$\mathrm{ord}_2(c_n) \ge n + 1$$

if $n = 2^\ell$, but we cannot apply 'the strongest wins' property in this case. Some numerical calculations listed below suggest that $\mathrm{ord}_2(c_n)$ may take a comparatively large value when n is a power of 2. Indeed, we will show that

$$\mathrm{ord}_2(c_n) \ge n + \frac{\log n}{\log 2} + 2$$

if $n = 2^\ell \ge 8$. To see this we can assume $\ell \ge 5$. Put $m = 2^{\ell-1}$ for brevity. We have

$$c_n = c_{n+m} - \sum_{k=1}^{m} \frac{2^{n+k}}{n+k}$$

$$= c_{n+m} - 2^n c_m + n2^n \sum_{k=1}^{m} \frac{2^k}{k(n+k)}.$$

n	$\mathrm{ord}_2(c_n)$	n	$\mathrm{ord}_2(c_n)$	n	$\mathrm{ord}_2(c_n)$	n	$\mathrm{ord}_2(c_n)$
1	1	11	10	21	22	31	27
2	2	12	12	22	21	32	40
3	2	13	12	23	21	33	33
4	5	14	12	24	27	34	34
5	8	15	12	25	25	35	34
6	5	16	22	26	26	36	37
7	5	17	17	27	26 ·	37	39
8	13	18	18	28	27	38	37
9	9	19	18	29	27	39	37
10	10	20	21	30	27	40	48

Table 19.1   The dyadic valuations of c_n for $1 \le n \le 40$.

Therefore

$$\text{ord}_2(c_n) \geq \min(\text{ord}_2(c_{n+m}), n + \text{ord}_n(c_m), n + \ell + \Delta)$$

$$\geq \min\left(n + m + 1 - \left[\frac{\log(n+m+1)}{\log 2}\right],\right.$$

$$\left. n + m + 1 - \left[\frac{\log(m+1)}{\log 2}\right], n + \ell + \Delta\right)$$

$$= \min(n + m + 1 - \ell, n + \ell + \Delta),$$

where

$$\Delta = \text{ord}_2\left(\sum_{k=1}^{m} \frac{2^k}{k(n+k)}\right).$$

Since $\text{ord}_2(n+k) = \text{ord}_2 k$ for $1 \leq k \leq m$, we have

$$\Delta \geq \min\left(\delta, \text{ord}_2\left(\frac{2^3}{3(n+3)}\right), \min_{5 \leq k \leq m} \text{ord}_2\left(\frac{2^k}{k(n+k)}\right)\right)$$

$$= \min\left(\delta, 3, \min_{5 \leq k \leq m} (k - 2\,\text{ord}_2 k)\right),$$

where

$$\delta = \text{ord}_2\left(\frac{2}{n+1} + \frac{2^2}{2(n+2)} + \frac{2^4}{4(n+4)}\right)$$

$$= \text{ord}_2\left(\frac{2^{2\ell} + 2^{\ell+2} + 2^{\ell-2} + 4}{(2^{\ell-2}+1)(2^{\ell-1}+1)(2^\ell+1)}\right) = 2,$$

because $\ell \geq 5$. Since $k - 2\,\text{ord}_2 k \geq k - 2\log k/\log 2 \geq 2$ for $k \geq 8$ and this holds even for $k = 5, 6$ and 7, we have $\Delta \geq 2$. This implies that $\text{ord}_2(c_n) \geq n + \ell + 2$ for $\ell \geq 5$, as required.

Chapter 20

Differential Equations

[Summary of Basic Points]

1. Let $f_k(x, y_1, \ldots, y_m)$ be a function defined on a domain $D \subset \mathbb{R} \times \mathbb{R}^m$ for $1 \leq k \leq m$. The problem to find a (local) solution $(y_1(x), \ldots, y_m(x))$ satisfying the following system of differential equations:

$$y_k'(x) = f_k(x, y_1(x), \ldots, y_m(x)), \quad 1 \leq k \leq m \tag{20.1}$$

in a neighborhood of a given point $(x_0, u_1, \ldots, u_m)$ in D is said to be the initial value problem or the Cauchy problem with initial condition $y_k(x_0) = u_k$, $1 \leq k \leq m$. The equation (20.1) can be written in vectorial form as

$$\boldsymbol{y}'(x) = \boldsymbol{f}(x, \boldsymbol{y}(x)). \tag{20.2}$$

2. It may be convenient to introduce the following norm $\|\cdot\|_1$ on $\mathbb{R}^m$:

$$\|\boldsymbol{y}\|_1 = \sum_{j=1}^{m} |y_j|$$

where $\boldsymbol{y} = (y_1, \ldots, y_m)$. This norm is equivalent to the Euclidean norm $\|\cdot\|$ in the sense that

$$\|\boldsymbol{y}\| \leq \|\boldsymbol{y}\|_1 \leq \sqrt{m} \, \|\boldsymbol{y}\|$$

holds for any $\boldsymbol{y} \in \mathbb{R}^m$. We say that the function f_k satisfies a Lipschitz condition on D provided that there exists a constant K satisfying

$$|f_k(x, \boldsymbol{y}) - f_k(x, \boldsymbol{z})| \leq K \|\boldsymbol{y} - \boldsymbol{z}\|_1$$

for any $(x, \boldsymbol{y})$ and $(x, \boldsymbol{z}) \in D$. Note that the right-hand side is independent of x; in other words, the inequality holding uniformly in x is required. In particular, any function satisfying a Lipschitz condition is continuous in $y_1, \ldots, y_m$.

3. If each function f_k is independent of x, then the system of equations (20.1) is called autonomous.

4. Assume that $\{f_k\}$ are continuous and satisfy Lipschitz conditions on a closed hypercube

$$D = \Big\{ (x, y_1, \ldots, y_m) \;\Big|\; |x - x_0| \le \alpha, |y_k - u_k| \le \beta, 1 \le k \le m \Big\}$$

for some positive constants α and β. Then there exists a unique solution of the equation (20.2) with initial condition $y_k(x_0) = u_k$, $1 \le k \le m$ on the interval

$$|x - x_0| \le \min\left(\alpha, \frac{\beta}{M} \right)$$

where

$$M = \max_{1 \le k \le m} \; \max_{(x,y) \in D} |f_k(x, y_1, \ldots, y_m)|.$$

This existence and uniqueness theorem of the initial value problem, also called Cauchy-Lipschitz theorem, is usually proved by either the Euler-Cauchy polygon method or Picard's successive approximation method or the contraction principle. Moreover, if D can be taken as $[x_0 - \alpha, x_0 + \alpha] \times \mathbb{R}^m$, then the same conclusion holds on the whole interval $[x_0 - \alpha, x_0 + \alpha]$.

5. The Wronskian of $n - 1$ times differentiable functions $g_1(x), g_2(x), \ldots, g_n(x)$ is the following determinant

$$W(x) = \begin{vmatrix} g_1 & g_2 & \cdots & g_n \\ g_1' & g_2' & \cdots & g_n' \\ \vdots & \vdots & \ddots & \vdots \\ g_1^{(n-1)} & g_2^{(n-1)} & \cdots & g_n^{(n-1)} \end{vmatrix}.$$

If $\{g_k(x)\}$ are linearly dependent on an interval I, then the Wronskian vanishes everywhere on I. Peano (1889) noticed that the Wronskian of $g_1(x) = x^2$ and $g_2(x) = x|x|$ vanishes everywhere but they are linearly independent in any neighborhood of the origin. However, if the Wronskian vanishes everywhere on I, then there exists a subinterval $J \subset I$ such that $\{g_k(x)\}$ are linearly dependent on J.

6. The differential equation (20.2) is said to be linear if each function f_k can be written in the form

$$f_k(x, y) = p_{k,1}(x)y_1 + \cdots + p_{k,m}(x)y_m + q_k(x) \tag{20.3}$$

where $p_{k,1}(x), \ldots, p_{k,m}(x), q_k(x)$ are given functions usually assumed to be continuous on an interval $[a, b]$. In such a case each f_k satisfies a Lipschitz condition on $D = [a, b] \times \mathbb{R}^m$, and therefore there exists a unique solution of (20.2) with any initial condition $y(x_0) = u \in \mathbb{R}^m$.

7. A linear differential equation of mth order

$$y^{(m)}(x) = p_1(x)y(x) + \cdots + p_m(x)y^{(m-1)}(x) + q(x) \qquad (20.4)$$

can be transformed into the form (20.2) if we substitute $y_k = y^{(k-1)}$ and

$$f_1(x, y_1, \ldots, y_m) = y_2$$
$$f_2(x, y_1, \ldots, y_m) = y_3$$
$$\vdots$$
$$f_{m-1}(x, y_1, \ldots, y_m) = y_m$$
$$f_m(x, y_1, \ldots, y_m) = p_1(x)y_1 + \cdots + p_m(x)y_m + q(x).$$

The equation (20.4) is said to be homogeneous if $q(x)$ vanishes everywhere on $[a, b]$. The Wronskian of m solutions $Y_1(x), \ldots, Y_m(x)$ of the homogeneous equation vanishes everywhere on $[a, b]$ if and only if they are dependent on $[a, b]$. It can be easily seen that $W'(x) = p_m(x)W(x)$, from which we have

$$W(x) = W(x_0) \exp\left(\int_{x_0}^{x} p_m(t)\, dt \right)$$

for any $x_0 \in [a, b]$. This is known as Liouville's formula.

8. All solutions of a given homogeneous linear differential equation of m-order forms an m-dimensional linear space V. Then the m solutions $Y_1(x), \ldots, Y_m(x)$ are called fundamental solutions if $\{Y_k(x)\}$ is a basis of V.

PROBLEM 20.1

Suppose that $a(x) \in \mathscr{C}(\mathbb{R})$ satisfies $m \le a(x) \le M$ for any $x \in \mathbb{R}$, where m and M are constants. Show that the first-order differential equation

$$y' = y(a(x) - y)$$

has a unique global solution $y(x)$ satisfying $m \le y(x) \le M$.

PROBLEM 20.2

Suppose that $a(x), b(x) \in \mathscr{C}(\mathbb{R})$ satisfy $a(x) > 0$ for any $x \geq 0$,

$$\int_0^\infty a(x)\,dx = \infty \quad \text{and} \quad \int_0^\infty \frac{b^2(x)}{a(x)}\,dx < \infty.$$

Prove that there exists a unique solution $y(x)$ of the first-order linear differential equation

$$y' = a(x)y + b(x)$$

satisfying

$$\int_0^\infty a(x)y^2(x)\,dx < \infty.$$

Show moreover that

$$\int_0^\infty a(x)y^2(x)\,dx \leq \int_0^\infty \frac{b^2(x)}{a(x)}\,dx.$$

PROBLEM 20.3

Suppose that $f(x) \in \mathscr{C}[0, \infty)$. Show that the initial value problem

$$\frac{y'}{e^y + e^{-y}} = f(x), \quad y(0) = 0$$

has a global solution on $[0, \infty)$ if and only if

$$-\frac{3\pi}{4} < \int_0^x f(t)\,dt < \frac{\pi}{4}$$

for any $x > 0$.

PROBLEM 20.4

Find all differentiable functions $f(x)$ defined on $\mathbb{R}$ satisfying

$$f(x + y) = f(x)f'(y) + f'(x)f(y)$$

for all $x, y \in \mathbb{R}$.

One may notice immediately that this is the trigonometric addition formula when $f(x) = \sin x$.

9. For $q(x) \in \mathscr{C}[a, b]$ we consider non-zero solutions of the second-order linear differential equation

$$y'' + q(x)y = 0.$$

Every zero of such a solution is clearly simple by the uniqueness of solutions. This means that any non-zero solution can be expressed by the polar coordinate system as

$$y(x) = r(x)\sin\theta(x),$$
$$y'(x) = r(x)\cos\theta(x).$$

We simply say that $\theta(x)$ is the angular function corresponding to $y(x)$. Then we have the following system of non-linear differential equations:

$$r'(x) = (1 - q(x))r(x)\sin\theta(x)\cos\theta(x),$$
$$\theta'(x) = \cos^2\theta(x) + q(x)\sin^2\theta(x). \tag{20.5}$$

Note that (20.5) is a non-linear first-order differential equation on the angular function, which satisfies a Lipschitz condition on $[a, b] \times \mathbb{R}$; indeed,

$$|f(x, y) - f(x, z)| \le 2\Big(1 + \max_{a \le x \le b}|q(x)|\Big)|y - z|$$

where $f(x, y) = \cos^2 y + q(x)\sin^2 y$. In particular, if $q(x)$ is positive on $[a, b]$, then every curve $(y(x), y'(x))$ rotates around the origin with at least a fixed angular speed, because

$$\theta'(x) \ge \min\Big(1, \min_{a \le x \le b} q(x)\Big). \tag{20.6}$$

❦ ✳ ❧

PROBLEM 20.5

Suppose that $q(x), Q(x) \in \mathscr{C}[a, b]$ satisfy $q(x) < Q(x)$ on $[a, b]$. Let $\theta(x)$ and $\Theta(x)$ be the angular functions corresponding to non-zero solutions of

$$y'' + q(x)y = 0 \quad and \quad y'' + Q(x)y = 0$$

respectively. Show that, if $\theta(a) = \Theta(a)$, then $\theta(x) < \Theta(x)$ for any $a < x \le b$.

PROBLEM 20.6

Suppose that $q(x) \in \mathscr{C}[a, b]$ satisfies $q(x) \le M$ on $[a, b]$ for some constant $M > 0$. Show that if a non-zero solution $y(x)$ of

$$y'' + q(x)y = 0$$

has two distinct zeros z_1, z_2 in $[a, b]$, then

$$|z_1 - z_2| \ge \frac{\pi}{\sqrt{M}}.$$

PROBLEM 20.7

Suppose that $q(x) \in \mathscr{C}[a, b]$ satisfies $q(x) > m$ on $[a, b]$ for some constant $m > 0$. Show that every non-zero solution of

$$y'' + q(x)y = 0$$

has at least one zero in (a, b) if $b - a \ge \pi/\sqrt{m}$.

From (20.6) one easily has

$$\theta(b) - \theta(a) = \int_a^b \theta'(x)\, dx > (b - a) \min(1, m).$$

This implies that if $b - a \ge \pi/\min(1, m)$, then the solution $y(x)$ has at least one zero in (a, b). However $\sqrt{m} \ge \min(1, m)$ holds always, so the problem asks a stronger result.

PROBLEM 20.8

Show that any solution $y(x)$ of the second-order linear differential equation

$$y'' + xy = 0$$

is bounded on the interval $[0, \infty)$.

PROBLEM 20.9

Let $y(x)$ be a non-zero solution of the second-order differential equation

$$y'' + xy = 0$$

and $\{z_n\}$ be the sequence of all positive roots of $y(x)$ arranged in ascending order. Show that

$$\lim_{n \to \infty} \frac{z_n}{n^{2/3}} = \left(\frac{3\pi}{2}\right)^{2/3}.$$

PROBLEM 20.10

Find all differentiable functions $f(x)$ defined on $(0, \infty)$ satisfying $f(1) = 1$ and

$$f'(x)f\left(\frac{1}{x}\right) = 1$$

for $x > 0$.

♦ ✸ ♦

Solutions for Chapter 20

SOLUTION 20.1

Note that the right-hand side is a quadratic function in y; hence it does not satisfy any Lipschitz condition globally in y. This means that a local solution may blow up at a finite x. For example, in the specific case $a(x) = m = M = 0$ every general solution with constant C:

$$y(x) = \frac{1}{x + C}$$

blows up at $x = -C$. In this case the only solution of the problem is a singular solution $y = 0$.

Without loss of generality we can assume that the interval $[m, M]$ does not contain 0; for otherwise the trivial solution $y = 0$ would satisfy the required condition. We can moreover assume that $0 < m \le M$; for otherwise consider $\tilde{y}(x) = -y(-x)$ and $\tilde{a}(x) = -a(x)$. Substituting $y = 1/w$ we get the first-order linear differential equation of the form

$$w' = 1 - a(x)w. \tag{20.7}$$

Clearly $y(x)$ is a global solution of the original equation if and only if $w = 1/y$ is a global solution of (20.7) with constant sign. We must seek any solution w with $1/M \le w \le 1/m$. Put

$$A(x) = \int_0^x a(s)\, ds.$$

Then the general solution of (20.7) is given by

$$w(x) = e^{-A(x)}\left(\int_0^x e^{A(s)}\, ds + C_0 \right)$$

with an arbitrary constant C_0. Since $A(x)$ diverges to $-\infty$ as $x \to -\infty$, the factor $e^{-A(x)}$ diverges to ∞. Hence we must take

$$C_0 = -\int_0^{-\infty} e^{A(s)}\, ds;$$

otherwise $w(x)$ would be unbounded. Hence

$$w(x) = e^{-A(x)} \int_{-\infty}^{x} e^{A(s)} \, ds = \int_{-\infty}^{x} \exp\left(-\int_{s}^{x} a(t) \, dt\right) ds.$$

Using the inequality $a(x) \geq m > 0$ we get

$$w(x) \leq \int_{-\infty}^{x} e^{-m(x-s)} \, ds = \int_{0}^{\infty} e^{-mt} \, dt = \frac{1}{m}.$$

Similarly we obtain $w(x) \geq 1/M$, as required. $\qquad\square$

REMARK. The first-order differential equation of the form

$$y' + p(x)y = q(x)y^n$$

is called the Bernoulli differential equation when $n \neq 1$. By the substitution $w = y^{1-n}$ it reduces to a first-order linear differential equation on w.

SOLUTION 20.2

Put

$$A(x) = \int_{0}^{x} a(s) \, ds \quad \text{and} \quad K = \int_{0}^{\infty} \frac{b^2(x)}{a(x)} \, dx < \infty.$$

The general solution is given by

$$y(x) = e^{A(x)} \left(\int_{0}^{x} b(s) e^{-A(s)} \, ds + C \right)$$

where C is an arbitrary constant. Since $A(x)$ diverges to ∞ as $x \to \infty$, we must take

$$C = -\int_{0}^{\infty} b(s) e^{-A(s)} \, ds; \tag{20.8}$$

otherwise $|y(x)|$ would tend to ∞ and would not satisfy

$$\int_{0}^{\infty} a(x) y^2(x) \, dx < \infty.$$

Note that, by the Cauchy-Schwarz inequality,

$$\left(\int_{0}^{\infty} |b(s)| \, e^{-A(s)} \, ds \right)^2 \leq \int_{0}^{\infty} \frac{b^2(s)}{a(s)} \, ds \cdot \int_{0}^{\infty} a(s) e^{-2A(s)} \, ds = \frac{K}{2}$$

in view of

$$\int_{0}^{\infty} a(s) e^{-2A(s)} \, ds = -\frac{1}{2} e^{-2A(s)} \Big|_{0}^{\infty} = \frac{1}{2}$$

and so the improper integral (20.8) converges absolutely.

We next show that the solution

$$y(x) = -e^{A(x)} \int_x^\infty b(s)e^{-A(s)}\, ds$$

satisfies the inequality

$$\int_0^\infty a(x)y^2(x)\, dx \le K.$$

In the same way as above we have

$$y^2(x) \le e^{2A(x)} \int_x^\infty \frac{b^2(s)}{a(s)}\, ds \cdot \int_x^\infty a(s)e^{-2A(s)}\, ds$$

$$= \frac{1}{2} \int_x^\infty \frac{b^2(s)}{a(s)}\, ds,$$

which converges to 0 as $x \to \infty$. Integrating the equation $ay^2 = yy' - by$ from 0 to L, we obtain

$$\sigma_L = \int_0^L a(x)y^2(x)\, dx = \frac{y^2(L) - y^2(0)}{2} - \int_0^L b(x)y(x)\, dx$$

$$\le \frac{y^2(L)}{2} + \int_0^L |b(x)y(x)|\, dx.$$

Since

$$\left(\int_0^L |b(x)y(x)|\, dx \right)^2 \le \sigma_L \int_0^L \frac{b^2(x)}{a(x)}\, dx \le K\sigma_L,$$

we get $\sigma_L \le \epsilon_L + \sqrt{K\sigma_L}$ where $\epsilon_L = y^2(L)/2$. Let σ^* be a unique positive solution of the equation $\sigma = \epsilon_L + \sqrt{K\sigma}$. Then

$$\sigma_L \le \sigma^* = \epsilon_L + \frac{K}{2} + \frac{\sqrt{K^2 + 4K\epsilon_L}}{2}$$

$$\le \epsilon_L + \frac{K}{2} + \frac{K + 2\epsilon_L}{2} = K + 2\epsilon_L.$$

Since ϵ_L converges to 0 as $L \to \infty$, we obtain

$$\int_0^\infty a(x)y^2(x)\, dx \le K,$$

as required.   □

SOLUTION 20.3

It is easily seen that $Y(x) = e^{y(x)}$ satisfies the following differential equation

$$\frac{Y'}{1 + Y^2} = f(x)$$

with $Y(0) = e^0 = 1$. Hence

$$Y(x) = \tan\left(\int_0^x f(t)\,dt + \frac{\pi}{4}\right)$$

becomes a solution on $[0, \infty)$ if and only if

$$\left|\int_0^x f(t)\,dt + \frac{\pi}{4}\right| < \frac{\pi}{2}$$

for any $x > 0$. $\qquad\square$

SOLUTION 20.4

If $f(y_0) \neq 0$ for some $y_0 \in \mathbb{R}$, then it follows from

$$f(x + y) = f(x)f'(y) + f'(x)f(y) \qquad (20.9)$$

that

$$\frac{f'(x+h) - f'(x)}{h} = \frac{1}{f(y_0)} \cdot \frac{f(x + y_0 + h) - f(x + y_0)}{h}$$
$$- \frac{f'(y_0)}{f(y_0)} \cdot \frac{f(x+h) - f(x)}{h}$$

for any x and $h \neq 0$, which implies that $f(x)$ is twice differentiable everywhere and satisfies

$$f'(x + y) = f'(x)f'(y) + f''(x)f(y) \qquad (20.10)$$

for any x, y. Of course, this holds even when $f(x)$ is identically zero. Repeating this argument we can conclude that $f \in \mathscr{C}^\infty(\mathbb{R})$.

Put $\alpha = f(0)$ and $\beta = f'(0)$ for brevity. Putting $x = y = 0$ in (20.9) we have $\alpha(1 - 2\beta) = 0$. We distinguish two cases, as follows.

Case (i) $\alpha \neq 0$.

We get $\beta = 1/2$ and the first-order differential equation $2\alpha f'(x) = f(x)$ by putting $y = 0$ in (20.9). If $\alpha = 0$, then we have a constant solution $f(x) = 0$. Otherwise, solving this equation with the initial condition $f(0) = \alpha$, we obtain

$$f(x) = \frac{1}{2c}e^{cx}$$

where $c = 1/(2\alpha)$ is any non-zero constant. Obviously this satisfies (20.9) for any x, y.

Case (ii) $\alpha = 0$.

We can assume that $f(y_0) \neq 0$ for some $y_0 \in \mathbb{R}$. Putting $x = y_0$ and $y = 0$ in (20.9) we get $(1 - \beta)f(y_0) = 0$ and so $\beta = 1$. It also follows from (20.10) that $f''(x)f(y) = f(x)f''(y)$; hence $f''(x) = Cf(x)$ on J and $f''(x) = C'f(x)$ on J', where C, C' are some constants and $J = (0, a)$, $J' = (-a', 0)$ are maximal intervals on which $f(x)$ does not vanish, respectively. Such intervals certainly exist, because $f'(0) = 1$. However, since $f'''(0) = C\beta = C'\beta$, we have $C = C'$ and the second-order differential equation $f''(x) = Cf(x)$ holds on $(-a', a)$. We distinguish three cases, as follows.

Case (a) $C = 0$.

Solving $f''(x) = 0$ with $f(0) = 0$, $f'(0) = 1$, we obtain $f(x) = x$. In this case $a = a' = \infty$ and clearly $f(x) = x$ satisfies (20.9) for any x, y.

Case (b) $C > 0$.

Put $C = c^2, c > 0$. Any solution of the equation $f''(x) = c^2 f(x)$ with $f(0) = 0$, $f'(0) = 1$ is

$$f(x) = \frac{e^{cx} - e^{-cx}}{2c} = \frac{1}{c}\sinh(cx).$$

In this case $a = a' = \infty$ and it clearly satisfies (20.9) for any x, y.

Case (b) $C < 0$.

Put $C = -c^2, c > 0$. Any solution of $f''(x) = -c^2 f(x)$ with $f(0) = 0$, $f'(0) = 1$ is

$$f(x) = \frac{e^{icx} - e^{-icx}}{2ic} = \frac{1}{c}\sin(cx). \tag{20.11}$$

In this case $a = a' = \pi/c$. Since $f'(x) = \cos(cx)$ on $(-\pi/c, \pi/c)$, we get $f'(\pi/c) = -1$. Putting $x = \pi/c, y = x$ in (20.10) we get $f'(x + \pi/c) + f'(x) = 0$; so, $f(x + \pi/c) + f(x) = 0$ for any x, because $f(\pi/c) = 0$. Hence $f(x)$ is a periodic function with period $2\pi/c$ and (20.11) is obviously a solution on $\mathbb{R}$.

Conversely it can be seen that the functions

$$0, \quad x, \quad \frac{1}{2c}e^{cx}, \quad \frac{1}{c}\sin(cx) \quad \text{and} \quad \frac{1}{c}\sinh(cx),$$

where c is any non-zero constant, are actually the solutions of the equation.   □

SOLUTION 20.5

Put $\theta(a) = \Theta(a) = \theta_0$ and $\sigma(x) = \Theta(x) - \theta(x)$; so, $\sigma(a) = 0$. Since

$$\sigma'(x) = \Theta'(x) - \theta'(x)$$
$$= \cos^2 \Theta(x) - \cos^2 \theta(x)$$
$$+ Q(x) \sin^2 \Theta(x) - q(x) \sin^2 \theta(x),$$

we have

$$\sigma'(a) = (Q(a) - q(a)) \sin^2 \theta_0 \geq 0.$$

We distinguish two cases as follows:

Case (i) $\sin \theta_0 \neq 0$.

Since $\sigma'(a) > 0$, one has $\sigma(x) > 0$ in a right neighborhood of a.

Case (ii) $\sin \theta_0 = 0$.

It follows from (20.5) that

$$\theta'(a) = \cos^2 \theta_0 + q(a) \sin^2 \theta_0 = 1,$$
$$\Theta'(a) = \cos^2 \theta_0 + Q(a) \sin^2 \theta_0 = 1.$$

Hence we have

$$\sin^2 \theta(x) = \sin^2(x - a + o(x - a)) = (x - a)^2 + o(x - a)^2$$

as $x \to a+$ and $\sin^2 \Theta(x)$ has also the same estimate; therefore

$$\sigma'(x) = (Q(a) - q(a))(x - a)^2 + o(x - a)^2$$

as $x \to a+$. Thus $\sigma(x) > 0$ in a right neighborhood of a.

We now show that $\sigma(x) > 0$ on $(a, b]$. Suppose, on the contrary, that $\sigma(c) = 0$ for some $c \in (a, b]$. We can assume that $\sigma(x) > 0$ on (a, c); so, clearly $\sigma'(c-) \leq 0$. Put $\theta(c) = \Theta(c) = \theta_1$. Since

$$\sigma'(c-) = \Theta'(c-) - \theta'(c-) = (Q(c) - q(c)) \sin^2 \theta_1 \geq 0,$$

we have $\sigma'(c-) = 0$ and hence $\sin \theta_1 = 0$. By the same argument as Case (ii) we conclude that

$$\sigma'(x) = (Q(c) - q(c))(x - c)^2 + o(x - c)^2$$

as $x \to c-$. This implies that $\sigma(x) < 0$ in a left neighborhood of c, which is a contradiction. $\qquad \square$

SOLUTION 20.6

We can assume that $z_1 < z_2, \theta(z_1) = 0$ and $\theta(z_2) = \pi$, where $\theta(x)$ is the angular function corresponding to the solution $y(x)$ of $y'' + q(x)y = 0$. For any $\epsilon > 0$ let $\Theta(x)$ be the angular function corresponding to the solution of

$$y'' + (M + \epsilon)y = 0 \tag{20.12}$$

with $\Theta(z_1) = \theta(z_1) = 0$. Then it follows from **PROBLEM 20.5** that $\Theta(x) > \theta(x)$ for any $z_1 < x \le b$. Since $\Theta(z_2) > \theta(z_2) = \pi$, there exists $w \in (z_1, z_2)$ with $\Theta(w) = \pi$.

On the other hand, we see that

$$\phi(x) = \sin\left(\sqrt{M + \epsilon}\,(x - z_1)\right)$$

is a solution of (20.12) satisfying $\phi(z_1) = 0$. Hence $\phi(x) = R(x) \sin \Theta(x)$ for some positive function $R(x)$ and $\phi(w) = R(w) \sin \Theta(w) = 0$. Since

$$\sqrt{M + \epsilon}\,(w - z_1) \ge \pi,$$

we have

$$z_2 - z_1 > w - z_1 \ge \frac{\pi}{\sqrt{M + \epsilon}};$$

so, $z_2 - z_1 \ge \pi/\sqrt{M}$, because ϵ is arbitrary. □

SOLUTION 20.7

We can assume that $b = a + \pi/\sqrt{m}$. Suppose, on the contrary, that $y(x) \ne 0$ on (a, b). Since the Wronskian

$$W(x) = \begin{vmatrix} y(x) & \sin(\sqrt{m}\,(x - a)) \\ y'(x) & \sqrt{m}\,\cos(\sqrt{m}\,(x - a)) \end{vmatrix}$$

satisfies

$$W'(x) = (q(x) - m)\,y(x)\sin(\sqrt{m}\,(x - a)),$$

we have

$$\mathrm{sgn}\,(W(b) - W(a)) = \mathrm{sgn}\,W'(x) = \mathrm{sgn}\,y(x)$$

on (a, b). However we have

$$W(b) - W(a) = -\sqrt{m}\,(y(a) + y(b)),$$

a contradiction. □

SOLUTION 20.8

We show that any solution $y(x)$ satisfies the inequality

$$y^2(x) \le y^2(1) + (y'(1))^2$$

for any $x > 1$. To see this, multiplying the differential equation by y' and integrating from 1 to x we have

$$\int_1^x y'(t)y''(t)\,dt + \int_1^x t\,y(t)y'(t)\,dt = 0.$$

It follows from integration by parts to the second integral that

$$(y'(t))^2 \Big|_1^x + ty^2(t)\Big|_1^x = \int_1^x y^2(t)\,dt;$$

therefore

$$xu(x) \le c + \int_1^x u(t)\,dt$$

where $u(x) = y^2(x)$ and $c = y^2(1) + (y'(1))^2$. Hence, putting

$$v(x) = \frac{1}{x}\int_1^x u(t)\,dt,$$

we get

$$v'(x) = \frac{u(x)}{x} - \frac{1}{x^2}\int_1^x u(t)\,dt \le \frac{c}{x^2}.$$

Integrating from 1 to x again, we have

$$v(x) \le c\int_1^x \frac{dt}{t^2} = c\left(1 - \frac{1}{x}\right),$$

which implies $u(x) \le c/x + v(x) \le c$. Thus $u(x)$ is bounded. $\qquad\square$

REMARK. Putting

$$y(x) = \sum_{n=0}^{\infty} a_n x^n,$$

we obtain a power series solution by solving the recursive formula

$$a_n = -\frac{a_{n-3}}{n(n-1)} \quad (n \ge 3).$$

with $a_2 = 0$. Hence we obtain

$$
a_n = \begin{cases}
\dfrac{(-1)^{n/3} a_0}{n(n-1)(n-3)(n-4)\cdots 3 \cdot 2} & (n \equiv 0 \pmod 3) \\[4mm]
\dfrac{(-1)^{[n/3]} a_1}{n(n-1)(n-3)(n-4)\cdots 4 \cdot 3} & (n \equiv 1 \pmod 3) \\[4mm]
0 & (n \equiv 2 \pmod 3)
\end{cases}
$$

which implies that the radius of convergence is ∞; in other words, every solution is real analytic.

SOLUTION 20.9

It follows from **PROBLEM 20.6** and **PROBLEM 20.7** that

$$
\frac{\pi}{\sqrt{z_{n+1}}} \leq z_{n+1} - z_n \leq \frac{\pi}{\sqrt{z_n}}
$$

for all n. Let z_{n_0} be the least zero of $y(x)$ containing in the interval $[(\pi/2)^{2/3}, \infty)$. Note that

$$
z_{n_0+1} \geq z_{n_0} + \frac{\pi}{\sqrt{z_{n_0} + \pi/\sqrt{z_{n_0}}}} > \pi^{2/3}.
$$

Putting $\zeta_n = 1/z_n$ for $n \geq n_0$ we have

$$
\phi(\zeta_n) \leq \zeta_{n+1} \leq \Phi(\zeta_n)
$$

where $\phi(x) = x/(1 + \pi x^{3/2})$ and $\Phi(x)$ is the inverse function of $f(x) = x/(1 - \pi x^{3/2})$. Clearly $\phi(x)$ and $\Phi(x)$ are strictly monotone increasing on $[0, (2/\pi)^{2/3}]$ and on $[0, \infty)$ respectively. Moreover $0 < \phi(x) < \Phi(x) < x$ on $[0, (2/\pi)^{2/3}]$. We see that

$$
\phi^n(\zeta_{n_0}) \leq \zeta_{n+n_0} \leq \Phi^n(\zeta_{n_0})
$$

for $n \geq 1$, where ϕ^n and Φ^n are the nth iterate of ϕ and Φ respectively. Since

$$
\phi(x) = x - \pi x^{5/2} + O(x^4) \quad \text{and} \quad \Phi(x) = x - \pi x^{5/2} + O(x^4)
$$

as $x \to 0+$, it follows from **PROBLEM 1.14** that

$$
\liminf_{n \to \infty} n^{2/3} \zeta_n \geq \lim_{n \to \infty} n^{2/3} \phi^n(\zeta_{n_0}) = \left(\frac{2}{3\pi}\right)^{2/3}
$$

and

$$\limsup_{n\to\infty} n^{2/3}\, \zeta_n \le \lim_{n\to\infty} n^{2/3}\Phi^n(\zeta_{n_0}) = \left(\frac{2}{3\pi}\right)^{2/3}.$$

This completes the proof. $\square$

SOLUTION 20.10

From the given relation we see that $f(x)$ must be a positive and strictly monotone increasing function on $(0, \infty)$. Since $f'(x) = 1/f(1/x)$, f is twice differentiable on $(0, \infty)$. Hence, differentiating $f(1/x) = 1/f'(x)$ we have

$$f'\left(\frac{1}{x}\right) = \frac{x^2 f''(x)}{f'^2(x)}.$$

Since $f'(1/x) = 1/f(x)$, we thus get the second-order non-linear differential equation

$$x^2 f(x) f''(x) = f'^2(x). \tag{20.13}$$

If $f(x)$ is a solution of (20.13), then $cf(x)$ is also a solution for any constant c. This suggests that $g(x) = \log f(x)$ may satisfy a simpler differential equation, and indeed, this new function $g(x)$ satisfies

$$g''(x) = \left(\frac{1}{x^2} - 1\right) g'^2(x).$$

Note that the above equation does not involve the term $g(x)$ itself and therefore can be solvable by integration. Since $g'(x) > 0$ on $(0, \infty)$ and $g'(1) = f'(1)/f(1) = 1$, we have

$$g'(x) = \frac{x}{x^2 - x + 1};$$

thus,

$$g(x) = \int_1^x \frac{t}{t^2 - t + 1}\, dt,$$

because $g(1) = 0$. Hence we get

$$f(x) = \sqrt{x^2 - x + 1}\, \exp\left(\frac{1}{\sqrt{3}} \arctan \frac{2x - 1}{\sqrt{3}} - \frac{\pi}{6\sqrt{3}}\right).$$

It is not hard to see that this unique solution of (20.13) with the initial conditions $f(1) = f'(1) = 1$ satisfies actually the given relation. $\square$

Bibliography

[**Books**]

Achieser, N. I. (1956). Theory of Approximation, Frederick Ungar Pub. Co., New York.
[Translated by C. J. Hyman from the author's lectures on approximation theory at Univ. Kharkov.]

Ahlfors, L. V. (1979). Complex Analysis, Third Edition, McGraw-Hill.

Apostol, T. M. (1957). Mathematical Analysis – A Modern Approach to Advanced Calculus, Addison-Wesley.

Artin, E. (1964). The Gamma Function, Holt, Rinehart and Winston, New York.
[Translated by M. Butler from 'Einführung in die Theorie der Gammafunktion', Hamburger Math. Einzelschriften, Verlag B. G. Teubner, Leipzig, 1931.]

Bass, J. (1966). Exercises in Mathematics : Simple and multiple integrals. Series of functions. Fourier series and Fourier integrals. Analytic functions. Ordinary and partial differential equations, Academic Press, New York-London.
[Translated by Scripta Technica, Inc. from 'Exercices de Mathématiques. Algèbre linéaire. Intégrales simples et multiples. Séries de fonctions. Séries et intégrales de Fourier. Fonctions analytiques. Equations différentielles et aux dérivées partielles. Calcul des probabilités', Masson et Cie, Éditeurs, Paris, 1965.]

Berndt, B. C. (1994). Ramanujan's Notebooks, Part IV, Springer-Verlag.

Bernoulli, Johannis (1697). Opera Omnia, Tom I, Edited by J. E. Hofmann, Georg Olms Verlagsbuchhandlung, Hildesheim, 1968 (in particular pp. 184–185).

Borwein, J. M. and Borwein, P. B. (1987). Pi and the AGM, Wiley Interscience Pub. (in particular p. 381).

Carathéodory, C. (1954). Theory of Functions of a Complex Variable, Chelsea, New York (in particular p. 119).
[Translated from 'Funktionentheorie I', Verlag Birkhäuser, Basel, 1950.]

Cauchy, A. L. (1841). Exercices d'analyse et de physique mathématique, Vol. 2, Bachelier, Paris (in particular p. 380).

Cheney, E. W. (1966). Introduction to Approximation Theory, McGraw-Hill Book Co., New York.

Delahaye, J.-P. (1997). Le fascinant nombre π, Pour la Science, Diffusion Belin, Paris (in particular p. 60).

Dieudonné, J. (1971). Infinitesimal Calculus, Hermann, Paris.
[Translated from 'Calcul infinitésimal', Hermann, Paris, 1968.]

Dini, U. (1892). Theorie der Functionen, Verlag B. G. Teubner, Leipzig (in particular pp. 148–150).
[Translated from 'Fondamenti per la teorica delle funzioni di variabili reali', Tipografia Nistri, Pisa, 1878.]

Fichtenholz, G. M. (1964). Differential- und Integralrechnung II, VEB Deutscher Verlag der Wissenschaften, Berlin.
[Translated from t'A Course of Differential and INtegral Calculus' II, Moskow 1958.]

Finch, S. R. (2003). Mathematical Constants, Encyclopedia of Math. and its Appl. 94, Cambridge Univ. Press.

Genocchi, A (1884). Calcolo differenziale e principii di calcolo integrale (pubblicato con aggiunte dal Dr. Giuseppe Peano), Fratelli Bocca, Torino (in particular p. 174).

Hardy, G. H. (1958). A Course of Pure Mathematics, Cambridge Univ. Press.

——— (1963). Divergent Series, 3rd ed., Oxford Univ. Press.

Hardy, G. H., Littlewood, J. E. and Pólya, G. (1934). Inequalities, Cambridge Univ. Press.

Hardy, G. H. and Wright, E. M. (1979). An Introduction to the Theory of Numbers, 5th ed., Oxford Univ. Press.

Hata, M. (2015). Neurons, A Mathematical Ignition, Series on Number Theory and Its Appl. 9, World Scientific.

Hobson, E. W. (1957). The Theory of Functions of a Real Variable and the Theory of Fourier's Series, Vol. II, Dover edition.

Kac, M. (1959). Statistical Independence in Probability, Analysis and Number Theory, The Carus Math. Monographs No. 12, John Wiley & Sons.

Kanemitsu, S. and Tsukada, H. (2007). Vistas of Special Functions, World Scientific Publishing Company.

Katok, S. (2007). *p*-adic Analysis: Compared with Real, Student Math. Library 37, Amer. Math. Soc.

Koblitz, N. (1977). *p*-adic Numbers, *p*-adic Analysis, and Zeta-Functions, Springer-Verlag New York Berlin Heidelberg.

Korevaar, J. (2004). Tauberian Theory – A Century of Developments, Springer-Verlag, Berlin.

Korobov, N. M. (1992). Exponential Sums and their Applications, Kluwer Acad. Pub.
[Translated from 'Trigonometrical Sums and their Applications', Nauka, Moscow, 1989.]

Lang, S. (1965). Algebra, Addison-Wesley.

le Lionnais, F. (1983). Les nombres remarquables, Hermann, Paris (in particular p. 22).

Levin, L. (1981). Polylogarithms and Associated Functions, North-Holland, New York (in particular p. 4).

Lyusternik, L.A. and Yanpol'skii, A.R. (1965). Mathematical Analysis, Functions, Limits, Series, Continued Fractions, Pergamon Press.
[Translated by D. E. Brown from the original book in Russian published in 1963.]

Niven, I. and Zuckerman, H. S. (1960). An Introduction to the Theory of Numbers, New York, John Wiley & Sons.

Pólya, G. and Szegö, G. (1972). Problems and Theorems in Analysis I, Springer-Verlag New York Berlin Heidelberg.
[Translated from 'Aufgaben und Lehrsätze aus der Analysis I', 4th ed., Heider-

berger Taschenbücher, Bd. 73, 1970.]

——— (1976). Problems and Theorems in Analysis II, Springer-Verlag New York Berlin Heidelberg.

[Translated from 'Aufgaben und Lehrsätze aus der Analysis II', 4th ed., Heiderberger Taschenbücher, Bd. 74, 1971.]

Staudt, K. G. C. von (1845). De Numeris Bernoullianis : Loci in Senatu Academico : Rite obtinendi Causa, Erlangae, typis Adolphi Ernesti Junge.

Stirling, J. (1730). Methodus Differentialis : sive tractatus de summatione et interpolatione serierum infinitarum, G. Bowyer, G. Strahan, London.

Wall, H. S. (1948). Analytic Theory of Continued Fractions, D. Van Nostrand Company, Inc., Princeton (in particular p. 331).

Zygmund, A. (1979). Trigonometric Series, Vol. 1, Cambridge Univ. Press.

[Papers]

Abel, N. H. (1826). Recherches sur la série $1 + \frac{m}{1}x + \frac{m(m-1)}{1\cdot 2}x^2 + \frac{m(m-1)(m-2)}{1\cdot 2\cdot 3}x^3 + \cdots$, *J. Reine Angew. Math.* **1**, pp. 311–339.

= Œuvres complètes de Niels Henrik Abel, Nouvelle Édition, Tome I, Christiania Grøndahl & Søn, 1881, pp. 219–250 (in particular p. 223).

Apéry, R. (1979). Irrationalité de $\zeta(2)$ et $\zeta(3)$, *Astérisque* **61**, pp. 11–13.

Apostol, T. M. (1973). Another elementary proof of Euler's formula for $\zeta(2n)$, *Amer. Math. Monthly* **80**, no. 4, pp. 425–431.

——— (1983). A proof that Euler missed: Evaluating $\zeta(2)$ the easy way, *Math. Intelligencer* **5**, pp. 59–60.

Arratia, A. (1999). Algunas maneras juveniles de evaluar $\zeta(2k)$, *Bol. Asoc. Mat. Venez.* **6**, no. 2, pp. 167–176.

Ayoub, R. (1974). Euler and the zeta function, *Amer. Math. Monthly* **81**, no. 10, pp. 1067–1086.

Barnes, E. W. (1899). The theory of the Gamma function, *Mess. Math.* **29**, pp. 64–128.

Bateman, P. T. and Diamond, H. G. (1996). A hundred years of prime numbers, *Amer. Math. Monthly* **103**, no. 9, pp. 729–741.

Berndt, B. C. (1975). Elementary evaluation of $\zeta(2n)$, *Math. Mag.* **48**, no. 3, pp. 148–154.

Bernstein, S. N. (1912a). Démonstration du théorème de Weierstrass fondée sur le calcul de probabilité, *Comm. Soc. Math. Kharkov* **13**, pp. 1–2.

——— (1912b). Sur l'ordre de la meilleure approximation des fonctions continues par des polynomes de degré donné, *Acad. Roy. Belgique Cl. Sci. Mém. Coll. in-4°* (2) **4**, pp. 1–103.

——— (1928). Sur les fonctions absolument monotones, *Acta Math.* **52**, pp. 1–66.

——— (1931). Sur les polynômes orthogonaux relatifs à un segment fini II, *J. Math. Pures Appl.* (9) **10**, pp. 219–286.

Beukers, F. (1978). A note on the irrationality of $\zeta(2)$ and $\zeta(3)$, *Bull. London Math. Soc.* **11**, pp. 268–272.

Beukers, F., Calabi, E. and Kolk, J. A. C. (1993). Sums of generalized harmonic series and volumes, *Nieuw Arch. Wisk.* (4) **11**, pp. 217–224.

Binet, J. (1839). Mémoire sur les intégrales définies eulériennes, et sur leur application à la théorie des suites, ainsi qu'à l'évaluation des fonctions des grands nombres, *J. École Roy. Polytech.* **16**, pp. 123–343. See also C. R. Acad. Sci. Paris, **9** (1839), pp. 39–45.

Bohman, H. (1952). On approximation of continuous and of analytic functions, *Ark. Mat.* **2**, no. 3, pp. 43–56.

Bohr, H. and Mollerup, J. (1922). Lærebog i matematisk Analyse (A textbook of mathematical analysis), III, Grænseprocesser, Jul. Gjellerups Forlag, København.

Borel, E. (1895). Sur quelques points de la théorie des fonctions, *Ann. Sci. École Norm. Sup.* (3) **12**, pp. 9–55 (in particular p. 44).

Boyd, D. W. (1969). Transcendental numbers with badly distributed powers, *Proc. Amer. Math. Soc.* **23**, no. 2, pp. 424–427.

Brown, G. and Koumandos, S. (1997). On a monotonic trigonometric sum, *Monatsh. Math.* **123**, pp. 109–119.

Caianiello, E. R. (1961). Outline of a theory of thought-processes and thinking machine, *J. Theoret. Biol.* **2**, pp. 204–235.

Callahan, F. P. (1964). Density and uniform density, *Proc. Amer. Math. Soc.* **15**, no. 5, pp. 841–843.

Carleman, T. (1922). Sur les fonctions quasi-analytiques, Proc. 5th Scand. Math. Congress, Helsingfors, Finland, pp. 181–196.
     = Édition complète des articles de Torsten Carleman, Litos Reprotryck, Malmö, 1960, pp. 199–214.

——— (1923). Über die Approximation analytischer Funktionen durch linear Aggregate von vorgegebenen Potenzen, *Ark. Mat. Astr. Fys.* **17**, no. 9, pp. 1–30.

——— (1927). Sur un théorème de Weierstrass, *ibid.* **20B**, no. 4, pp. 1–5.

Carleson, L. (1954). A proof of an inequality of Carleman, *Proc. Amer. Math. Soc.* **5**, no. 6, pp. 932–933.

Carlitz, L. (1961). A recurrence formula for $\zeta(2n)$, *Proc. Amer. Math. Soc.* **12**, no. 6, pp. 991–992.

Carlson, F. (1935). Une inégalité, *Ark. Mat. Astr. Fys.* **25B**, no. 1, pp. 1–5.

Catalan, E. (1875). Sur la constante d'Euler et la fonction de Binet, *J. Math. Pures Appl.* (3) **1**, pp. 209–240.

Cesàro, E. (1888). *Nouv. Ann. Math.* (3) **17**, p. 112.

——— (1893). Sulla determinazione assintotica delle serie di potenze, *Atti R. Accad. Sc. Fis. Mat. Napoli* (2) **7**, pp. 187–195.
     = Ernesto Cesàro, Opere Scelte, Vol. 1, Parte seconda, Ed. Cremonese, Roma, 1965, pp. 397–406.

——— (1906). Fonctions continues sans dérivée, *Arch. Math. Phys.* (3) **10**, pp. 57–63.

Chebyshev, P. L. (1852). Mémoire sur les nombres premiers, Œuvres de P. L. Tchebychef, Tome I, Chelsea Pub. Co., New York, 1961, pp. 51–70.

——— (1854). Théorie des mécanismes connus sous le nom de parallélogrammes, Œuvres de P. L. Tchebychef, Tome I, Chelsea Pub. Co., New York, 1961, pp. 111–143.

——— (1859). Sur l'interpolation dans le cas d'un grand nombre de données fournies par les observations, Œuvres de P. L. Tchebychef, Tome I, Chelsea Pub. Co., New York, 1961, pp. 387–469.

——— (1881). Sur les fonctions qui s'écartent peu de zéro pour certaines valeurs de la variable, Œuvres de P. L. Tchebychef, Tome II, Chelsea Pub. Co., New York, 1961, pp. 335–356.

Chen, M.-P. (1975). An elementary evaluation of $\zeta(2m)$, *Chinese J. Math.* **3**, no. 1, pp. 11 –15.

Choe, B. R. (1987). An elementary proof of $\sum_{n=1}^{\infty} 1/n^2 = \pi^2/6$, *Amer. Math. Monthly* **94**, no. 7, pp. 662–663.

Choi, J. and Rathie, A. K. (1997). An evaluation of $\zeta(2)$, *Far East J. Math. Sci.* **5**, no. 3, pp. 393–398.

Choi, J., Rathie, A. K. and Srivastava, H. M. (1999). Some hypergeometric and other evaluations of $\zeta(2)$ and allied series, *Appl. Math. Comput.* **104**, no. 2-3, pp. 101–108.

Clausen, T. (1840). Lehrsatz aus einer Abhandlung über die Bernoullischen Zahlen, *Astronomische Nachrichten* **17**, no. 406, pp. 351–352.

Davis, P. J. (1959). Leonhard Euler's integral : a historical profile of the Gamma function, *Amer. Math. Monthly* **66**, no. 10, pp. 849–869.

de Boor, C. and Schoenberg, I. J. (1976). Cardinal interpolation and spline functions VIII. The Budan-Fourier theorem for splines and applications, Lect. Notes in Math., 501, pp. 1–77.

de la Vallée Poussin, Ch.-J. (1896). Recherches analytiques sur la théorie des nombres premiers. Première partie: La fonction $\zeta(s)$ de Riemann et les nombres premiers en général, suivi d'un Appendice sur des réflexions applicables à une formule donnée par Riemann, *Ann. Soc. Sci. Bruxelles* (deuxième partie) **20**, pp. 183–256.
= Charles-Jean de La Vallée Poussin Collected Works, Vol. I, Acad. Roy. Belgique, Circolo Mat. Palermo, 2000, pp. 223–296.

de Rham, G. (1957). Sur un exemple de fonction continue sans dérivée, *Enseign. Math.* (2) **3**, pp. 71–72.

Denquin, C. (1912). Sur quelques séries numériques, *Nouv. Ann. Math.* (4) **12**, pp. 127– 135.

Diamond, H. G. (1982). Elementary methods in the study of the distribution of prime numbers, *Bull. Amer. Math. Soc.* (N.S.) **7**, no. 3, pp. 553–589.

Duncan, J. and McGregor, C. M. (2003). Carleman's inequality, *Amer. Math. Monthly* **110**, no. 5, pp. 424–431.

Duffin, R. J. and Schaeffer, A. C. (1941). A refinement of an inequality of the brothers Markoff, *Trans. Amer. Math. Soc.* **50**, no. 3, pp. 517–528.

Elkies, N. D. (2003). On the sums $\sum_{k=-\infty}^{\infty}(4k + 1)^{-n}$, *Amer. Math. Monthly* **110**, no. 7, pp. 561–573.

Erdös, T. (1949). On a new method in elementary number theory which leads to an elementary proof of the prime number theorem, *Proc. Nat. Acad. Sci. USA* **35**, pp. 374 –384.

Estermann, T. (1947). Elementary evaluation of $\zeta(2k)$, *J. London Math. Soc.* **22**, pp. 10– 13.

Faber, G. (1907). Einfaches Beispiel einer stetigen nirgends differentiierbaren Funktion, *Jber. Deutsch. Math. Verein.* **16**, pp. 538–540.

——— (1910). Über die Orthogonalfunktionen des Herrn Haar, *Jber. Deutsch. Math. Verein.* **19**, pp. 104–112.

Fejér, L. (1900). Sur les fonctions bornées et intégrables, *C. R. Acad. Sci. Paris* **131**, pp. 984 –987.

———— (1910). Über gewisse Potenzreihen an der Konvergenzgrenze, *S.-B. math.-phys. Akad. Wiss. München* **40**, no. 3, pp. 1–17.

= Leopold Fejér Gesammelte Arbeiten, Band I, Birkhäuser, 1970, pp. 573–583.

———— (1925). Abschätzungen für die Legendreschen und verwandte Polynome, *Math. Z.* **24**, pp. 285–298.

Fekete, M. (1923). Über die Verteilung der Wurzeln bei gewissen algebraischen Gleichungen mit ganzzahligen Koeffizienten, *Math. Z.* **17**, pp. 228–249 (in particular p. 233).

Franklin, F. (1885). Proof of a theorem of Tchebycheff's on definite integrals, *Amer. J. Math.* **7**, no. 4, pp. 377–379.

Gibbs, J. W. (1899). letter to the editor, Fourier Series, *Nature* **59**, p. 200 and p. 606.

Giesy, D. P. (1972). Still another elementary proof that $\sum 1/k^2 = \pi^2/6$, *Math. Mag.* **45**, no. 3, pp. 148–149.

Goldscheider, F. (1913). *Arch. Math. Phys.* **20**, pp. 323–324.

Goldstein, L. J. (1973). A history of the prime number theorem, *Amer. Math. Monthly* **80**, no. 6, pp. 599–615.

Gronwall, T. H. (1912). Über die Gibbssche Erscheinung und die trigonometrischen Summen $\sin x + \dfrac{1}{2}\sin 2x + \cdots + \dfrac{1}{n}\sin nx$, *Math. Ann.* **72**, pp. 228–243.

———— (1913). Über die Laplacesche Reihe, *ibid.* **74**, pp. 213–270.

———— (1918). The gamma function in the integral calculus, *Ann. Math.* (2) **20**, no. 2, pp. 35–124.

Grosswald, E. (1980). An unpublished manuscript of Hans Rademacher, *Historia Math.* **7**, no. 4, pp. 445–446.

Hadamard, J. (1896). Sur la distribution des zéros de la fonction $\zeta(s)$ et ses conséquences arithmétiques, *Bull. Soc. Math. France* **24**, pp. 199–220.

———— (1914). Sur le module maximum d'une fonction et de ses dérivées, *ibid.* **42**, pp. 68–72.

= Œuvres de Jacques Hadamard, Tome I, Edition du C. N. R. S., Paris, 1968, pp. 379–382.

Hardy, G. H. (1912). Note on Dr. Vacca's series for γ, *Quart. J. Math.* **43**, pp. 215–216.

———— (1914). Sur les zéros de la fonction $\zeta(s)$ de Riemann, *C. R. Acad. Sci. Paris* **158**, pp. 1012–1014.

Hardy, G. H. and Littlewood, J. E. (1914). Tauberian theorems concerning power series and Dirichlet's series whose coefficients are positive, *Proc. London Math. Soc.* (2) **13**, pp. 174–191.

———— (1914). Some problems of Diophantine approximation, *Acta Math.* **37**, pp. 155–190.

Harper, J. D. (2003). Another simple proof of $1 + 1/2^2 + 1/3^2 + \cdots = \pi^2/6$, *Amer. Math. Monthly* **110**, no. 6, pp. 540–541.

Hata, M. (1982). Dynamics of Caianiello's equation, *J. Math. Kyoto Univ.* **22**, no. 1, pp. 155–173.

———— (1995). Farey fractions and sums over coprime pairs, *Acta Arith.* **70**, no. 2, pp. 149–159.

Hausdorff, F. (1925). Zum Hölderschen Satz über $\Gamma(x)$, *Math. Ann.* **94**, pp. 244–247.

Hecke, E. (1921). Über analytische Funktionen und die Verteilung von Zahlen mod. eins, *Abh. Math. Sem. Hamburg Univ.* **1**, pp. 54–76.

= Erich Hecke Mathematische Werke, Göttingen, 1959, pp. 313–335.

Hofbauer, J. (2002). A simple proof of $1 + 2^{-2} + 3^{-2} + \cdots = \pi^2/6$ and related identities, *Amer. Math. Monthly* **109**, no. 2, pp. 196–200.

Hölder, O. (1887). Ueber die Eigenschaft der Gammafunction keiner algebraischen Differentialgleichung zu genügen, *Math. Ann.* **28**, pp. 1–13.

Holme, F. (1970). En enkel beregning av $\sum_{k=1}^{\infty} 1/k^2$, *Nordisk Mat. Tidskr.* **18**, pp. 91–92.

Hovstad, R. M. (1972). The series $\sum_{k=1}^{\infty} 1/k^{2p}$, the area of the unit circle and Leibniz' formula, *Nordisk Mat. Tidskr.* **20**, pp. 92–98.

Hua, L.-G. (1965). On an inequality of Opial, *Sci. Sinica* **14**, pp. 789–790.

Jackson, D. (1911). Über eine trigonometrische Summe, *Rend. Circ. Math. Palermo* **32**, pp. 257–262.

Jensen, J. L. W. V. (1906). Sur les fonctions convexes et les inégalités entre les values moyennes, *Acta Math.* **30**, pp. 175–193.

——— (1916). An elementary exposition of the theory of the Gamma function. Authorized translation from the Danish by T. H. Gronwall, *Ann. Math.* (2) **17**, no. 3, pp. 124–166.

Ji, C.-G. and Chen, Y.-G. (2000). Euler's formula for $\zeta(2k)$, proved by induction on k, *Math. Mag.* **73**, no. 2, pp. 154–155.

Kakeya, S. (1914). On approximate polynomials, *Tôhoku Math. J.* **6**, pp. 182–186.

Kalman, D. (1993). Six ways to sum a series, *College Math. J.* **24**, no. 5, pp. 402–421.

Karamata, J. (1930). Über die Hardy-Littlewoodschen Umkehrungen des Abelschen Stätigkeits-satzes, *Math. Z.* **32**, pp. 319–320.

Katznelson, Y. and Stromberg, K. (1974). Everywhere differentiable, nowhere monotone, functions, *Amer. Math. Monthly* **81**, no. 4, pp. 349–354.

Kempner, A. J. (1914). A curious convergent series, *Amer. Math. Monthly* **21**, no. 2, pp. 48–50.

Khintchine, A. (1924). Über einen Satz der Wahrscheinlichkeitsrechnung, *Fund. Math.* **6**, pp. 9–20.

Kimble, G. (1987). Euler's other proof, *Math. Mag.* **60**, no. 5, p. 282.

Klamkin, M. S. (1970). A comparison of integrals, *Amer. Math. Monthly* **77**, no. 10, p. 1114; ibid. **78**, no. 6, pp. 675–676.

Kluyver, J. C. (1900). Der Staudt-Clausen'sche Satz, *Math. Ann.* **53**, pp. 591–592.

——— (1928). Über Reihen mit positiven Gliedern, *J. London Math. Soc.* **3**, pp. 205–211.

Knopp, K. and Schur, I. (1918). Über die Herleitung der Gleichung $\sum_{n=1}^{\infty} 1/n^2 = \pi^2/6$, *Arch. Math. Phys.* (3) **27**, pp. 174–176.

Kolmogorov, A. N. (1939). On inequalities for suprema of consecutive derivatives of an arbitrary function on an infinite interval (Russian), *Uchenye Zapiski Moskov. G. Univ. Mat.* **30**, no. 3, pp. 3–13.

= Amer. Math. Soc. Transl., Ser. 1, Vol. 2, 1962, pp. 233–243.

= Selected Works of A. N. Kolmogorov, Vol. 1, ed. V. Tikhomirov, Kluwer, 1991, pp. 277–290.

Kolmogorov, A.N. (1926). Une série de Fourier-Lebesgue divergente partout, *C. R. Acad. Sci. Paris* **183**, pp. 1327–9.

Köpcke, A. (1887). Ueber Differentiirbarkeit und Anschaulichkeit der stetigen Functionen, *Math. Ann.* **29**, pp. 123–140.

——— (1889). Ueber eine durchaus differentiirbare, stetige Function mit Oscillationen in jedem Intervalle, *ibid.* **34**, pp. 161–171.

———— (1890). Nachtrag zu dem Aufsatze „Ueber eine durchaus differentiirbare, stetige Function mit Oscillationen in jedem Intervalle" (Annalen, Band XXXIV, pag. 161 ff.), *ibid.* **35**, pp. 104–109.

Korevaar, J. (1982). On Newman's quick way to the prime number theorem, *Math. Intelligencer* **4**, no. 3, pp. 108–115.

Korkin, A. N. and Zolotareff, E. I. (1873). Sur une certain minimum, *Nouv. Ann. Math.* (2) **12**, pp. 337–355.

Korovkin, P. P. (1953). On convergence of linear positive operators in the space of continuous functions (Russian), *Dokl. Akad. Nauk SSSR* **90**, no. 6, pp. 961–964.

Kortram, R. A. (1996). Simple proofs for $\sum_{k=1}^{\infty} 1/k^2 = \pi^2/6$ and $\sin x = x \prod_{k=1}^{\infty} (1 - x^2/(k\pi)^2)$, *Math. Mag.* **69**, no. 2, pp. 122–125.

Koumandos, S. (2001). Some inequalities for cosine sums, *Math. Inequal. Appl.* **4**, no. 2, pp. 267–279.

Kuhn, S. (1991). The Derivative à la Carathéodory, *Amer. Math. Monthly* **98**, no. 1, pp. 40–44.

Kummer, E. E. (1847). Beitrag zur Theorie der Function $\Gamma(x) = \int_0^{\infty} e^{-v} v^{x-1} dv$, *J. Reine Angew. Math.* **35**, pp. 1–4.

Kuo, H.-T. (1949). A recurrence formula for $\zeta(2n)$, *Bull. Amer. Math. Soc.* **55**, pp. 573–574.

Landau, E. (1908). Über die Approximation einer stetigen Funktionen durch eine ganze rationale Funktion, *Rend. Circ. Mat. Palermo* **25**, pp. 337–345.

———— (1913). Einige Ungleichungen für zweimal differentiierbare Funktionen, *Proc. London Math. Soc.* (2) **13**, pp. 43–49.

———— (1915). Über die Hardysche Entdeckung unendlich vieler Nullstellen der Zetafunktion mit reellem Teil $\frac{1}{2}$, *Math. Ann.* **76**, pp. 212–243.

———— (1934). Über eine trigonometrische Ungleichung, *Math. Z.* **37**, p. 36.

Landsberg, G. (1908). Über Differentiierbarkeit stetiger Funktionen, *Jber. Deutsch. Math. Verein.* **17**, pp. 46–51.

Lerch, M. (1888). Ueber die Nichtdifferentiirbarkeit gewisser Functionen, *J. Reine Angew. Math.* **103**, pp. 126–138.

———— (1903). Sur un point de la théorie des fonctions génératrices d'Abel, *Acta Math.* **27**, pp. 339–352.

Levinson, N. (1966). On the elementary proof of the prime number theorem, *Proc. Edinburgh Math. Soc.* (2) **15**, pp. 141–146.

———— (1969). A motivated account of an elementary proof of the prime number theorem, *Amer. Math. Monthly* **76**, no. 3, pp. 225–245.

Lebesgue, H. (1898). Sur l'approximation des fonctions, *Bull. Sci. Math.* **22**, pp. 278–287.

Littlewood, J. E. (1910). The converse of Abel's theorem on power series, *Proc. London Math. Soc.* (2) **9**, pp. 434–448.

Macdonald, I. G. and Nelsen, R. B. (1979). E2701, *Amer. Math. Monthly* **86**, no. 5, p. 396.

Malmstén, C. J. (1847). Sur la formule $hu'_x = \Delta u_x - \frac{h}{2} \Delta u'_x + \frac{B_1 h^2}{1.2} \cdot \Delta u''_x - \frac{B_2 h^4}{1...4} \Delta u_x^{IV} +$ etc, *J. Reine Angew. Math.* **35**, no. 1, pp. 55–82.

Markov, A. A. (1889). On a problem of D. I. Mendeleev (Russian), *Zap. Imp. Akad. Nauk, St. Petersburg* **62**, pp. 1–24.

Markov, V. A. (1892). On functions deviating the least from zero on a given interval (Russian), Depart. Appl. Math. Imperial St. Petersburg Univ.
= Über Polynome, die in einem gegebenen Intervalle möglichst wenig von Null abweichen, *Math. Ann.* **77**, pp. 213–258, 1916.

Matsuoka, Y. (1961). An elementary proof of the formula $\sum_{k=1}^{\infty} 1/k^2 = \pi^2/6$, *Amer. Math. Monthly* **68**, no. 5, pp. 485–487.

Mehler, F. G. (1872). Notiz über die Dirichlet'schen Integralausdrücke für die Kugelfunction $P^n(\cos \vartheta)$ und über eine analoge Integralform für die Cylinderfunction $J(x)$, *Math. Ann.* **5**, pp. 141–144.

Mertens, F. (1875). Ueber die Multiplicationsregel für zwei unendliche Reihen, *J. Reine Angew. Math.* **79**, pp. 182–184.

Michelson, A. A. (1898). letter to the editor, Fourier Series, *Nature* **58**, pp. 544–545.

Mirkil, H. (1956). Differentiable functions, formal power series, and moments, *Proc. Amer. Math. Soc.* **7**, no. 4, pp. 650–652.

Mittag-Leffler, G. (1900). Sur la représentation analytique des fonctions d'une variable réelle, *Rend. Circ. Mat. Palermo* **14**, pp. 217–224.

Möbius, A. F. (1832). Über eine besondere Art von Umkehrung der Reihen, *J. Reine Angew. Math.* **9**, no. 2, pp. 105–123.

Moore, E. H. (1897). Concerning transcendentally transcendental functions, *Math. Ann.* **48**, pp. 49–74.

Müntz, Ch. H. (1914). Über den Approximationssatz von Weierstraß, Mathematische Abhandlungen Hermann Amandus Schwarz zu seinem fünfzigjährigen Doktorjubiläum am 6 August 1914 gewidmet von Freunden und Schülern, Berlin, J. Springer, pp. 303–312 [There is a textually unaltered reprint by Chelsea Pub. Co. in 1974].

Neville, E. H. (1951). A trigonometrical inequality, *Proc. Cambridge Philos. Soc.* **47**, pp. 629–632.

Newman, D. J. (1980). Simple analytic proof of the prime number theorem, *Amer. Math. Monthly* **87**, no. 9, pp. 693–696.

Nikolić, A. (2002). Jovan Karamata (1902–1967), *Novi Sad J. Math.* (1) **32**, pp. 1–5.

Niven, I. (1947). A simple proof that π is irrational, *Bull. Amer. Math. Soc.* **53**, p. 509.

Okada, Y. (1923). On approximate polynomials with integral coefficients only, *Tôhoku Math. J.* **23**, pp. 26–35.

Opial, Z. (1960). Sur une inégalité, *Ann. Polon. Math.* **8**, pp. 29–32.

Osler, T. J. (2004). Finding $\zeta(2p)$ from a product of sines, *Amer. Math. Monthly* **111**, no. 1, pp. 52–54.

Ostrowski, A. (1919). Neuer Beweis des Hölderschen Satzes, daß die Gammafunktion keiner algebraischen Differentialgleichung genügt, *Math. Ann.* **79**, pp. 286–288.

——— (1925). Zum Hölderschen Satz über $\Gamma(x)$, *ibid.* **94**, pp. 248–251.

Pál, J. (1914). Zwei kleine Bemerkungen, *Tôhoku Math. J.* **6**, pp. 42–43.

Papadimitriou, I. (1973). A simple proof of the formula $\sum_{k=1}^{\infty} k^{-2} = \pi^2/6$, *Amer. Math. Monthly* **80**, no. 4, pp. 424–425.

Peano, G. (1889). Sur le déterminant Wronskien, *Mathesis: recueil mathématique* **9**, pp. 75–76, 110–112.

Pereno, I. (1897). Sulle funzioni derivabili in ogni punto ed infinitamente oscillati in ogni intervallo, *Giorn. di Mat.* **35**, pp. 132–149.

Picard, É. (1891). Sur la représentation approchée des fonctions, *C. R. Acad. Sci. Paris* **112**, pp. 183–186.

Pisot, Ch. (1938). La répartition modulo 1 et les nombres algébriques, *Ann. Sc. Norm. Sup. Pisa* (2) **7**, pp. 205–248.

——— (1946). Répartition (mod 1) des puissances successives des nombres réels, *Comment. Math. Helv.* **19**, pp. 153–160.

Pólya, G. (1911). *Nouv. Ann. Math.* (4) **11**, pp. 377–381.

——— (1926). Proof of an inequality, *Proc. London Math. Soc.* (2) **24**, p. vi in 'Records of Proceedings at Meetings'.

——— (1931). *Jber. Deutsch. Math. Verein.* **40**, p. 81.

Pringsheim, A. (1900). Ueber das Verhalten von Potenzreichen auf dem Convergenzkreise, *S.-B. math.-phys. Akad. Wiss. München* **30**, pp. 37–100.

Raabe, J. L. (1843). Angenäherte Bestimmung der Factorenfolge 1.2.3.4.5...$n = \Gamma(1+n) = \int x^n e^{-x} dx$, wenn n eine sehr grosse Zahl ist, *J. Reine Angew. Math.* **25**, pp. 146–159.

——— (1844). Angenäherte Bestimmung der Function $\Gamma(1+n) = \int_0^\infty x^n e^{-x} dx$, wenn n eine ganze, gebrochene, order incommensurable sehr grosse positive Zahl ist, *ibid.* **28**, pp. 10–18.

Rademacher, H. (1922). Einige Sätze über Reihen von allgemeinen Orthogonalfunktionen, *Math. Ann.* **87**, pp. 112–138.

Rado, R. (1934). A new proof of a theorem of v. Staudt, *J. London Math. Soc.* (2) **9**, pp. 85–88.

Redheffer, R. (1967). Recurrent inequalities, *Proc. London Math. Soc.* (3) **17**, pp. 683–699.

Riemann, G. F. B. (1859). Ueber die Anzahl der Primzahlen unter einer gegebenen Grösse, Bernhard Riemann's Gesammelte Mathematische Werke und Wissenschaftlicher Nachlass, Zweite Auflage, Leipzig, 1892, pp. 145–153.
   [English Translation: Collected Papers Bernhard Riemann, Kendrick Press, 2004, pp. 135–143.]

Robbins, N. (1999). Revisiting an old favorite: $\zeta(2m)$, *Math. Mag.* **72**, no. 4, pp. 317–319.

Rogosinski, W. (1955). Some elementary inequalities for polynomials, *Math. Gazette* **39**, pp. 7–12.

Rogosinski, W. and Szegö, G. (1928). Über die Abschnitte von Potenzreihen, die in einem Kreise beschränkt bleiben, *Math. Z.* **28**, pp. 73–94.

Rosenthal, A. (1953). On functions with infinitely many derivatives, *Proc. Amer. Math. Soc.* **4**, no. 4, pp. 600–602.

Runge, C. (1885a). Zur Theorie der eindeutigen analytischen Functionen, *Acta Math.* **6**, pp. 229–245.

——— (1885b). Über die Darstellung willkürlicher Functionen, *ibid.* **7**, pp. 387–392.

Russell, D. C. (1991). Another Eulerian-type proof, *Math. Mag.* **64**, no. 5, p. 349.

Schauder, J. (1928). Eine Eigenschaft des Haarschen Orthogonalsystems, *Math. Z.* **28**, pp. 317–320.

Schlömilch, O. (1844). Einiges über die Eulerschen Integrale der zweiten Art, *Arch. Math. Phys.* (2) **4**, pp. 167–174.

Schönbeck, J. (2004). Thomas Clausen und die quadrierbaren Kreisbogenzweiecke, *Centaurus* **46**, pp. 208–229.

Schwering, K. (1899). Zur Theorie der Bernoulli'schen Zahlen, *Math. Ann.* **52**, pp. 171–173.

Selberg, A. (1949). An elementary proof of the prime-number theorem, *Ann. Math.* **50**, no. 2, pp. 305–313.

Skau, I. and Selmer, E. S. (1971). Noen anvendelser av Finn Holmes methode for beregning av $\sum_{k=1}^{\infty} 1/k^2$, *Nordisk Mat. Tidskr.* **19**, pp. 120–124.

Stäckel, P. (1913). *Arch. Math. Phys.* (3) **13**, p. 362.

Stark, E. L. (1969). Another proof of the formula $\sum 1/k^2 = \pi^2/6$, *Amer. Math. Monthly* **76**, no. 5, pp. 552–553.

――― (1972). A new method of evaluating the sums of $\sum_{k=1}^{\infty}(-1)^{k+1}k^{-2p}$, $p = 1, 2, 3, \ldots$ and related series, *Elem. Math.* **27**, no. 2, pp. 32–34.

――― (1974). The series $\sum_{k=1}^{\infty} k^{-s}$, $s = 2, 3, 4, \cdots$, once more, *Math. Mag.* **47**, no. 4, pp. 197–202.

Staudt, K. G. C. von (1840). Beweis eines Lehrsatzes, die Bernoullischen Zahlen betreffend, *J. Reine Angew. Math.* **21**, pp. 372–374.

Steinhaus, H. (1920). Sur les distances des points des ensembles de mesure positiv, *Fund. Math.* **1**, pp. 93–104.

Stieltjes, T. J. (1876). De la représentation approximative d'une fonction par une autre [traduction de la brochure imprimée à Delft en 1876], Œuvres complétes de Thomas Jan Stieltjes, Tome I, Noordhoff, Groningen, Netherlands, 1914, pp. 11–20.

――― (1890). Sur les polynômes de Legendre, Œuvres complétes de Thomas Jan Stieltjes, Tome II, Noordhoff, Groningen, Netherlands, 1914, pp. 236–252.

Szegö, G. (1934). *Jber. Deutsch. Math. Verein.* **43**, pp. 17–20.

――― (1948). On an inequality of P. Turán concerning Legendre polynomials, *Bull. Amer. Math. Soc.* **54**, pp. 401–405.

Takagi, T. (1903). A simple example of the continuous function without derivative, *Proc. Phys.-Math. Soc. Japan* (IIs) **1**, pp. 176–177.

Tatuzawa, T. and Iseki, K. (1951). On Selberg's elementary proof of the prime-number theorem, *Proc. Japan Acad.* **27**, pp. 340–342.

Tauber, A. (1897). Ein Satz aus der Theorie der unendlichen Reihen, *Monatsh. Math.* **8**, pp. 273–277.

Titchmarsh, E. C. (1926). A series inversion formula, *Proc. London Math. Soc.* (2) **26**, pp. 1–11.

Tsumura, H. (2004). An elementary proof of Euler's formula for $\zeta(2m)$, *Amer. Math. Monthly* **111**, no. 5, pp. 430–431.

Turán, P. (1950). On the zeros of the polynomials of Legendre, *Časopis Pěst. Mat. Fys.* **75**, pp. 113–122.

    = Collected Papers of Paul Turán, Vol. 1, Edited by Paul Erdös, Akad. Kiadó, Budapest, 1990, pp. 531–540.

Underwood, R. S. (1928). An expression for the summation $\sum_{m=1}^{n} m^p$, *Amer. Math. Monthly* **35**, no. 8, pp. 424–428.

Vacca, G. (1910). A new series for the Eulerian constant, *Quart. J. Pure Appl. Math.* **41**, pp. 363–368.

van der Waerden, B. L. (1930). Ein einfaches Beispiel einer nicht-differenzierbaren stetigen Funktion, *Math. Z.* **32**, pp. 474–475.

Verblunsky, S. (1945). On positive polynomials, *J. London Math. Soc.* **20**, pp. 73–79.

Vijayaraghavan, T. (1940). On the fractional parts of the powers of a number (I), *J. London Math. Soc.* **15**, pp. 159–160.

――― (1941). ― (II), *Proc. Cambridge Philos. Soc.* **37**, pp. 349–357.

――― (1942). ― (III), *J. London Math. Soc.* **17**, pp. 137–138.

――― (1948). ― (IV), *J. Indian Math. Soc.* **12**, pp. 33–39.

Volterra, V. (1897). Sul principio de Dirichlet, *Rend. Circ. Mat. Palermo* **11**, pp. 83–86.

von Mangoldt, H. (1895). Zu Riemanns Abhandlung „Ueber die Anzahl der Primzahlen unter einer gegebenen Grösse", *J. Reine Angew. Math.* **114**, no. 3-4, pp. 255–305.

Walsh, J. L. (1923). A closed set of normal, orthogonal functions, *Amer. J. Math.* **45**, no. 1, pp. 5–24.

Weierstrass, K. (1856). Über die Theorie der analytischen Facultäten, *J. Reine Angew. Math.* **51**, pp. 1–60.

――― (1885). Über die analytische Darstellbarkeit sogenannter willkürlicher Funktionen reeller Argumente, *S.-B. Königl. Akad. Wiss. Berlin*, pp. 633–639 and pp. 789–805. = Mathematische Werke von Karl Weierstrass, Band III, Berlin, Mayer & Müller, 1903, pp. 1–37.

Weyl, H. (1916). Über die Gleichverteilung von Zahlen mod. Eins, *Math. Ann.* **77**, pp. 313–352.

Wielandt, H. (1952). Zur Umkehrung des Abelschen Stetigkeitssatzes, *Math. Z.* **56**, no. 2, pp. 206–207.

William Lowell Putnam Mathematical Competition (1970). A-4, *Amer. Math. Monthly* **77**, no. 7, p. 723 and p. 725.

Williams, G. T. (1953). A new method of evaluating $\zeta(2n)$, *Amer. Math. Monthly* **60**, no. 1, pp. 19–25.

Williams, K. S. (1971). On $\sum_{n=1}^{\infty} (1/n^{2k})$, *Math. Mag.* **44**, no. 5, pp. 273–276.

Wright, E. M. (1952). The elementary proof of the prime number theorem, *Proc. Roy. Soc. Edinburgh Sect. A* **63**, pp. 257–267.

――― (1954). An inequality for convex functions, *Amer. Math. Monthly* **61**, no. 9, pp. 620–622.

Yaglom, A. M. and Yaglom, I. M. (1953). An elementary derivation of the formula of Wallis, Leibniz and Euler for the number π (Russian), *Uspehi Mat. Nauk* (N.S.) **8**, no. 5 (57), pp. 181–187.

Young, W. H. (1909). On differentials, *Proc. London Math. Soc.* (2) **7**, pp. 157–180.

――― (1912). On a certain series of Fourier, *ibid.* (2) **11**, pp. 357–366.

Zagier, D. (1997). Newman's short proof of the prime number theorem, *Amer. Math. Monthly* **11**, no. 8, pp. 705–708.

Index

Printed in the United States
By Bookmasters